Business Success Through Innovation

An Insider's Guide to the World of United States Patents

Business Success Through Innovation:

An Insider's Guide to the World of United States Patents

****

Copyright Notice

****

ISBN: 978-0-9819052-0-4

Library of Congress Control No. 2008931952

Printed in the United States of America

www.isopatent.com

Book Dedication

This book is dedicated to inventors everywhere, who move the world forward through innovation.

Acknowledgement

We would like to thank Amanda, Meredithe, Anna and our kind and loving families, friends and colleagues for their wonderful support.

TABLE OF CONTENTS

Introduction

We wrote this book to provide an insider's guide to the world of United States patents and to explain how savvy inventors and corporations make innovation a key component of their business plans.

This book is divided into two parts. In Part I, *Generating Wealth Through Innovation*, we explain what a patent is, what distinguishes a valuable patent from a worthless one and how to develop wealth through patents using the same techniques employed by large, successful businesses. Patents are no longer the sole domain of geeks, nerds, engineers or patent attorneys. They are, and should be treated as, strategic business tools – financial assets that, when properly conceived and written, can significantly add to the bottom line.

It is estimated that over $500 billion in licensing revenue was generated by U.S. entities in 2005.[1] The bulk of that was realized by a relatively small group of large businesses – IBM, Texas Instruments and Qualcomm to name a few – which already understand the patent system and use it to generate wealth.[2] There are also start-up businesses and entrepreneurs that generate fortunes from patents. Without patents there would likely be no fortunes since competitors could freely copy their innovations.

These businesses and entrepreneurs are the insiders that already know what this book explains, namely, that patents can lead to enormous wealth. Simply by placing words on paper (often accompanied by drawings), filing the resulting document as a patent application and obtaining a patent, you can monopolize a market segment defined by your innovation. You can then license or sell the patent, or sell products and services protected by the patent. Only your imagination, hard work, and at least a general understanding of how to manufacture your innovation limit the wealth you can obtain through patents.[3]

In Part II, *A Statistical Breakdown of Who Is Patenting What in the United States*, we present a statistical analysis of United States patents. The United States issues more patents than any other nation, which is not surprising since the United States is the world's largest marketplace and has strong patent laws.[4] The analysis shows which businesses (U.S. and foreign) and universities obtain the most United States patents, how they rank in terms of issued patents, how each of the fifty United States ranks, and the technology categories in which most patent activity occurs.

You will discover for yourself the massive patent portfolios being developed by large businesses, thereby protecting market segments for their own exclusive use. You may also learn what your competitors and potential competitors are patenting, which could help you formulate your own game plan for innovation.

PART I

Generating Wealth Through Innovation

Chapter 1
Innovate for Success

Innovation is the New Business Model

The United States was once a manufacturing powerhouse. But, that time has passed. Today, industries such as automotive, steel and apparel have moved overseas or south of the border to countries such as China, India and Mexico, with significantly lower wages than the United States.

Nearly 80% of the U.S. economy is now service based.[5] This loss of manufacturing jobs does not, however, signal economic decline, but rather economic *transition*. The United States still has by far the largest Gross Domestic Product ("GDP") of any country at nearly $14 trillion.[6] That is three times larger than Japan (which ranks second), and four times larger than either Germany or China.[7]

How do countries like the United States, without cheap labor, maintain their economic edge? Through continuous research and development. An assembly line of innovations.[8] And, just as importantly, protecting (and thereby controlling) the innovations through legal mechanisms, particularly patents, properly valuing the innovations and profiting from them.

In many industries it is no longer important to be a manufacturing source because manufacturing is a commodity input provided at a commodity price. Instead, it is important to be the source of, and control, the *innovation*. By controlling the innovation you can control the product pipeline, from manufacturing, to distribution to sales. This is the new business model for nations with mature economies. Conference rooms and white boards have replaced factories and heavy equipment.

Thomas Jefferson once said "The price of liberty is constant vigilance." Today, the price of economic liberty, and our economic future, is constant investment in research and development ("R&D") and constant innovation. If we cease to innovate we will stagnate and decline.

The United States Government understands the value of innovation, which is why it has enacted some of the strongest intellectual property[9] laws in the world. When Congress passed the Industrial Espionage Act to guard against the theft of proprietary economic information it recognized that the United States produces the vast majority of intellectual property in the world, and that in the ten-year period between 1982-1992 the *"intangible assets"* of U.S. mining and manufacturing companies *grew from about 38% of their value to about 62% of their value, that the value of these intangible assets would only continue to grow, and* <u>*that a piece of information can be as valuable as a factory*</u>:

> *The United States produces the vast majority of the intellectual property in the world. This includes patented inventions, copyrighted material, and proprietary economic information...*
>
> *As this Nation moves into the high-technology, information age, the value of these intangible assets will only continue to grow.*
>
> *In a world where a nation's power is now determined as much by economic strength as by armed might, we cannot afford to neglect to protect our intellectual property. Today, a piece of information can be as valuable as a factory is to a business.*[10]

American business also understands the value of innovation. In 2006, the United States spent an estimated $344 billion on R&D. That was significantly more than Japan, China or the entire European Union spent:[11]

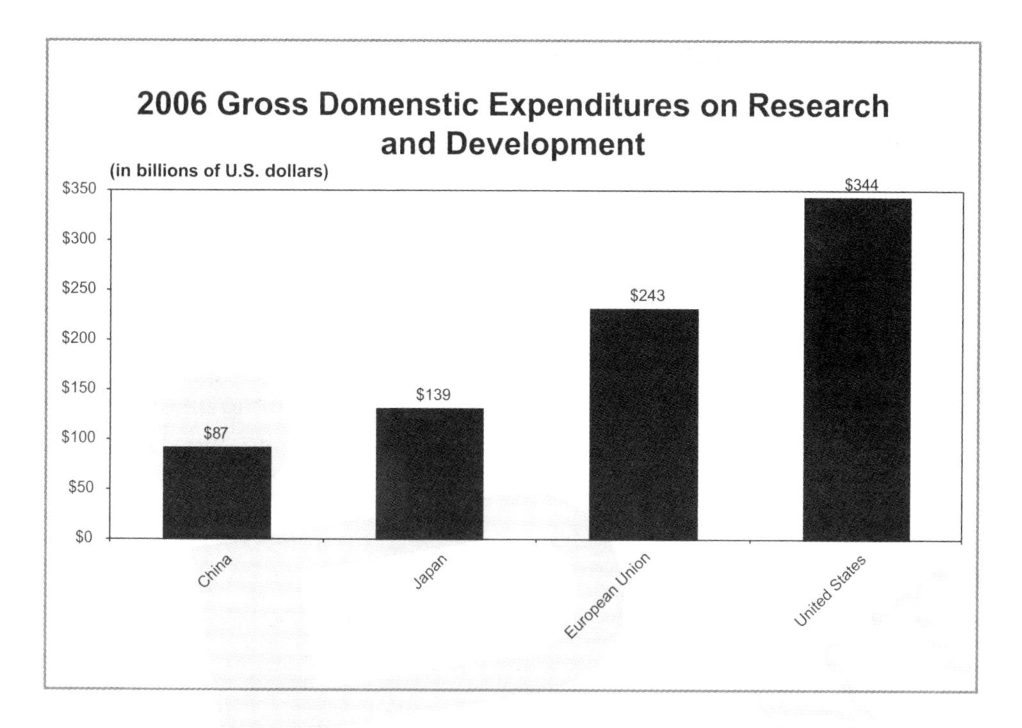

It is estimated that about 75% of the value of publicly-traded United States companies is now in intangible assets, and that in 2005 the total value of United States intellectual property was more than $5 trillion, which is greater than the entire GDP of any other country.[12] But, United States business is still losing an estimated $1 trillion by under utilizing patent assets alone.[13] Any business not already generating wealth through innovation should realize that it may be missing significant opportunities and implement an innovation plan.

Key points to remember:

1. The United States has by far the world's largest GDP at nearly $14 trillion.

2. The United States issues more patents than any other nation.

3. The United States spends more money on R&D than Japan, China, or the entire European Union.

4. The majority of the value of United States' manufacturing and mining companies is in intangible assets rather than plant or equipment.

5. The United States has strong patent laws.

6. Innovation is crucial for countries that cannot viably compete in commodity manufacturing.

Chapter 2
What is a Patent?

A Document that Uses Words and Drawings to Protect Your Innovation

Most people know the word "patent" and link it to protecting innovations, but few fully understand what a patent really is. A patent is a document, usually between about five to twenty pages in length, which uses words (and often drawings) to describe a piece of "intellectual property." Like a piece of land, the value of this intellectual property is based on its size and location. Generally speaking, the larger the "size" of a patent (referred to as the "scope" or "breadth" of the patent) and the better its "location" (i.e., the inherent value of the innovation protected by the patent), the more valuable the patent. Of course, the more valuable the patent, the more likely you are to realize positive financials from sales of products or services protected by the patent and/or the sale or licensing of the patent.

In the United States a patent is issued by the United States Patent and Trademark Office ("USPTO") and provides rights for twenty years from the date the application for the patent was originally filed.[14]

The most common, and usually by far the most valuable, type of patent is a "utility" patent, which is what most people think of when they refer to patents and which is the focus of this book.[15] A utility patent protects new processes, machines, articles of manufacture, or compositions of matter (such as chemical compounds), or any new and useful improvement of an existing product, process or composition of matter.[16] Utility patents comprise about 86 percent of all United States patents.[17]

Key points to remember:

1. A patent is a document issued (in the United States) by the USPTO.

2. A patent uses words (and often drawings) to describe a piece of intellectual property that protects a market segment into which your innovation falls.

3. A patent provides rights for twenty years from the date the application for the patent was originally filed.

4. A utility patent is by far the most common type of patent and usually the most valuable.

Chapter 3
What Can Be Patented?

"Anything Under the Sun That Is Made by Man."[18]

The United States has perhaps the most versatile patent system in the world. Essentially anything invented by Man can potentially be patented, and that includes products, processes, or compositions of matter.[19] Some examples of things falling within the ambit of patentable subject matter are: daytime planners, brooms, games, methods of providing insurance, software programs, rubber compounds, metal alloys, methods of determining a stock price, vacuum cleaners, pharmaceutical compounds, laser sights, tools, weapons, mixtures of soil and plant parts, methods of conducting a sweepstakes, pumps, back scrubbers, shoes, heart valves, bone screws, medical instruments, pacemakers, heated pet beds, and methods of detecting cancer cells. Things for which you cannot obtain a patent include scientific principles, unaltered, naturally-occurring substances, and methods that (1) are not tied to a particular machine or apparatus, and (2) do not transform an article into a different state or thing.

An important point for people conducting business in the United States to understand is that methods of *providing services*, often called "business methods," can be patented in the United States.[20] Examples of business methods are methods for providing insurance, sales contests, marketing promotions, or determining stock prices.

Some countries have more limited patentable subject matter than the

United States and do not permit, for example, patents for business methods, surgical procedures, or methods of playing games (although the physical game structure can be patented).

Certain business methods may be patented in countries other than the United States, but it depends upon the country, the nature of the business method and the manner in which the patent application is prepared. While not discussed in detail here, your attorney should understand how to prepare your application for each of the countries in which you desire patent protection. He or she cannot simply leave foreign patent issues to a foreign attorney (often referred to as a "foreign associate") because the United States patent application prepared by your U.S. attorney forms the basis of your foreign patent rights.[21]

In addition to falling within the ambit of patentable subject matter, your innovation must also be "new" to be patented. In the patent world "new" means "novel" and "non-obvious." "Novel" means that nothing exactly like your innovation was publicly known anywhere in the world prior to your invention date.[22] "Non-obvious" means that a hypothetical skilled person in the technological field to which your innovation pertains, with knowledge of all publicly-available information in the world that predates your innovation, and faced with the same problem as you solved with your innovation would *not* have viewed your innovation as one possible solution to the problem.[23] If this standard sounds subjective and complicated, that's because it is. Attorneys and inventors sometimes spend years and thousands of dollars arguing with the USPTO over whether an innovation is obvious and unpatentable, or non-obvious and patentable.[24]

So, when considering whether an innovation is worthy of patent protection, do not overlook anything new and valuable, even if at first blush the innovation seems like a simple improvement or something you would not normally consider patenting – such as a sales contest or method for valuing securities.

Remember, too, that in the United States patent protection is not limited to highly scientific subject matter such as computer components or pharmaceutical compounds, or to breakthrough innovations such as the transistor or integrated circuit. In fact, most patents are for innovations that represent incremental technological steps forward.

In summary, if you have a valuable innovation, do not assume it cannot be protected. Ask your attorney.

Key points to remember:

1. Methods of providing services, such as marketing or sales methods, providing insurance or determining stock prices, can potentially be patented in the United States. Do not limit your thinking of patentable innovations to highly scientific subject matter such as computer components or pharmaceuticals.

2. Improvements are also patentable and often valuable. Do not limit your thinking of what is patentable to only breakthrough innovations.

3. Things for which you cannot obtain a patent include scientific principles, unaltered, naturally-occurring substances, and methods that (a) are not tied to a particular machine or apparatus, and (b) do not transform an article into a different state or thing.

4. Certain methods of providing services may be patented in countries other than the United States, but that depends upon the country, the nature of the services and the manner in which the patent application is prepared.

Chapter 4
How Many United States Patents Are Issued?

Over Eight Million Since 1790, but Obtaining a Patent is Becoming More Difficult

Since 1790, the United States has issued over 8 million patents. From 1998 through 2007, the United States issued slightly less than 1.8 million patents, which equates to a ten-year annual average of almost 180,000 patents. Of those 1.8 million United States patents, 53 percent were granted to persons, companies, universities or governmental agencies based in the United States. The remaining 47 percent were issued to entities or persons from other countries.[25]

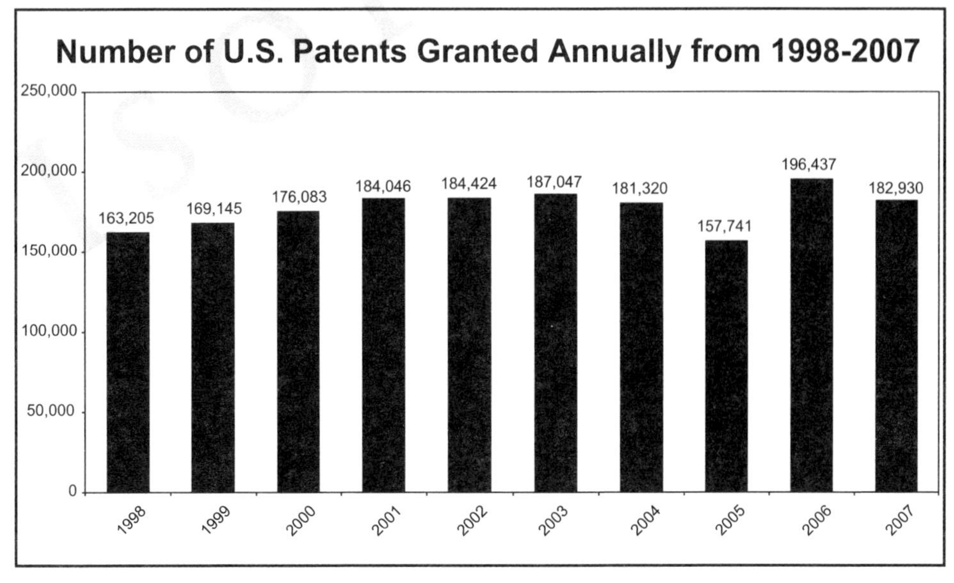

While it took 75 years (from 1836 to 1911) to issue one million patents after the advent of the current patent numbering system, it took only six years (from 2002 through 2007) to increase the number of U.S. patents by more than a million.[26]

Despite the high number of patents being issued, it has become more difficult to obtain a United States patent. Between 2000 and 2007, the patent procurement rate in the United States, which is a measure of the number of patent applications that ultimately issue as patents, declined by more than 20 percentage points, from 72 percent to 51 percent.[27]

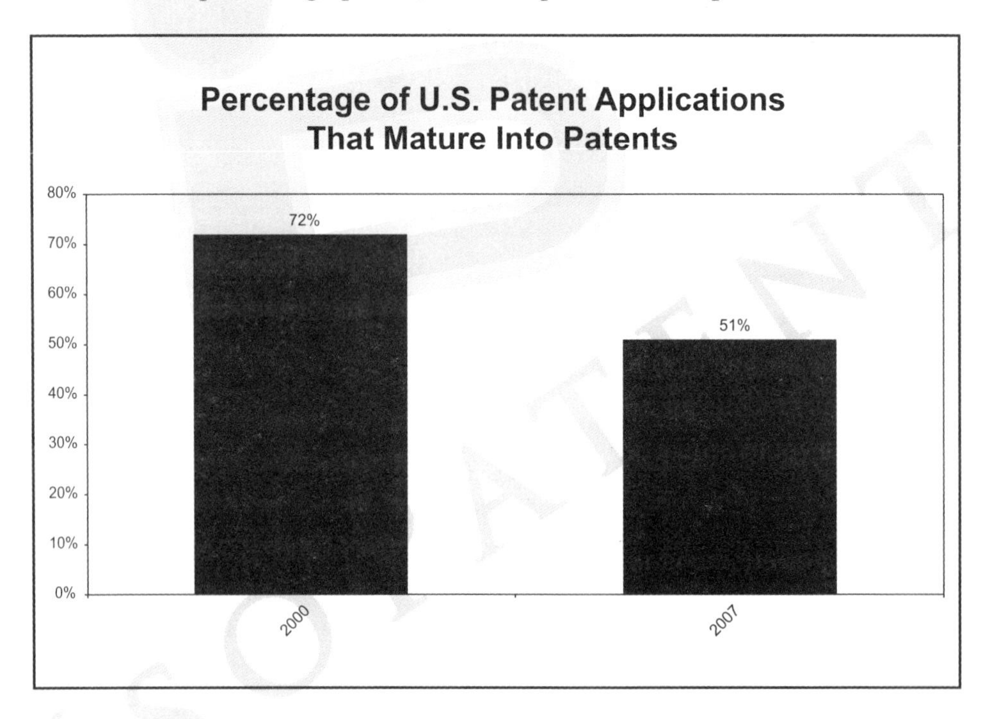

Consequently, you or your patent attorney must be knowledgeable and persistent in order to push your patent application[28] through the USPTO *without* modifying the application's scope in a way that can render the resulting patent worthless. The goal is not simply to obtain a patent. The goal is to generate wealth by obtaining a patent with broad scope that monopolizes a valuable market segment.

Key points to remember:

1. Over eight million U.S. patents have been issued.
2. On average, nearly 180,000 U.S. patents were issued annually between 1998 and 2007.
3. The patent procurement rate in the U.S. dropped from 72% in 2000 to 51% in 2007.
4. Your goal should be to secure a patent with broad scope that monopolizes a valuable market segment.

Chapter 5
How Patents Generate Wealth

By Protecting Your Innovation a Patent Increases Its Value

A patent creates a legal barrier preventing entry into the market segment defined by your innovation. The patent owner has the right to operate exclusively within that segment and to stop any trespass (called an "infringement") into the segment, which means the patent owner can exclude others from making, using, selling, offering to sell or importing products or services covered by the patent *regardless* of whether the patent owner ever provides the products or services. That is the power of a patent – its mere existence monopolizes a market segment.[29] You can generate profits without ever providing products or services, or dealing with vendors, customers, governmental regulations or employees.

A patent's barrier to entry provides many benefits:

1. If you choose to manufacture your innovation (or have it manufactured), it gives you time to establish manufacturing, marketing and sales channels;

2. You can potentially charge premium prices within the market segment protected by the patent;

3. You have a tangible asset to attract business partners, investors and potential buyers to your product/service or company; and/or

4. You can generate income by simply licensing or selling the patent, and hence your barrier to entry, to another.

Without patent protection there is no legal barrier to entry and others are free to copy your innovation.[30]

The more market segments covered by a patent, and the larger those market segment(s) are, the more valuable the patent. If your patent fails to capture the entire scope of your innovation, competitors will be free to practice the scope not protected. This is often called “designing around” a patent. If a design-around option is discovered, competitors can use it to practice your innovation without infringing your patent, as illustrated in the following diagram:

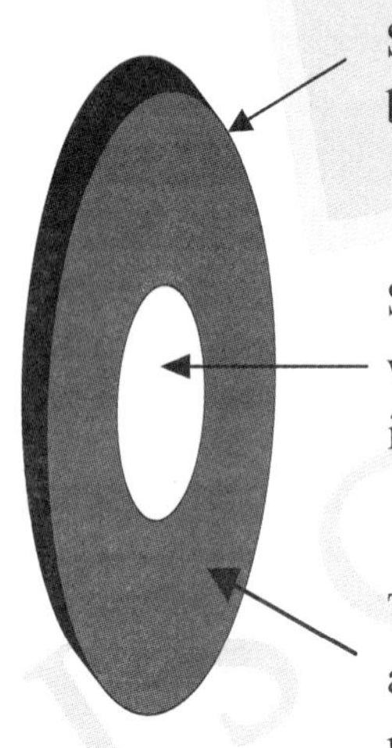

Scope of the innovation that could have been protected by the original inventor’s patent.

Scope of the innovation often protected by a patent, which is just the specific example conceived by the inventor.

The scope of the innovation left for others to practice and/or potentially patent, but that could have been protected by the original inventor’s patent.

As shown above, patents are sometimes too narrow in scope, which means they fail to protect the entire innovation and leave design-around options. Inventors and attorneys often patent just the example of the innovation conceived by the inventor rather than broadening its scope to protect the entire innovation and market segment(s) it defines. Additionally, there is usually little or no input from other business disciplines to broaden the patent scope. The result is a patent worth significantly less than it could have been and perhaps numerous lost marketing or licensing opportunities.

An Example of Failing to Protect the Entire Scope of an Innovation

Assume that your innovation is the chair, which is an improvement over a stool. The improvement lies in providing a back and thereby providing back support. While developing the chair you conceive what you believe to be an optimal version. This optimal version has a seat, four legs attached to the seat, and a back fastened to the rear edge of the seat and extending upwardly from the seat. You take the innovation to your patent attorney who prepares a patent covering it as follows:

> *A chair having a horizontal seat, four parallel legs attached to and extending downwardly from the seat, and a vertical back connected to the seat and extending upwardly therefrom.*

Sounds perfect; it defines your innovation to a "T," right? Not really. It merely defines the specific version that you described to your attorney. It fails to capture the entire innovative scope and leaves numerous design-around options.

For example, your competitors could make a chair with fewer than four legs, with legs that are not parallel, with a back that is not connected to the seat, with a seat that is not horizontal, or with a back that is not vertical. If the inventor and attorney had thought carefully about the innovation, its function and market applications, and placed themselves in the shoes of a competitor trying to design around a patent covering the innovation, they would have attempted to protect *any* device that provides back support and on which you can sit above the floor, rather than just the specific version of the innovation the inventor described to the attorney.

This book does not delve into the specific aspects of patent preparation that make a patent valuable. But, based on experience and anecdotal evidence, many patents have significantly less value than they should. Once a patent is issued by the USPTO a maintenance fee must be paid 3½ years, 7½ years and 11½ years after the issue date. If the maintenance fee is not paid the patent lapses, leaving others free to copy the innovation the patent had once protected.[31] About 80% of patent maintenance fees are paid at the 3½ year date, 60% at the 7½ year date and 40% at the 11½ year date.[32]

Thus, nearly 20% of patent owners, only 3½ years after patent issuance, do not believe the patent is worth a maintenance fee of just $490 to keep it from lapsing.[33] The solution to this problem is to put in the effort and thought to create patents that capture and monopolize broad, valuable market segments, leaving few or no design-around options.

Incorporate Ideas from All Disciplines to Optimize Patent Strategy

You can expand an innovation beyond the inventor's original concept through interdisciplinary brainstorming involving sales, marketing, finance, engineering, manufacturing, product development and legal personnel. When brainstorming think about target markets, the manner in which the innovation functions, how it will be sold and packaged, what spare parts will be utilized with the innovation, and other products, systems or services with which the innovation will be used. If possible and economically practical, patent the combinations, systems, spare parts, and various methods with which the innovation may be utilized. Patent an expanded scope to capture broad market segments.

Sometimes one or a few people provide input from multiple business disciplines, particularly in a small business where a few people wear many hats. But, in too many cases the perspectives from multiple disciplines are not factored into the innovation or patent strategy. The result is usually a weaker patent and lower potential profits plus a lack of organizational knowledge of innovation strategy, the organization's patent portfolio and how to use patents as profitable business tools.

The Four Key Factors for Valuing Innovations

Innovations, particularly those protected by patents, can create enormous wealth. But that wealth is largely dependent on the following four factors:

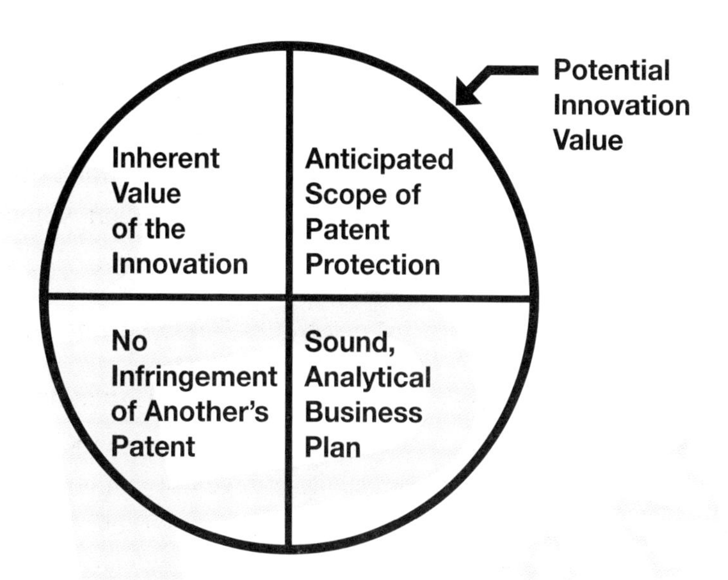

You should be familiar with these factors if you hope to profit from your own innovation or plan to invest in an innovation or start-up business. A brief explanation of these factors follows:

1. **Inherent Value of the Innovation.** The inherent value of the innovation with patent protection and without patent protection. Is the innovation a product or service that the market wants? If so, what profits can it generate?

2. **Anticipated Scope of Patent Protection.** The scope of the patent or patent application covering the innovation. Does it cover the entire scope of the innovation, or does it leave design-around options for competitors? The broader the scope, the more valuable the patent and the innovation.

3. **No Infringement of Another's Patent.** With or without a patent, can the innovation be provided to the market without infringing the pre-existing rights of another? If your innovation infringes another's pre-existing patent rights you

may never get it to market unless you license or buy the pre-existing rights or license/sell your innovation to the owner of the pre-existing rights.

4. **Sound, Analytical Business Plan.** A sound, analytical, quantitative business plan that explains both at a high level and in detail the business case for the innovation. Without such a plan you are unlikely to realize the full profit potential for your innovation.

Without any one of the above, you may have a difficult time attracting investors or business partners to, or profiting from, your innovation. The application of these principles is discussed further in Chapter 6: *The Innovation Decision Grid*, and Chapter 8: *You Need a Sound Business Plan*.

Key points to remember:

1. Patents have one purpose and that is to generate wealth. They should be considered valuable business assets.

2. Your underlying innovation must have inherent value.

3. Without patent protection competitors can legally copy your innovation.[34]

4. Without *broad* patent protection competitors will design around your patent and still be free to copy your innovation or aspects of it.

5. Your patent application should capture the entire, protectable scope of your innovation.

6. Your patent application should be prepared using a multi-disciplinary approach.

7. Always consult *The Four Key Factors for Valuing Innovations* when determining an innovation's potential value.

Chapter 6
The Innovation Decision Grid

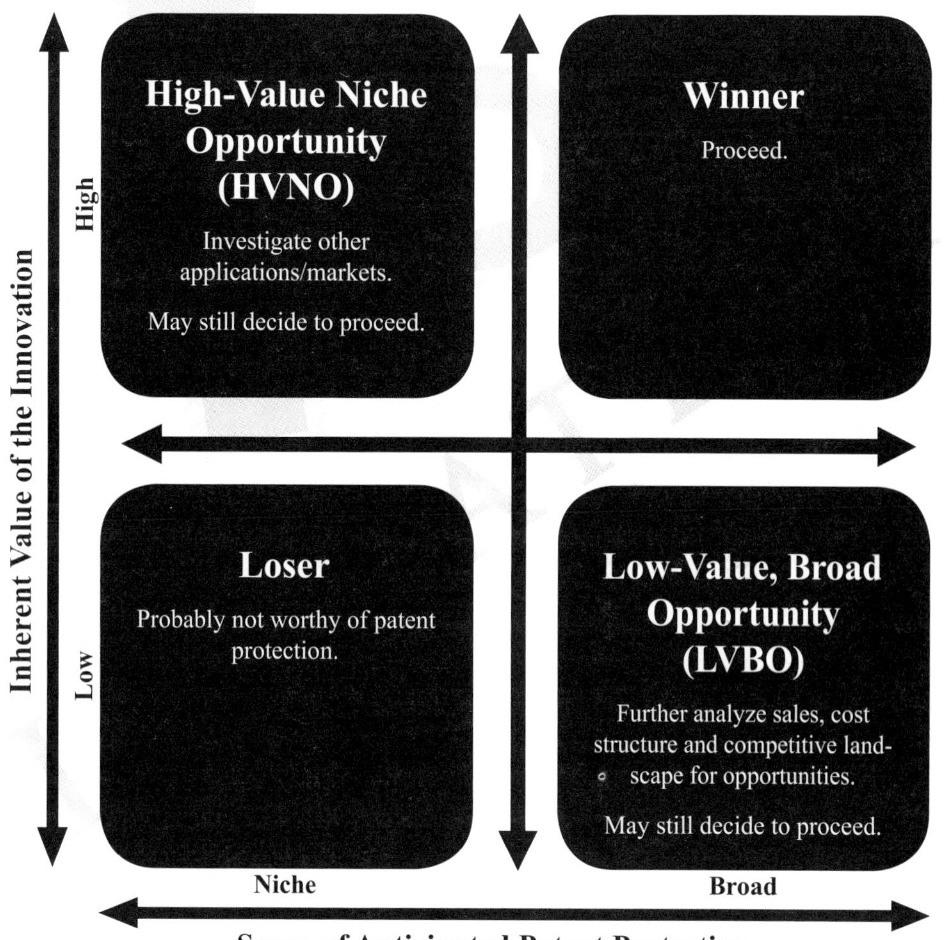

The *Innovation Decision Grid* is a simple matrix for estimating the success of an innovation by evaluating the first two factors of *The Four Key Factors for Valuing Innovations*: (1) *Inherent Value of the Innovation*, which should take into account things such as market size, start-up costs, anticipated market acceptance, and substitute products, and (2) *Anticipated Scope of Patent Protection*, which should take into account patent scope (after innovation brainstorming) and relevant pre-existing products, services and patents. Using the *Grid*, you can place an innovation into one of four Quadrants: Winner, High-Value Niche Opportunity ("HVNO"), Low-Value Broad Opportunity ("LVBO") or Loser.

Before or while evaluating your innovation using these factors, you should also analyze the third of *The Four Key Factors for Valuing Innovations* – namely, whether making or selling your innovation could infringe another's patent. If it could, you may decide either not to proceed at all, to modify your innovation to design around the other's patent, or to attempt to purchase or license the other's patent.

If your innovation falls within the Winner Quadrant you should move forward with patent protection and a business plan (the fourth of *The Four Key Factors for Valuing Innovations*) to capitalize on the innovation.

If an innovation falls within the HVNO Quadrant, it may be a valuable opportunity depending upon the size of the niche market and anticipated potential profits. But before proceeding, brainstorm to determine if the innovation can be broadened to expand the protected market segment. If so, and the cost/benefit analysis is favorable, file patent applications to protect the expanded market segment rather than just the original niche segment.

Next is the LVBO Quadrant, into which great ideas with exceptional breadth, but only limited present market potential, fall. With interdisciplinary brainstorming, you may discover more existing markets for the innovation and move it into the Winner Quadrant, or you may decide to gamble and file one or more patent applications to protect the innovation hoping that it will be valuable in the future. That is how some "patent trolls," discussed later in this book, made fortunes. They out-thought large, established businesses, guessed which innovations would be valuable five to ten years down the road, and filed patent applications to protect the innovations. Later, they extracted royalties as businesses entered the market segments protected by their patents.

The Loser Quadrant represents inherently low-value concepts with little chance of expansion and little chance of obtaining broad patent protection. Small, incremental enhancements to existing products may fit into this category. Being in the Loser Quadrant does not mean that the innovation should not be implemented; it means that the innovation is probably not a candidate for patent protection since the resulting patent would likely be of little value. The cost is simply not worth the benefit.

Key points to remember:

1. Use a quantitative approach, such as *The Innovation Decision Grid*, to determine whether an innovation is worth the cost of patent protection, or worth pursuing at all, with or without patent protection.

2. Always determine whether an innovation can be broadened to obtain greater scope.

3. Analyze others' pre-existing patent rights before proceeding with your innovation.

Chapter 7
A Twelve-Step Approach to Generating Wealth Through Innovation

Innovation Should Be a Quantitative Business Function

The following twelve steps represent one possible approach to successful innovation. Given your competitive situation, organizational parameters and the nature of the particular innovation, these steps may be performed in any logical sequence, and steps may be added or deleted. The key is to quantify and treat innovation and patent protection as any other business function, with an eye towards generating wealth.

A Twelve-Step Approach to Successful Innovation

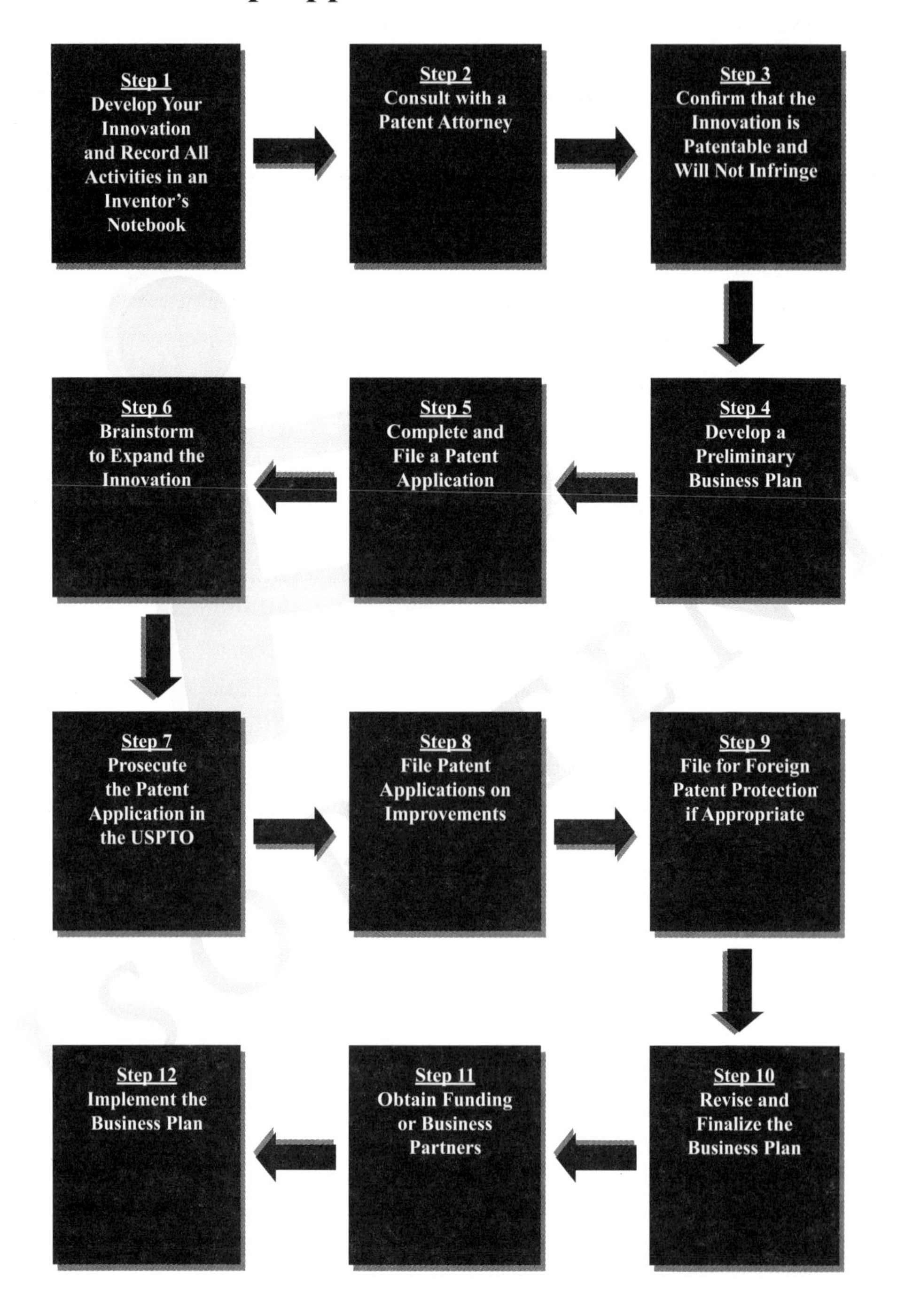

Step 1: <u>Develop Your Innovation and Record All Activities in an Inventor's Notebook</u>. This may comprise a number of sub-steps, and

should begin by identifying unmet market needs. A good source for identifying unmet market needs, and related opportunities for innovation, is your customers or potential customers. By soliciting customer input related to problems or desired product features, you can often develop related innovations. Unmet market needs can also be identified through formal customer surveys conducted by an in-house or outside marketing research group. Such a survey must be carefully developed and directed to the appropriate audience. It should be designed to minimize obvious customer responses such as "lower the price" or "provide overnight delivery free of charge," unless you strike upon a realistic way to provide such a benefit in a manner that cannot quickly be copied, or outdone, by your competitors.

Obtaining customer input to develop innovations may sound evident, but is often overlooked, done half-heartedly or too infrequently. Businesses are sometimes prone to maintain the status quo and ignore customer problems. Deflecting problems and stressing the strengths of existing products and/or settling for a low market share may be a necessary short-term solution since innovations cannot be developed, tested and brought to market overnight. Long-term failure to identify or address customer needs, however, is a formula for handing competitors your market share on a silver platter. Remember our statement in Chapter 1 of this book: *"If we cease to innovate we will stagnate and decline."* Businesses can no longer have a "business as usual" attitude. Be proactive, innovate and get ahead of the curve.

Developing your innovation is where multi-disciplinary brainstorming comes into play. Your sales force, customer service department, manufacturing personnel, product development group, marketing personnel, engineers and others should participate. Each likely has a unique perspective of your customers, competitors and your organization's capabilities. You may wish to offer incentives, such as vacation days, free dinners and/or bonuses to encourage innovation. The best approach to innovation should be carefully conceived and modified as necessary given your business objectives.

We encourage multi-disciplinary brainstorming and believe a team approach yields the best results. But, as *Some Real-World Success Stories* in Chapter 9 shows, many valuable innovations are developed by just one or a few people. They often innovate without marketing departments, R&D staffs or statistical breakdowns of customer surveys. This aptitude of gifted innovators does not weigh against, but reinforces, the multi-disciplinary team approach advocated here. For example, the gifted innovator in your organization might be a salesperson or customer service representative, and not the engineer or marketer typically responsible for product development. Or, the gifted innovator may be an engineer who is normally assigned the task of producing the marketing department's concepts, but not called upon to participate in developing innovations. Further, different team members may prove to be key innovators for different projects.

When developing your innovation, each person on the innovation team should keep a written log in a bound notebook, such as the *Inventor's Notebook*, which is shown in the back of this book. All efforts towards developing or improving the innovation should be recorded, dated and witnessed by a person who is not a coinventor. The purpose of the log is to establish invention dates because the United States grants patent rights to the first to invent rather than the first to file for patent protection.[35]

In summary, when developing an innovation, first identify unmet market needs. Then capitalize on the unmet needs using a multi-disciplinary team approach and properly record your development activities.

Step 2: <u>Consult with a Patent Attorney</u>. This should be done early and is necessary to complete steps 2, 3, 5, 7, 8, and 9. Quality, sophisticated legal advice can streamline the innovation process, generate additional ideas for expanding the innovation, save money and generate greater profits in the long run.

Step 3: <u>Confirm that the Innovation is Patentable and Will Not Infringe</u>. Confirm that your innovation is likely to be patented and estimate the probable scope of patent protection. Remember the second factor in *The Four Key Factors for Valuing Innovations*: *Anticipated Scope of Patent Protection*. Without meaningful patent protection your innovation will likely have less value, so you should be aware of the potential scope of patent protection before proceeding. If relevant, determine that your innovation will not infringe another's patent rights. If your innovation is likely to infringe another's pre-existing patent, the innovation should be modified to a non-infringing alternative or abandoned unless the pre-existing patent can be purchased or licensed. Some innovations, however, are meant to be improvements on existing patents and have value despite falling within the scope of another's patent. For example, the innovation could be licensed or sold to the owner of the pre-existing patent, or the pre-existing patent may expire soon.

Step 4: Develop a Preliminary Business Plan. This can be formal (recommended) or informal and should include all aspects of a business plan, as discussed in Chapter 8.

Step 5: Complete and File a Patent Application. Use the techniques discussed throughout this book to capture one or more broad, valuable market segments, rather than to simply obtain a patent. Hire a sophisticated, experienced patent attorney to provide you with sound advice. Be leery of attorneys who are vague, do not focus on expanding your innovation's scope, or who are overly enthusiastic about your innovation and eager to push you into filing a patent application.

Step 6: Brainstorm to Expand the Innovation. Using the techniques taught in this book, and the knowledge you gained as a result of the patent searches required for Step 3, attempt to expand the scope of the innovation using interdisciplinary brainstorming.

Step 7: Prosecute the Patent Application in the USPTO. This is a lengthy and detailed subject not discussed here. As previously stated, the key to successful patent prosecution is to move your application through the USPTO without surrendering valuable scope that could render the resulting patent worthless.

Step 8: File Patent Applications on Improvements. As more information becomes available, such as through interdisciplinary brainstorming, the scope of the innovation may be expanded to include additional concepts. Such concepts are often called "improvements."

New patent applications could be filed to cover improvements as the project matures.

Step 9: <u>File for Foreign Patent Protection if Appropriate</u>. Weigh the costs and benefits of foreign patent protection.[36] Decide if and where to file foreign patent applications based on an analysis of those countries that (1) could be potential markets or contain potential competitors, *and* (2) are likely to enforce your patent.

Step 10: <u>Revise and Finalize the Business Plan</u>. Based on information received from prosecution of your United States and (if applicable) foreign patent applications, you will have a better understanding of the scope of patent protection you are likely to receive and thus a better understanding of the legal barrier to entry that your patent(s) will provide. Your business plan can then be updated to focus on the features and benefits of your product or service that are likely to be protected. Your marketing strategy, competitive analysis and sales and profit projections should be updated accordingly.

Step 11: <u>Obtain Funding or Business Partners</u>. Secure necessary funds, find manufacturing and/or distribution sources or attempt to license or sell your patent application or patent. Your actions at this stage depend on your business strategy.

Step 12: <u>Implement the Business Plan</u>. This is an extension of Step 11. Once your patent applications are pending and you have received the required funding, implement the business plan. Adjust it as you go to account for new information, challenges and opportunities. If your plan is to sell or license your patent rights, the project may have been completed at Step 11.

In summary, these twelve steps represent one possible approach to developing innovations, protecting the innovations and generating wealth. Regardless of the approach you use, the key is to use the same discipline and care for innovation as for other business functions. Remember, innovation and patents are not scientific oddities solely for the engineering or legal department; they are valuable business tools whose sole purpose is to generate profit. They should be part of an overall business strategy and may even be the cornerstone of the strategy.

Key points to remember:

1. Make innovation a key business function, like manufacturing or sales.

2. Implement a program to generate wealth through innovations. The twelve-step approach outlined here may be used, or a hybrid of this approach, or any approach suitable for your business.

3. Consider offering employee incentives to foster innovation.

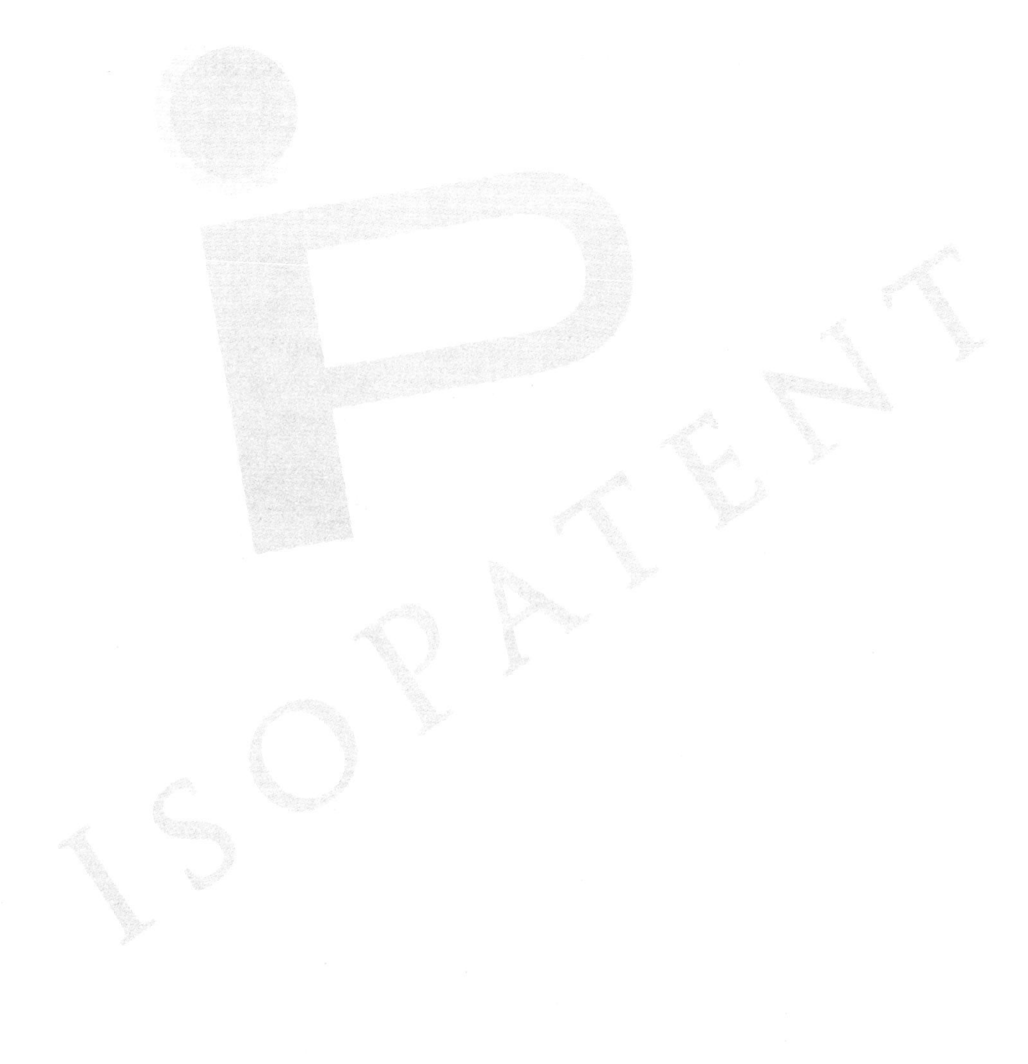

Chapter 8
You Need a Sound Business Plan

Without One You Likely Will Never Realize the Potential for Your Innovation

One common misconception is that wealth is automatically generated merely by filing a patent application or obtaining a patent. That is not the case. Even with a great innovation and great patent, the world will *not* beat a path to your door.[37] Regardless of whether you plan to manufacture, license, or sell your innovation, you need a sound business plan, clearly explaining how to generate wealth. Even if the innovation is to be developed and manufactured internally, a business plan acts as a project roadmap with a timeline, goals, costs and sales estimates, and assignment of responsibilities.

The business plan should be concise, disciplined, comprehensive, go from the big picture to the micro level, and include detailed, well-reasoned factual support for your assumptions. Expect to be competing for available resources and to have no more than a 60-minute presentation in which to convince a potential investor (inside or outside of your organization) to commit to your innovation. Poor organization, unsupported assumptions or rambling generalities in response to pointed questions is likely to elicit a “no” from potential investors.

The business plan should include a discussion of each of the following:

1. Market(s) for your innovation;

2. Target audience(s);
3. Competitors, competitive products and selling prices, and competitive patents;
4. Positioning of your innovation vis-à-vis the competitive products;
5. Manufacturing, distribution, sales and overhead costs related to your innovation;
6. Selling price of your innovation, and whether there will be different selling prices in different markets;
7. Start-up costs related to your innovation (including things such as tooling, plant, equipment and office space);
8. Timing on getting your innovation to market;
9. Key employees or business partners required (such as independent sales organizations, manufacturers or distributors);
10. Barriers to entry, including your patent protection or anticipated protection; and
11. Yearly sales and profit projections.

A typical business plan usually includes the following sections in this order:

Executive Summary

The executive summary is generally one or two pages and includes the critical components of the business plan. It should succinctly explain your innovation, marketing strategy, the financial opportunity the innovation represents, financial obligations (e.g., whether you need an investor and, if so, how much money you need), and the annual profits you expect to make over a five-year period.

The Current Market Situation

The current market situation serves as a foundation for the expected results of the plan. To prepare it properly you must gather, analyze and summarize data concerning market size and demographics, key customers, competitors, competitive products and market trends. This section should also analyze barriers to entry and include a discussion of competitors' patents.

Objectives

Objectives explain the business goal in terms of timing, market share and financials, and should include anticipated revenues, costs, margins and cash flows. Market share, quality rates, brand awareness and customer satisfaction objectives may also be included. Like running a race, in which your strategy would be different if you were running a 100-yard dash, a mile or a marathon, different business objectives dictate different business strategies.

Your objectives should be quantifiable, measurable, include a specific timeline and be aggressive, but realistic and not pie-in-the-sky. If the objectives are unrealistic you may become disenchanted with the project over time, or lose credibility with business partners and investors.

Strategy

The strategy is the game plan for achieving the objectives. It explains how your innovation will be positioned and used to generate profits, and dictates the action plan.

Action Plan

The action plan includes the tactical steps for carrying forth the strategy. It is generally written in a project management format identifying the person(s) or group(s) responsible for each aspect of the business plan, and a timeline for completion of each action step.

Financial Analysis

The financial analysis contains detailed financial information supporting the profitability of the plan and profit projections over time. The financial analysis should include a break-even analysis, a cash flow analysis, an income statement and a balance sheet by year over a five-year period.

Exit Strategy

Many start-up companies looking for outside investors (such as "angel investors" or "venture capitalists") have an "exit strategy," which is the anticipated way for the founders and early-stage investors to cash out and leave the business. Usually the exit strategy is to sell the business, either through an outright sale or sale of all or part of the business' assets. The sale is usually planned for some critical point in the business' life, such as when a particular sales target or market share is achieved, or when key technology is perfected.

Key points to remember:

1. Without a sound business plan you may never realize the full potential of your innovation.

2. Your business plan should be professional and thorough. It is a reflection of you, your innovation, and your business.

3. Go from the high-level, explaining objectives, strategy, costs, sales and profits, to the fine detail, including credible sources backing your assumptions.

4. Be ready to discuss your patent protection or other barriers to entry that give you a sustainable, competitive advantage.

5. When preparing and presenting your business plan, have an objective in mind and directly ask for the funding or business partnership sought.

Chapter 9
Some Real-World Success Stories

Innovation Works and Can Lead to Enormous Wealth

IBM is Awarded the Most Patents

In 2007, for the 15th consecutive year, IBM was awarded more U.S. patents than any other company, with a total of 3,125. With over 40,000 total patents, IBM generates over $1 billion annually in licensing revenues, which is a 3,000% increase from the $30 million it generated in 1990. Patent licensing is an integral part of its business strategy and the company actively encourages innovation.[38]

Xerox's Goal to Increase Patents Issued by One-Third

In 2005, Xerox announced a goal of increasing the number of patents it received annually by more than 30% over a three-year period. In 2007 Xerox was awarded 584 U.S. utility patents, which represented a 31-percent increase over its 2005 total.[39]

Qualcomm Pays $1 Billion for SnapTrack

Qualcomm touts its large portfolio of U.S. and foreign patents and applications and has one of the most widely licensed patent portfolios in the world.

In 2000 Qualcomm paid $1 billion for Snaptrack, a small San Jose company, to acquire wireless position location technology. In its January,

2000 press release Qualcomm stated, "Snaptrack's portfolio of 50 issued or pending patents are *critical* to wireless assisted GPS." Qualcomm apparently felt that Snaptrack's 50 issued or pending applications (not a particularly large patent portfolio) would help generate profits, through licensing and/or selling innovations falling within the scope of the patents, which would more than offset the $1 billion purchase price.[40]

In 2007, Qualcomm's revenues from patent licensing were more than $3 billion, representing over one-third of Qualcomm's total revenues.[41]

Takeda Pays $8.8 Billion for Millenium

In April 2008, Japan's largest drug company, Takeda Pharmaceutical, paid $8.8 billion for Millenium Pharmaceuticals. The deal represented a 53-percent premium over Millennium's stock price. One industry analyst called the deal "expensive" and claimed that it showed biotechnology companies "are valued more greatly by strategic buyers than they are by investors."[42]

We reviewed Millenium's U.S. patent portfolio at the time of the acquisition. Millenium had over 500 issued U.S. patents and over 500 pending, published U.S. applications. We assume Millenium had a large non-U.S. patent portfolio, as well. This large patent portfolio, with the potential to generate greater-than-normal profits from the company's innovations, was likely one of the reasons strategic buyers placed a high value on the company.

European Survey of Patent Value

In 2006, the European Commission published a comprehensive study of the value of patents entitled: "Study on Evaluating the Knowledge Economy – What are Patents Actually Worth?" The median value of the patents produced was €300,000 (about US $380,000 at the time of printing this book), with the average patent value being three million euros or about $3.8 million in U.S. dollars.[43] Those figures represent an enormous profit opportunity.

Velcro®

Velcro fasteners were conceived by Swiss inventor George de Mestral. The idea came about when, after a walk, de Mestral found burrs entangled in his clothing and he noticed the burrs had hooks that caused them to hold. The word "Velcro" comes from a combination of the words "velour" (French for "velvet") and "crochet" (French for "hooks"). In 1952, de Mestral filed his first patent application and, in 1955, established Velcro Industries. Eventually, his company sold more than 180 million feet of Velcro annually. He later sold his rights and today Velcro Industries has 160 patents and makes hundreds of various fasteners.[44]

M&Ms®

Forrest Mars developed, patented and marketed M&Ms candies and built the company now known as M&M/Mars, Inc.

In 1940, Mars developed M&Ms based on candies he had seen in Europe and took his innovation to the Hershey Corporation. There, he proposed an 80-20 partnership to Hershey whereby he would be the 80-percent partner. The M&Ms product name was based on the first initials of the last names of Mars and the president of Hershey at the time, William Murrie. The candy was patented in 1941.

The candies and later M&M/Mars were a tremendous economic success. Mars also patented a line of vending units that used an electronic recognition system instead of the typical weight drop mechanism. At the time of his death in 1999, Mars had an estimated net worth of $4 billion. As of 2007, M&M/Mars employed about 30,000 people worldwide with annual sales of more than $20 billion.[45]

Google

The Google page rank methodology was patented by Larry Page and Sergey Brin. The two men first met in 1995 at Stanford University. Today they are among the wealthiest people in the world. Their U.S. Patent No. 6,285,999 uses a series of algorithms to determine the number and relevancy of links and website content to rank web pages in a search.[46] Google began operations in 1998, and in 2007 had more than 16,000 employees and annual revenues of over $16 billion.

Intermittent Windshield Wipers

Robert Kearns invented the intermittent windshield wiper and in 1967 received the first of over 30 patents on his innovation. In 1969, Ford introduced intermittent wipers without a license from Kearns. He successfully sued Ford and collected over $10 million. He later collected over $20 million from Chrysler.[47]

Nike Airsole® Shoes

M. Frank Rudy designed the Nike Airsole shoe. The sole is hollow and pressurized with air and a small amount of inert gas. Rudy claimed that the development process took ten years and 23 unsuccessful attempts to sell his invention before striking a deal with Nike.[48] U.S. Patent Nos. 4,219,945, 4,936,029 and 6,013,340 are among the patents covering Nike's shoes.

Procter & Gamble's Strategy and the Swiffer®

Procter & Gamble Co. (P&G) had a goal to create a unique cleaning tool. So, P&G collaborated with another business called Continuum to develop a new cleaning technology. After several iterations, Continuum and P&G's personnel developed a way to "entrain" dirt in either a mop or a broom. Entrainment is an easier, neater way to clean as opposed to sweeping or mopping, because the entrainment cleaning device immediately captures dirt.

The first product launched, called the "Swiffer," utilized the technology and had strong test market results. After a successful national rollout, P&G launched the SwifferWet, followed by the SwifferWetJet. The products were a tremendous success for P&G, producing sales of $200 million in 2000.[49] By 2005, sales of the Swiffer line of products exceeded $2 billion.[50] The Swiffer is covered by one or more of United States Patent Nos. 6,305,046, 6,484,346, 6,651,290 and 6,561,354.[51]

P&G recognized that critical innovation was increasingly being done by smaller companies and individuals and set a goal to acquire 50% of its innovations from outside the company. In 2006, the company announced that by utilizing outside sources its innovation costs dropped while its innovation success rate doubled.[52]

Crocs®

The Crocs story began when three friends from Boulder, CO went sailing in the Caribbean. One of them had purchased a foam clog (a type of shoe) in Canada for the trip. He was so happy with the clog that the three of them decided to start a new business in 2003 to market it, and they purchased the manufacturer in 2004. Sales of the shoes, now known as "Crocs," increased from $1.2 million in 2003 to almost $850 million in 2007. The Crocs Company attributes much of its success to its proprietary processes, materials and intellectual property, and there are numerous utility and design patents protecting the shoes.[53]

The story does not end there. A stay-at-home mom opted to use clay and rhinestones to make charms that would fit securely in the holes of her family's numerous pairs of Crocs. She and her husband ultimately sold their business to Crocs for $10 million, plus a potential $10 million more depending upon earnings for the charms.[54]

Paul Winchell and the Artificial Heart

Paul Winchell is best known as a show business ventriloquist, but he obtained 30 patents during his life, including patents for a disposable razor, a retractable fountain pen and battery-heated gloves. Perhaps his best-known invention is an artificial heart, which he patented in 1963. Winchell donated his patent to a team of researchers led by Dr. Robert Jarvik and some of Winchell's basic principles were included in the Jarkvik-7, the first artificial heart to be successfully placed in a human, Barney Clark, in 1982.[55]

Thomas Edison and the Electric Light

Thomas Edison did not invent the light bulb. What he did invent was a practical light bulb, thereby improving on a concept that had been around for about 50 years at the time. While today most people focus on Edison's discovery of the proper filament that would work in a light bulb, Edison also invented the systems necessary for using electric lights.

The first public demonstration of Edison's electric lighting system was in December 1879, when the Menlo Park laboratory complex, where Edison worked, was electrically lighted. Edison then essentially created the electric industry from scratch.

On September 4, 1882, the first commercial power station, located on Pearl Street in lower Manhattan, went into operation providing electricity in a one-square-mile area. By the end of the 1880s, electric power stations were operating in many U.S. cities.

The success of the electric light brought Edison incredible fame and wealth. The various electric companies bearing his name were brought together in 1889 to form Edison General Electric. When Edison General Electric merged with its leading competitor, Thompson-Houston in 1892, "Edison" was dropped from the name, and the company became simply General Electric, which today is one of the world's largest corporations.[56]

The Integrated Circuit

Two separate inventors, unaware of each other's activities, invented the integrated circuit at nearly the same time.

Jack Kilby, an engineer, started working for Texas Instruments in 1958. A year earlier, research engineer Robert Noyce had co-founded the Fairchild Semiconductor Corporation.

The integrated circuit (also known as a microchip) placed the previously separated transistors, resistors, capacitors and all the connecting wiring onto a single semiconductor (or "chip"). Kilby originally used germanium and Noyce silicon for the semiconductor material.

In 1959, Kilby and Noyce both applied for patents and each received patents at about the same time. After years of legal battles the patents were cross licensed, allowing the respective businesses to manufacture integrated circuits. Today the annual market for integrated circuits is estimated at over $1 trillion.

Kilby holds over sixty patents. In 1970 he was awarded the National Medal of Science. Noyce holds sixteen patents and went on to found Intel Corporation in 1968.[57]

Monopoly®

The board game Monopoly was an improvement over other similar popular games related to renting or owning properties. In the early 1930s, Charles Darrow played such a game at a friend's house. Unemployed during the Great Depression, Darrow created his own real estate game modeled on Atlantic City, New Jersey. He made numerous innovations to his game, including color-coding the properties and deeds, and allowing the properties to be bought as well as rented. His game, *Monopoly*, was introduced in 1933 and patented in 1935.

Darrow's game, which he made at home and sold for $4 each, quickly became a success. By 1934, the game had become so successful that Darrow could no longer handle the demand himself. He presented the game to executives at Parker Brothers, but they rejected it, citing *fifty-two* design flaws. However, after hearing of the massive orders for the 1934 Christmas season, Parker Brothers changed its mind and acquired the rights. In 1936, *Monopoly* was the best selling game in America. That same year, Darrow retired a wealthy man. Today, over 200 million *Monopoly* games have been sold worldwide. Quite a testament to determination and belief in oneself despite being rejected by established industry experts.[58]

Stanley Mason

Stanley Mason invented dozens of items that Americans use every day and received over 50 patents during his life time. Mason's first invention (sold to his friends) was a clothespin fishing lure he designed when he was seven. Mason called himself an inventor of ordinary, everyday products. His inventions include the squeezable ketchup bottle, granola bars, heated pizza boxes, heatproof plastic microwave cookware, dental floss dispensers, and instant splints and casts for broken limbs.

In 1973, Mason founded Simco, Inc., which has successfully developed products for many Fortune 500 companies.[59]

Liquid Paper®

Bette Nesmith Graham, mother of Michael Nesmith of The Monkees, created Liquid Paper, which is still used extensively today. Graham, a high-school graduate and single mother, worked as a secretary. By the 1950s electric typewriters were becoming popular and Graham found typos difficult to erase and thought it would be easier to simply cover them. So she developed a formula using tempera water-based paint, tinting it to match her employer's stationery, and experimented with it using a watercolor brush to paint over mistakes.

The formula worked and soon other secretaries at her company began asking for it. Graham began mixing batches of the formula in her kitchen and called it "Mistake Out."

She soon started the Mistake Out Company in her home, mixing the formula and employing her son, Michael, and his friends. She became engrossed in her own business and was eventually fired from her secretarial job. She then devoted all of her efforts to Mistake Out.

Graham patented her formula and renamed it "Liquid Paper" in 1958. By 1969, Liquid Paper was manufacturing a million bottles of product annually. In 1979, Graham sold her company to the Gillette Corporation for $47.5 million.[60]

"Patent Trolls" and Jerome Lemelson

According to Norwegian fable, a "troll" is an ogre that lives under a bridge and charges people a fare to allow them to cross the bridge. The term "patent troll" was coined by large businesses forced to pay individual inventors or small businesses a fare (in the form of a patent licensing fee) to "cross the bridge," which in patent parlance means allowing the large company to manufacture, sell or use products falling within the scope of the patent troll's patent. Patent trolls usually do not manufacture or sell products, but simply look for opportunities to assert their patent rights and extract a licensing fee. While large businesses may not like this practice, the patent troll is simply enforcing the legal rights granted by the patent.

Some patent trolls use the cost of litigation to force large organizations to pay licensing fees. As an example, a patent troll might charge a licensing fee of $100,000 knowing that it would cost the large organization $500,000 or more to challenge the troll's patent in court. Thus, the large organization often settles to stave off the cost of litigation and not because the troll has strong patent rights. In that case, the patent troll may receive an unjust windfall just by working the legal system.

Regardless of your personal feelings, the troll's tactics are entirely legal. Large businesses employ similar tactics by creating huge patent portfolios with a labyrinth of overlapping and complex protections that would be extremely expensive, if not virtually impossible, to challenge. Others are then forced to license the large business' portfolio rather than battle it out in court. That is simply part of understanding the patent system and using it to generate wealth, and there is nothing stopping small businesses or individuals from utilizing similar strategies.

Jerome Lemelson (July 18, 1923 – October 1, 1997) was one of the most prolific American inventors in history, and was perhaps the most well known patent troll. His inventions included VCRs, bar code scanners, fax machines, camcorders, ATM's, dolls and toy masks. Over 600 patents have been issued to Lemelson.

Many large organizations considered Lemelson to be the ultimate abuser of the patent system because (to our knowledge) he never manufactured anything. He merely extracted profits from businesses providing products or services falling within the scope of his patents. Over $1 billion has been made by licensing and selling Lemelson's patents.[61]

So like him or loathe him, Lemelson used innovation, imagination and the patent system to create a fortune.

Chapter 10
Twelve Interesting Facts About Patents and Patent Value

Some Key Information for Businesses and Entrepreneurs

1. The number one business for obtaining patents in the United States and the world is IBM.[62]

2. The largest category in the United States for being awarded patents is not any single business, university or governmental agency, but *individual inventors*, who accounted for over 14,000 of all United States patents issued in 2007.[63]

3. Not all patents are valuable. The value of a patent, or group of patents, varies widely depending on the inherent value of the underlying innovation and the skill, imagination and care of the persons (especially the attorney) preparing and prosecuting the patent application(s). The inclusion or exclusion of even small amounts of information can mean the difference between a worthless piece of paper and a patent worth millions.

4. Patents can have a greater impact on the value of start-ups and small companies than they do on larger companies. Sometimes patents, or patent applications, represent the majority of the value of a small business.

5. You should not publicly disclose your innovation prior to filing a patent application. Doing so will bar you from obtaining patent protection in most of the world and start the running of a one-year grace period in the United States, Canada and Mexico.[64] Additionally, a public disclosure before filing an application is an invitation for the unscrupulous to steal your innovation.

6. A United States patent provides protection only in the United States. International patent protection is also available on a country-by-country basis. Even if you file for and obtain a "regional patent," such as a European patent, which can potentially cover as many as 38 European countries, the European patent must be registered in each European country in which you would like patent protection.[65]

7. You do not have to build a working prototype of your innovation to obtain a patent.

8. A pending patent application confers no right that can be enforced against others. You must ultimately obtain an issued patent to have enforceable rights.

9. Pending patent applications can be, and frequently are, sold.

10. Obtaining a patent does not mean you are free to manufacture or sell the innovation covered by the patent. As explained previously in *The Four Key Factors for Valuing Innovations*, you must first ensure that making, selling, offering to sell, importing or using the innovation does not infringe the patent rights of another. Consult a good patent attorney before embarking on a new project.

11. More patents are usually better than one. Like large, patent-savvy businesses, you would prefer to create a labyrinth of protection that is difficult and expensive to navigate, thus creating a large barrier to entry and increasing the value of your patent portfolio. It may also be more difficult to challenge

the "enforceability"[66] of multiple patents, and multiple patent applications (with relatively few claims per application) are sometimes easier and less expensive to move through the USPTO.

12. Do not be lulled into a false sense of security by filing a document called a provisional patent application ("PPA").[67] A PPA only potentially protects your innovation to the extent it (a) thoroughly describes the innovation, and (b) to ultimately obtain protection in some foreign jurisdictions, includes statements (similar to claims in a utility patent application) of various specific examples of the innovation. Many PPAs are prepared in haste with little thought about capturing broad patent scope or foreign patent protection.

PART II

A Statistical Breakdown of Who is Patenting What in the United States

Chapter 11
An Insider's Guide to the World of United States Patents

Great Competitive Information for Businesses and Entrepreneurs

We hope you will find the following statistics informative. They summarize who is obtaining patents, the technological categories in which most patents are issued and where opportunities may exist. They also suggest that there is a positive correlation between patents and income level, and that many large, profitable businesses are heavily focused on obtaining patents.

Whether you are an individual interested in profiting from innovation or a business looking for a competitive edge, we believe you will find this information helpful in formulating a game plan for business success through innovation.

Key points in the following statistics:

1. The top ten businesses in obtaining United States patents are IBM, Samsung Electronics, Canon Kabushiki Kaisha, Matsushita Electric Industrial Co., Intel, Microsoft, Toshiba, Micron Technology, Hewlett-Packard, and Sony.

2. The top ten universities in obtaining United States patents are the University of California, Massachusetts Institute of

Technology, California Institute of Technology, Stanford University, University of Texas, University of Wisconsin, Johns Hopkins University, University of Michigan, University of Florida, and Columbia University.

3. The top technology categories in which United States patents are granted include:

   - Drug, Bio-Affecting and Body Treating Compositions;
   - Semiconductor Device Manufacturing: Process;
   - Active Solid-State Devices;
   - Telecommunications;
   - Multiplex Communications;
   - Chemistry: Molecular Biology and Microbiology;
   - Computer Graphics Processing and Selective Visual Display Systems;
   - Organic Compounds;
   - Multicomputer Data Transferring; and
   - Static Information Storage and Retrieval.

4. Japanese businesses and individuals receive about 20% of all United States patents.

5. Businesses and individuals in California receive by far the most patents of any state in the United States, with a total of 22,601 issued in 2007. The next closest states are Texas and New York, each with slightly more than one-fourth of the patents issued to California residents.

6. Idaho received the most patents per capita of any state at about 93 patents per 100,000 residents.

Chapter 12
GDP of Selected Countries and Overall United States Patent Data[68]

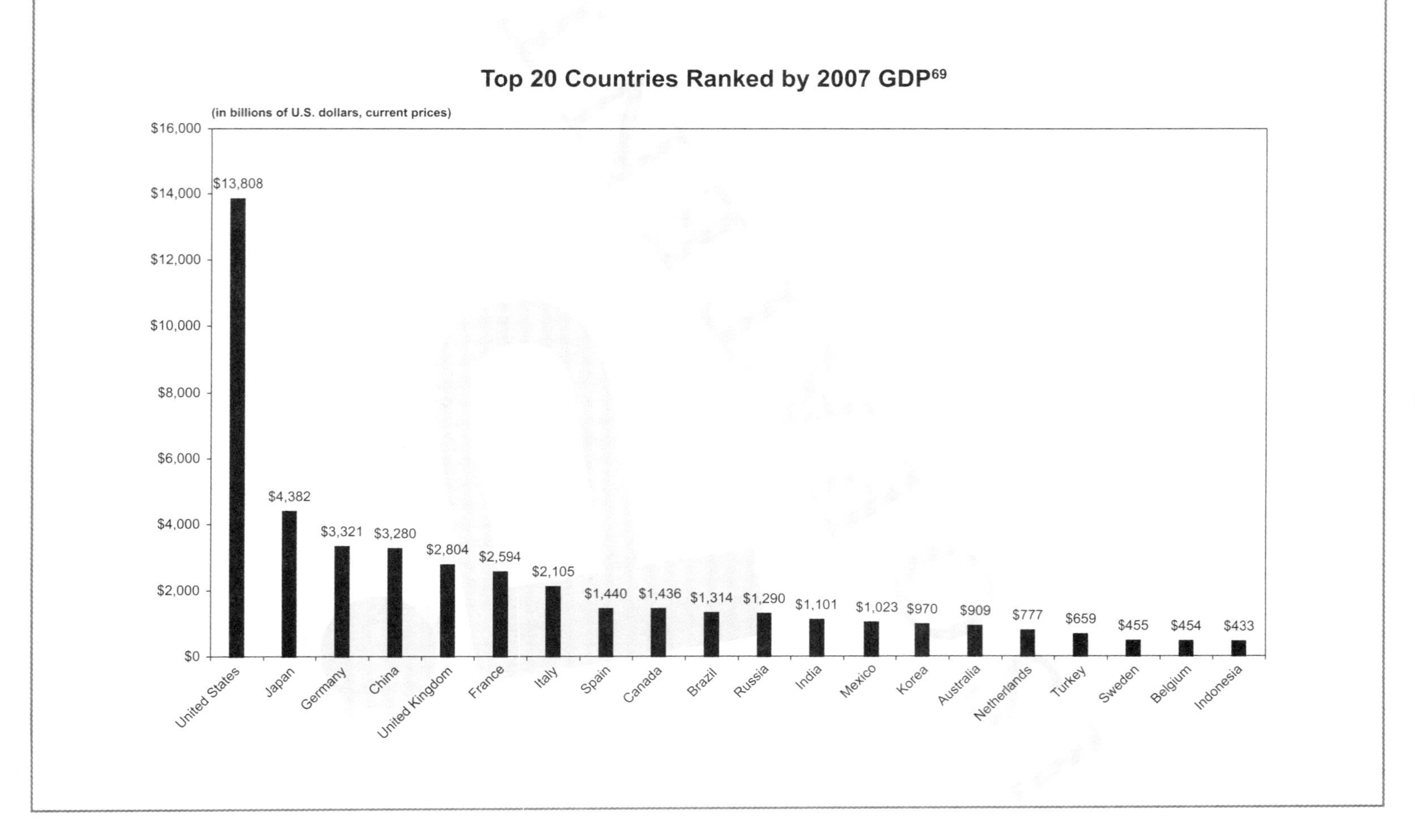
Top 20 Countries Ranked by 2007 GDP[69]
(in billions of U.S. dollars, current prices)
$16,000
$14,000
$12,000
$10,000
$8,000
$6,000
$4,000
$2,000
$0
$13,808
$4,382
$3,321
$3,280
$2,804
$2,594
$2,105
$1,440
$1,436
$1,314
$1,290
$1,101
$1,023
$970
$909
$777
$659
$455
$454
$433
United States
Japan
Germany
China
United Kingdom
France
Italy
Spain
Canada
Brazil
Russia
India
Mexico
Korea
Australia
Netherlands
Turkey
Sweden
Belgium
Indonesia

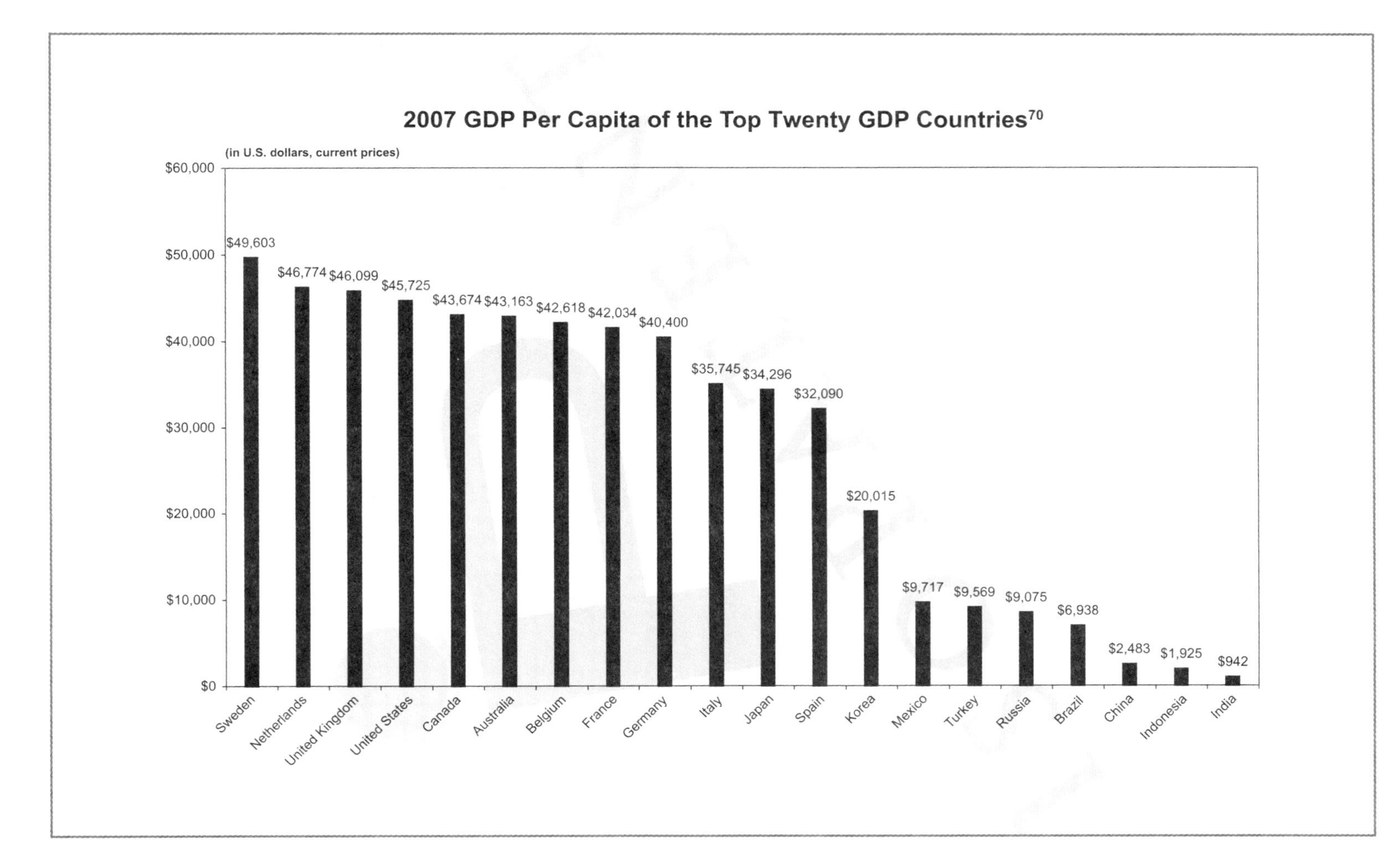

2007 GDP Per Capita of the Top Twenty GDP Countries70
(in U.S. dollars, current prices)
$60,000
$50,000
$40,000
$30,000
$20,000
$10,000
$0
$49,603
$46,774
$46,099
$45,725
$43,674
$43,163
$42,618
$42,034
$40,400
$35,745
$34,296
$32,090
$20,015
$9,717
$9,569
$9,075
$6,938
$2,483
$1,925
$942
Sweden
Netherlands
United Kingdom
United States
Canada
Australia
Belgium
France
Germany
Italy
Japan
Spain
Korea
Mexico
Turkey
Russia
Brazil
China
Indonesia
India

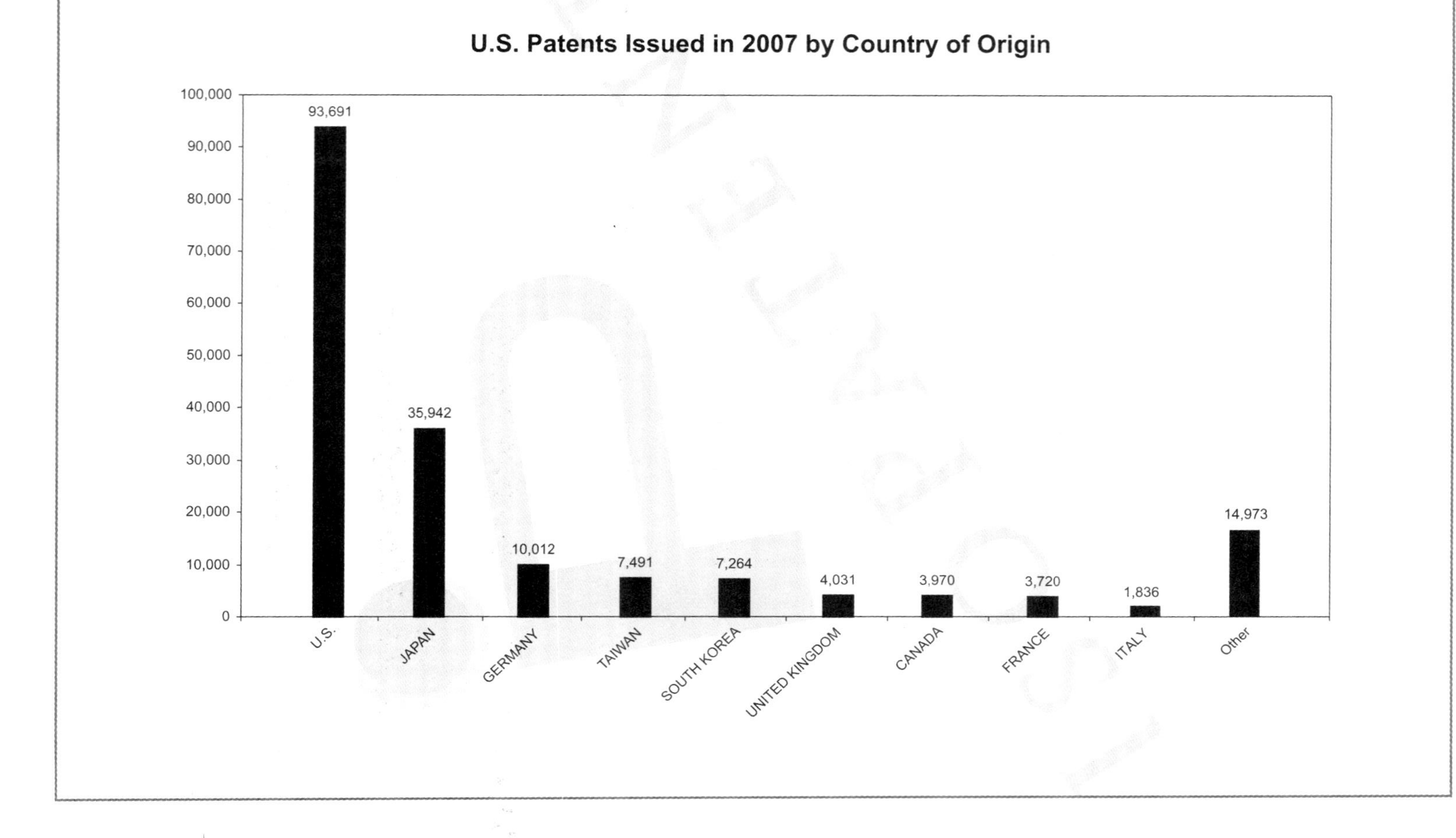
U.S. Patents Issued in 2007 by Country of Origin
100,000
90,000
80,000
70,000
60,000
50,000
40,000
30,000
20,000
10,000
0
93,691
35,942
10,012
7,491
7,264
4,031
3,970
3,720
1,836
14,973
U.S.
JAPAN
GERMANY
TAIWAN
SOUTH KOREA
UNITED KINGDOM
CANADA
FRANCE
ITALY
Other

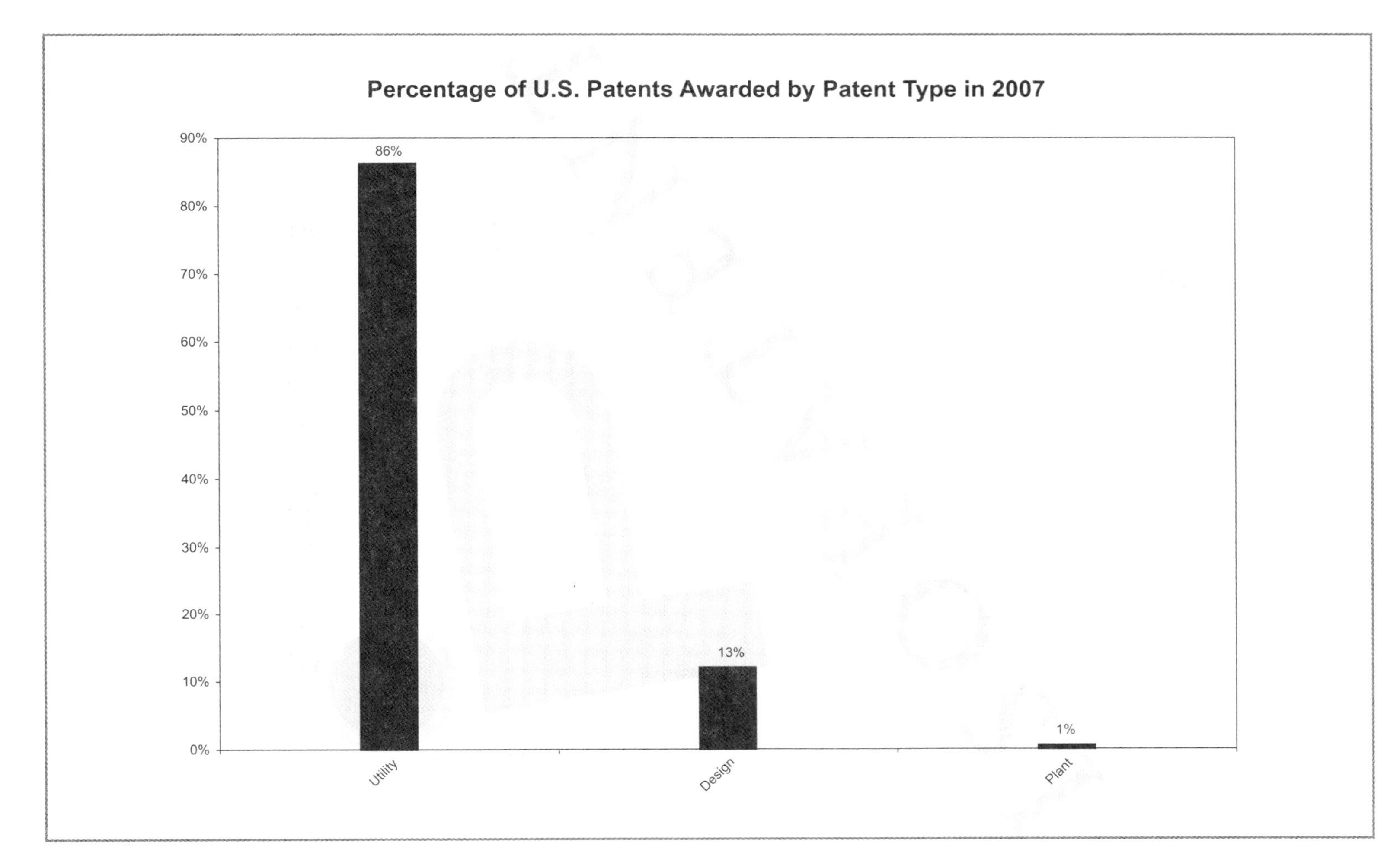
Percentage of U.S. Patents Awarded by Patent Type in 2007
90%
80%
70%
60%
50%
40%
30%
20%
10%
0%
86%
13%
1%
Utility
Design
Plant

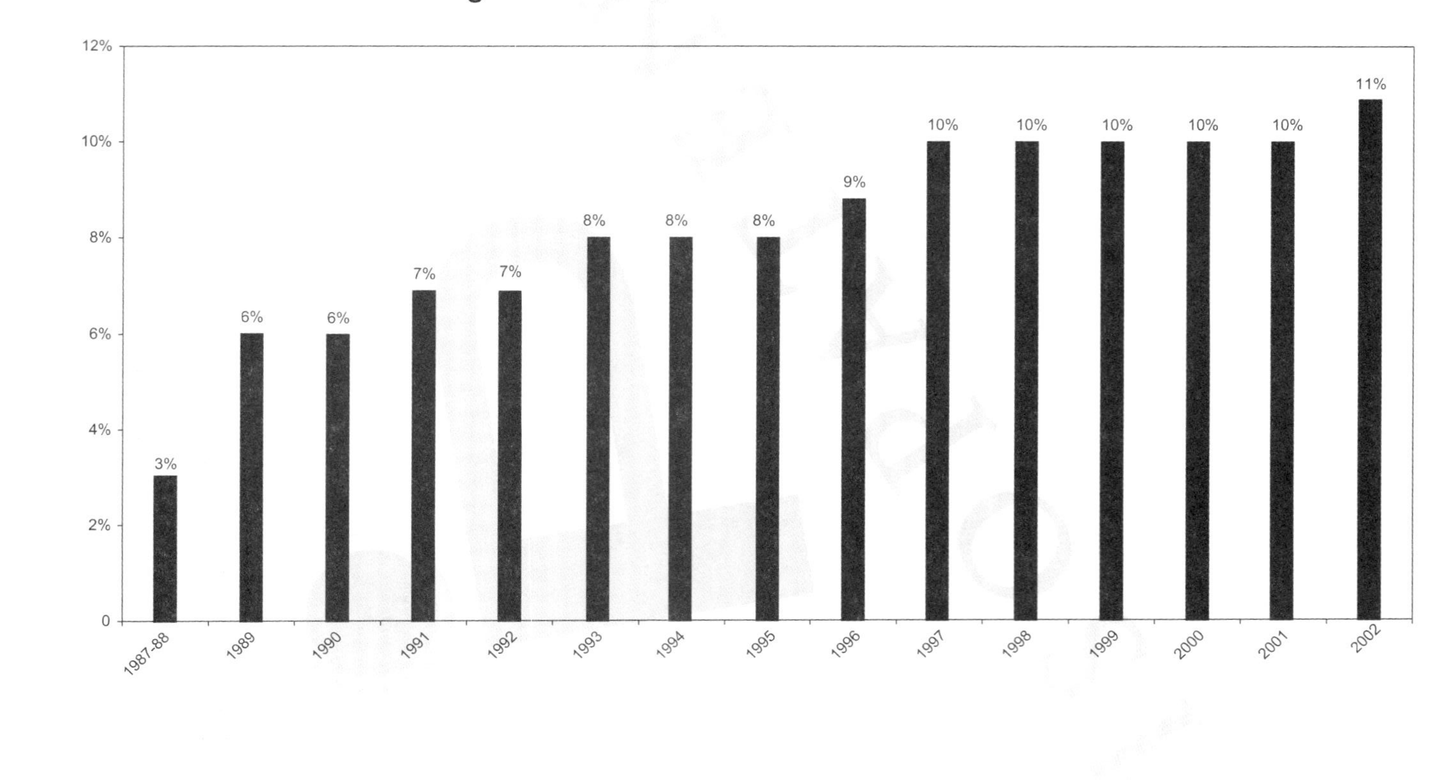

Percentage of U.S. Patents with at least One Woman Inventor
12%
10%
8%
6%
4%
2%
0
3%
6%
6%
7%
7%
8%
8%
8%
9%
10%
10%
10%
10%
10%
11%
1987-88
1989
1990
1991
1992
1993
1994
1995
1996
1997
1998
1999
2000
2001
2002

Chapter 13
The Top Ten in United States Patents

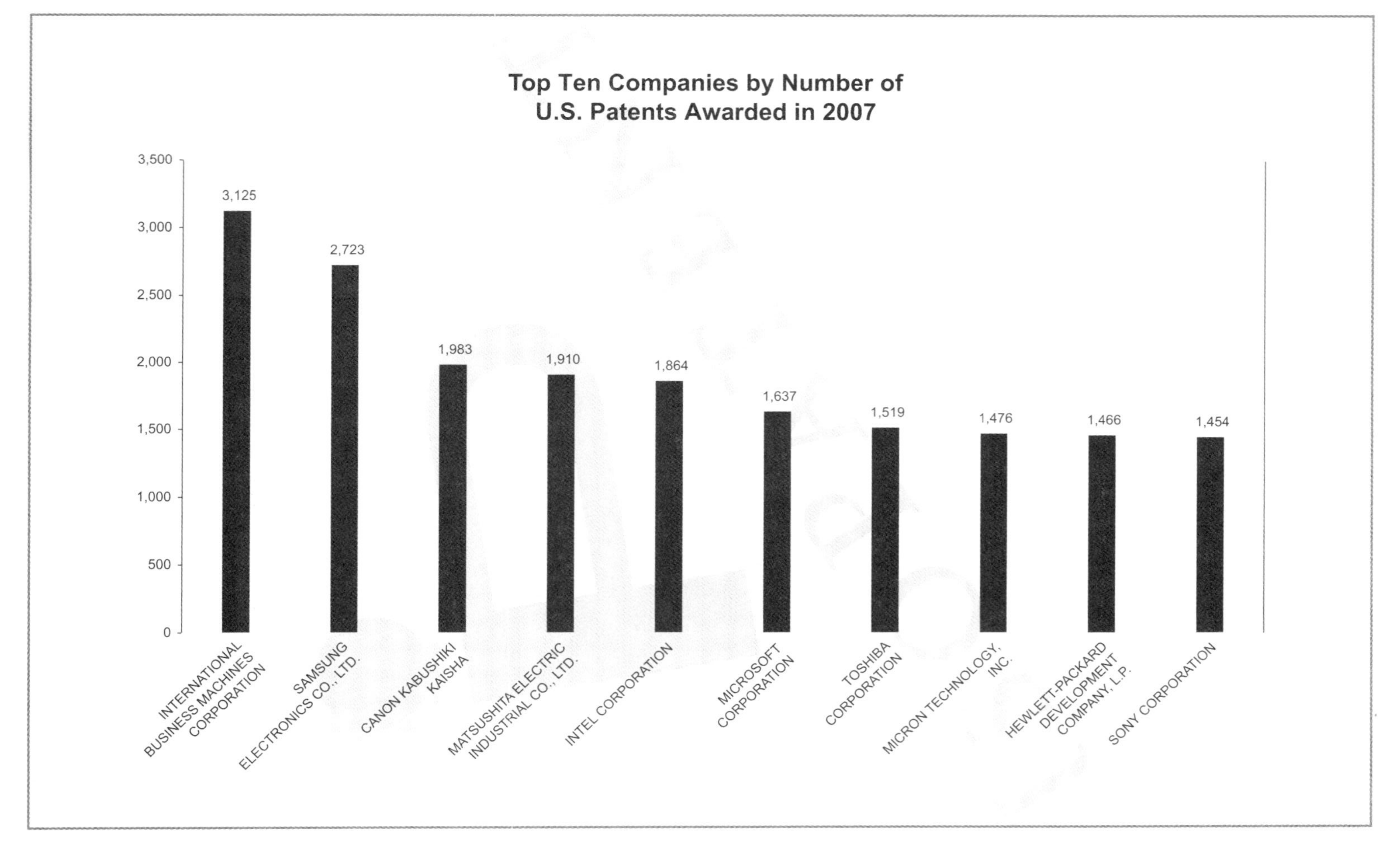

Top Ten Companies by Number of U.S. Patents Awarded in 2007
3,500
3,000
2,500
2,000
1,500
1,000
500
0
3,125
2,723
1,983
1,910
1,864
1,637
1,519
1,476
1,466
1,454
INTERNATIONAL BUSINESS MACHINES CORPORATION
SAMSUNG ELECTRONICS CO., LTD.
CANON KABUSHIKI KAISHA
MATSUSHITA ELECTRIC INDUSTRIAL CO., LTD.
INTEL CORPORATION
MICROSOFT CORPORATION
TOSHIBA CORPORATION
MICRON TECHNOLOGY, INC.
HEWLETT-PACKARD DEVELOPMENT COMPANY, L.P.
SONY CORPORATION

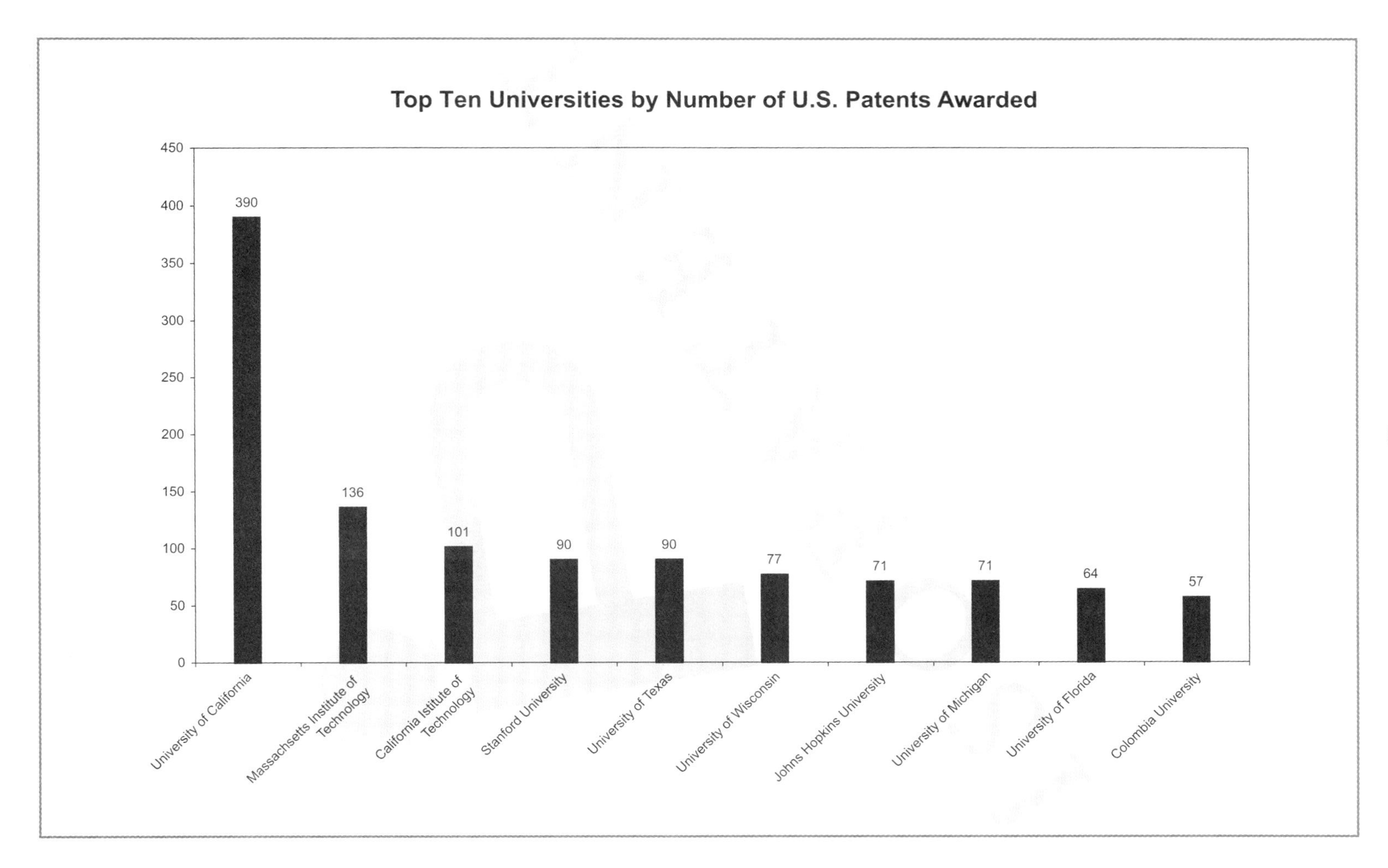

Top Ten Universities by Number of U.S. Patents Awarded
450
400
350
300
250
200
150
100
50
0
390
136
101
90
90
77
71
71
64
57
University of California
Massachsetts Institute of Technology
California Istitute of Technology
Stanford University
University of Texas
University of Wisconsin
Johns Hopkins University
University of Michigan
University of Florida
Colombia University

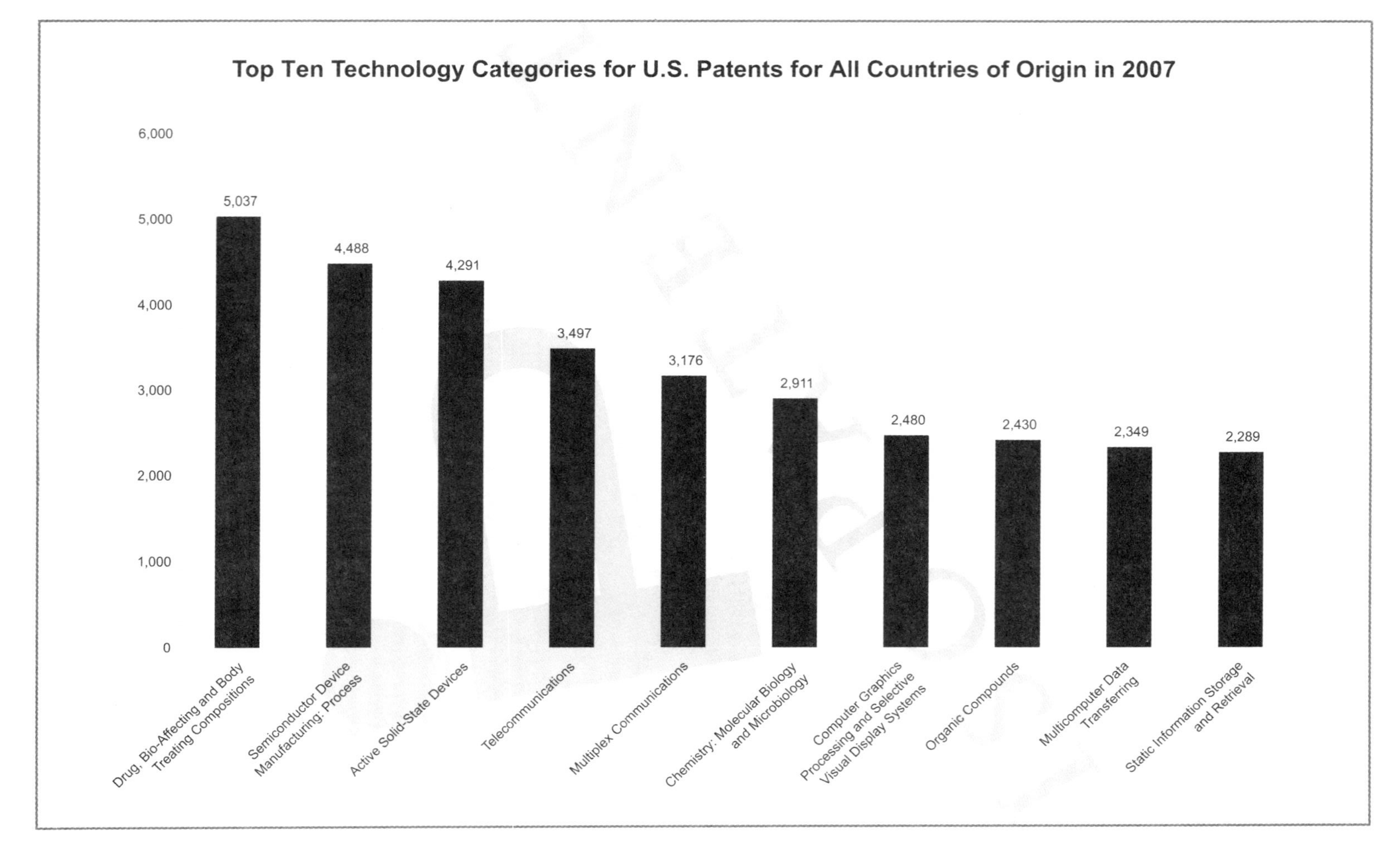

Top Ten Technology Categories for U.S. Patents for All Countries of Origin in 2007
6,000
5,000
4,000
3,000
2,000
1,000
0
5,037
4,488
4,291
3,497
3,176
2,911
2,480
2,430
2,349
2,289
Drug, Bio-Affecting and Body Treating Compositions
Semiconductor Device Manufacturing: Process
Active Solid-State Devices
Telecommunications
Multiplex Communications
Chemistry: Molecular Biology and Microbiology
Computer Graphics Processing and Selective Visual Display Systems
Organic Compounds
Multicomputer Data Transferring
Static Information Storage and Retrieval

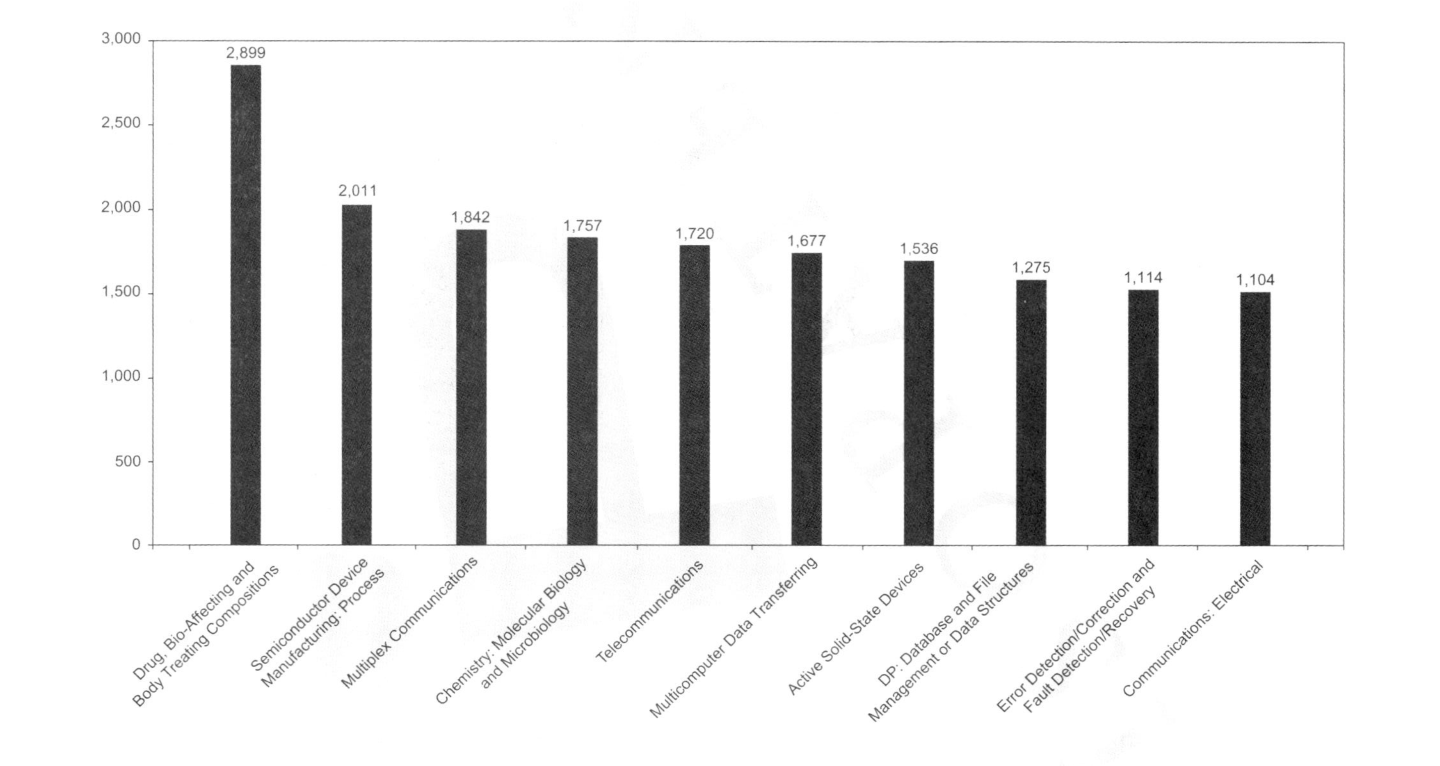

Top Ten Technology Categories for U.S. Patents for United States Country of Origin in 2007
3,000
2,500
2,000
1,500
1,000
500
0
2,899
2,011
1,842
1,757
1,720
1,677
1,536
1,275
1,114
1,104
Drug, Bio-Affecting and Body Treating Compositions
Semiconductor Device Manufacturing: Process
Multiplex Communications
Chemistry: Molecular Biology and Microbiology
Telecommunications
Multicomputer Data Transferring
Active Solid-State Devices
DP: Database and File Management or Data Structures
Error Detection/Correction and Fault Detection/Recovery
Communications: Electrical

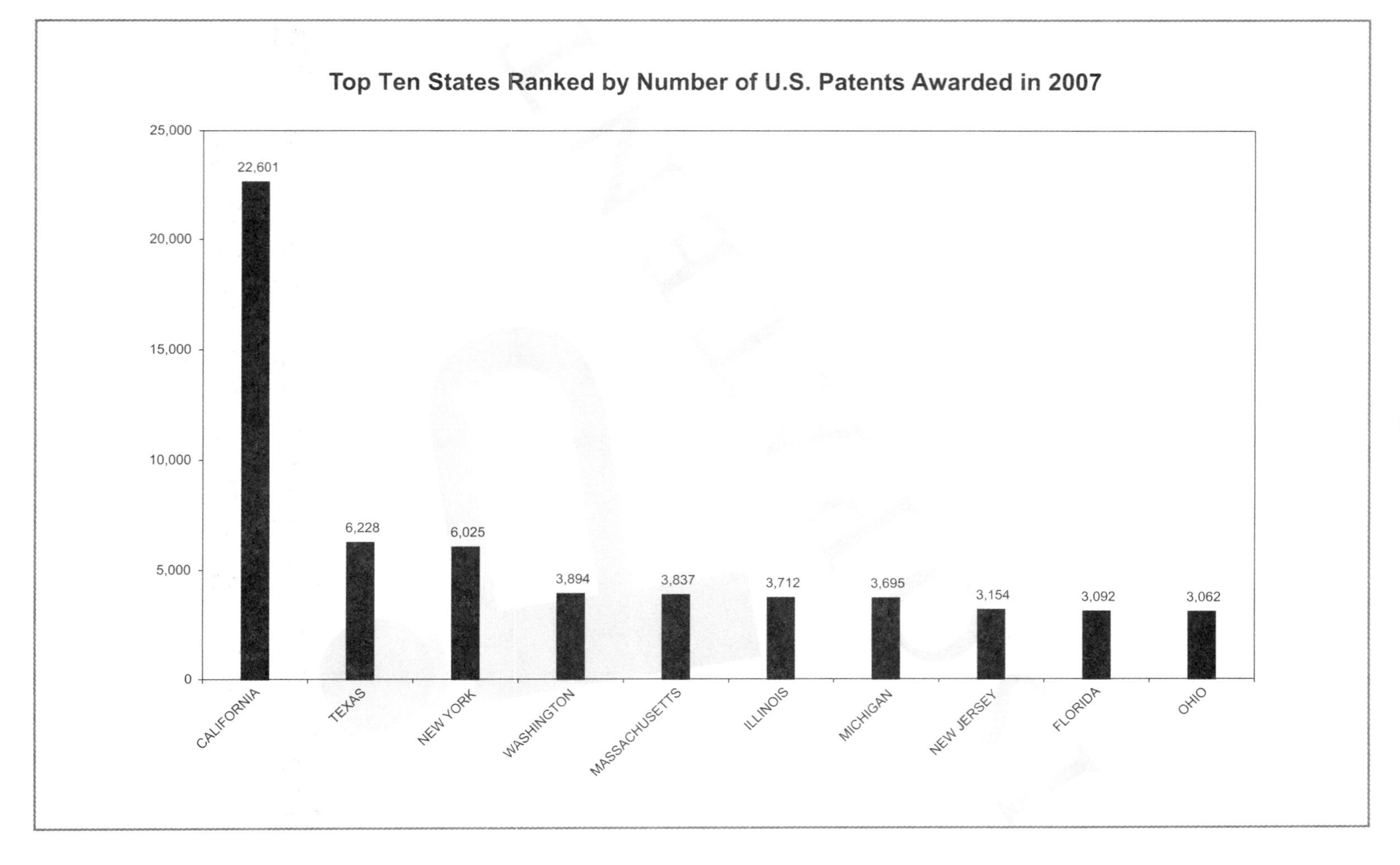

Top Ten States Ranked by Number of U.S. Patents Awarded in 2007
25,000
20,000
15,000
10,000
5,000
0
22,601
6,228
6,025
3,894
3,837
3,712
3,695
3,154
3,092
3,062
CALIFORNIA
TEXAS
NEW YORK
WASHINGTON
MASSACHUSETTS
ILLINOIS
MICHIGAN
NEW JERSEY
FLORIDA
OHIO

Chapter 14
Top United States Patent Recipients by Technology Category

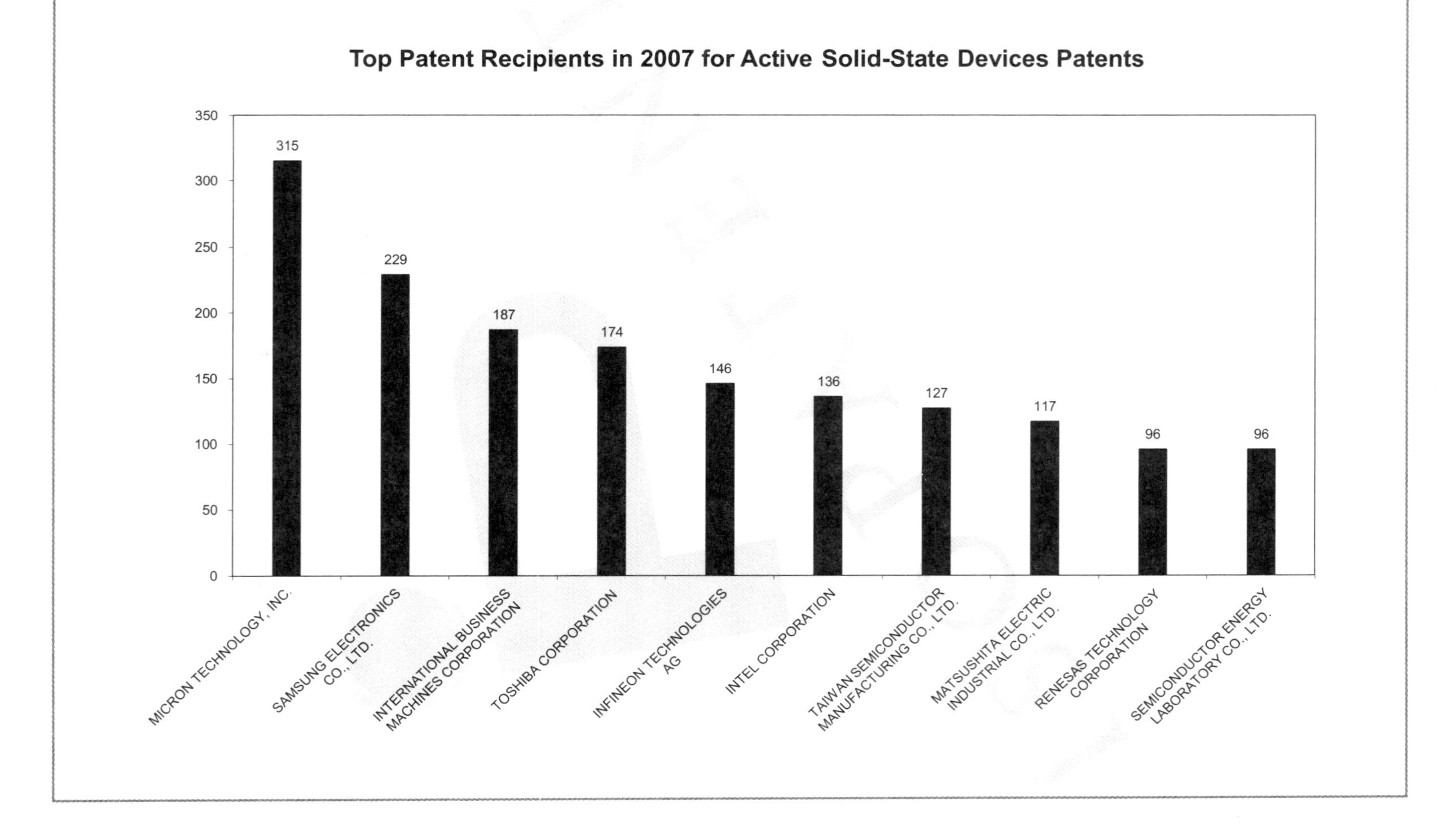
Top Patent Recipients in 2007 for Active Solid-State Devices Patents
350
300
250
200
150
100
50
0
315
229
187
174
146
136
127
117
96
96
MICRON TECHNOLOGY, INC.
SAMSUNG ELECTRONICS CO., LTD.
INTERNATIONAL BUSINESS MACHINES CORPORATION
TOSHIBA CORPORATION
INFINEON TECHNOLOGIES AG
INTEL CORPORATION
TAIWAN SEMICONDUCTOR MANUFACTURING CO., LTD.
MATSUSHITA ELECTRIC INDUSTRIAL CO., LTD.
RENESAS TECHNOLOGY CORPORATION
SEMICONDUCTOR ENERGY LABORATORY CO., LTD.

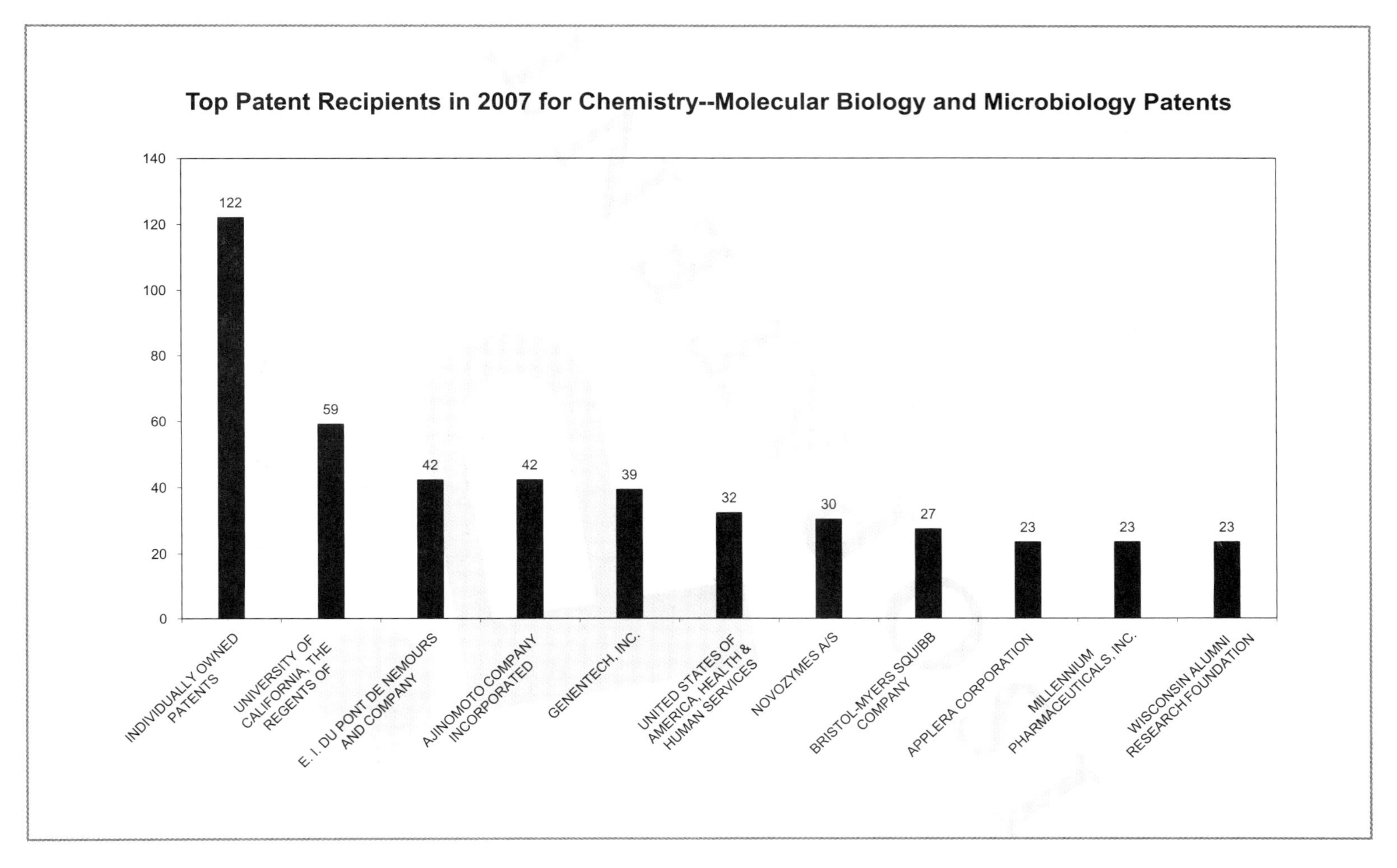

Top Patent Recipients in 2007 for Chemistry--Molecular Biology and Microbiology Patents
140
120
100
80
60
40
20
0
122
59
42
42
39
32
30
27
23
23
23
INDIVIDUALLY OWNED PATENTS
UNIVERSITY OF CALIFORNIA, THE REGENTS OF
E. I. DU PONT DE NEMOURS AND COMPANY
AJINOMOTO COMPANY INCORPORATED
GENENTECH, INC.
UNITED STATES OF AMERICA, HEALTH & HUMAN SERVICES
NOVOZYMES A/S
BRISTOL-MYERS SQUIBB COMPANY
APPLERA CORPORATION
MILLENNIUM PHARMACEUTICALS, INC.
WISCONSIN ALUMNI RESEARCH FOUNDATION

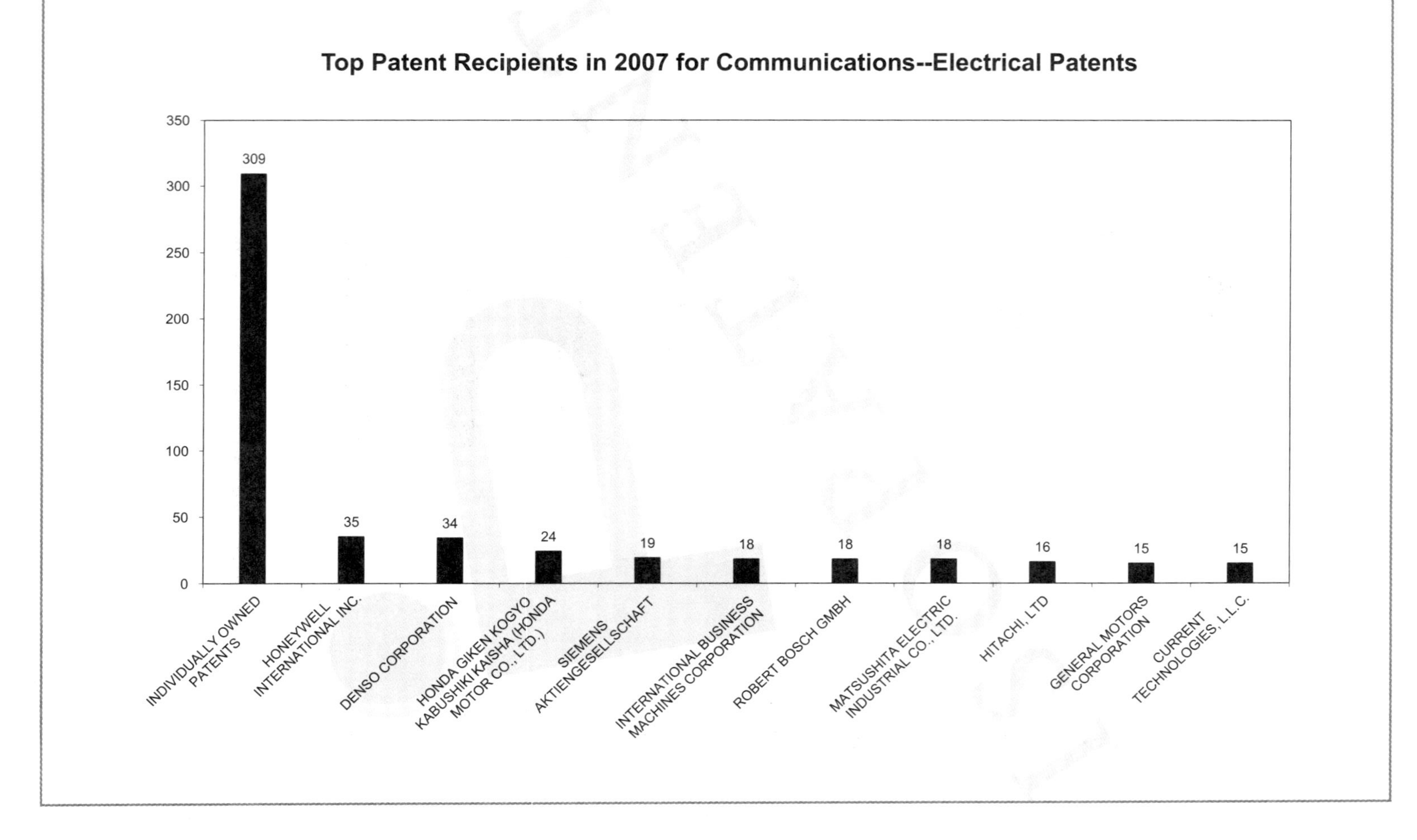

Top Patent Recipients in 2007 for Communications--Electrical Patents
350
300
250
200
150
100
50
0
309
35
34
24
19
18
18
18
16
15
15
INDIVIDUALLY OWNED PATENTS
HONEYWELL INTERNATIONAL INC.
DENSO CORPORATION
HONDA GIKEN KOGYO KABUSHIKI KAISHA (HONDA MOTOR CO., LTD.)
SIEMENS AKTIENGESELLSCHAFT
INTERNATIONAL BUSINESS MACHINES CORPORATION
ROBERT BOSCH GMBH
MATSUSHITA ELECTRIC INDUSTRIAL CO., LTD.
HITACHI, LTD
GENERAL MOTORS CORPORATION
CURRENT TECHNOLOGIES, L.L.C.

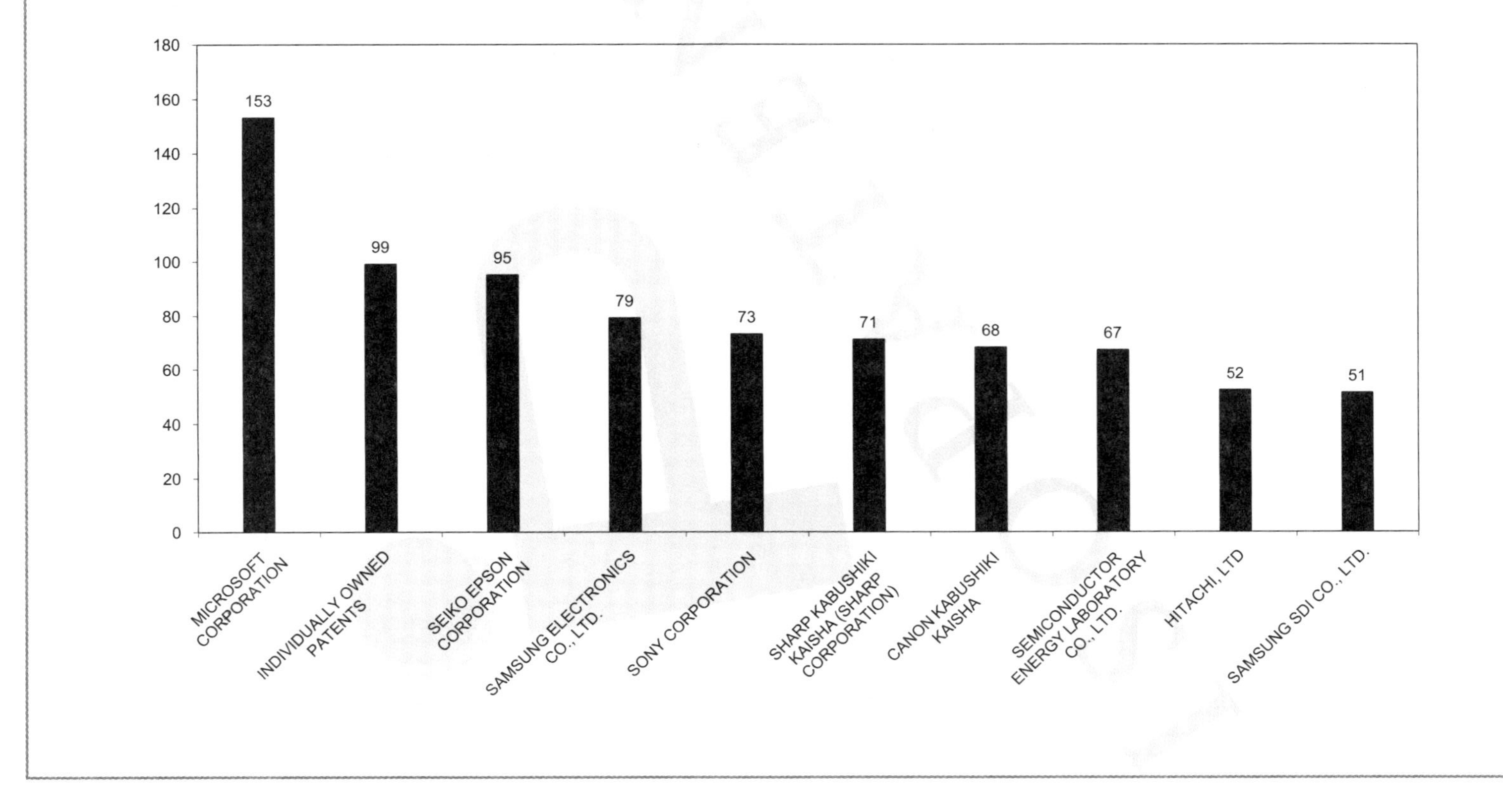
Top Patent Recipients in 2007 for Computer Graphics Processing and Selective Visual Display Systems Patents
180
160
140
120
100
80
60
40
20
0
153
99
95
79
73
71
68
67
52
51
MICROSOFT CORPORATION
INDIVIDUALLY OWNED PATENTS
SEIKO EPSON CORPORATION
SAMSUNG ELECTRONICS CO., LTD.
SONY CORPORATION
SHARP KABUSHIKI KAISHA (SHARP CORPORATION)
CANON KABUSHIKI KAISHA
SEMICONDUCTOR ENERGY LABORATORY CO., LTD.
HITACHI, LTD
SAMSUNG SDI CO., LTD.

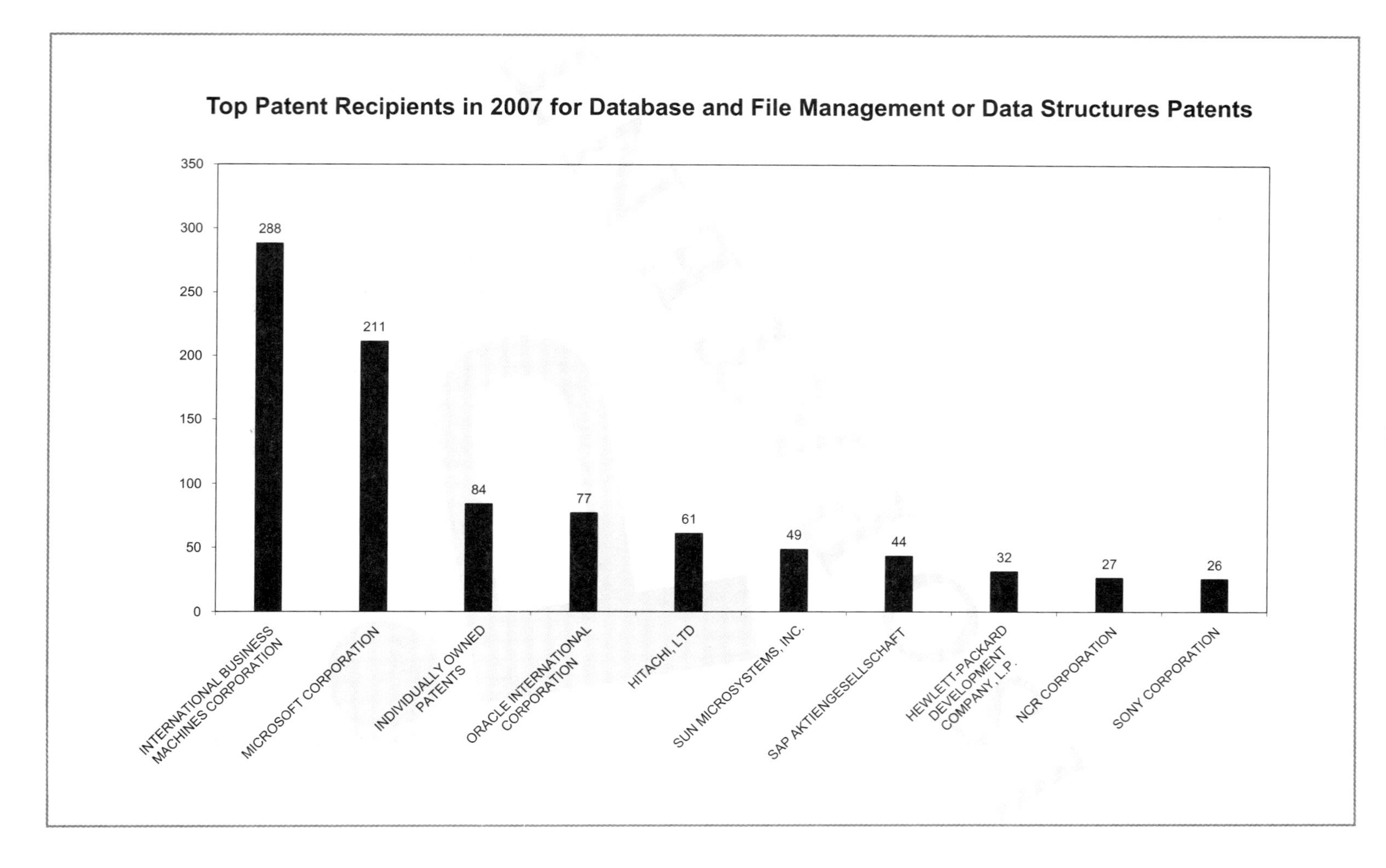

Top Patent Recipients in 2007 for Database and File Management or Data Structures Patents
350
300
250
200
150
100
50
0
288
211
84
77
61
49
44
32
27
26
INTERNATIONAL BUSINESS MACHINES CORPORATION
MICROSOFT CORPORATION
INDIVIDUALLY OWNED PATENTS
ORACLE INTERNATIONAL CORPORATION
HITACHI, LTD
SUN MICROSYSTEMS, INC.
SAP AKTIENGESELLSCHAFT
HEWLETT-PACKARD DEVELOPMENT COMPANY, L.P.
NCR CORPORATION
SONY CORPORATION

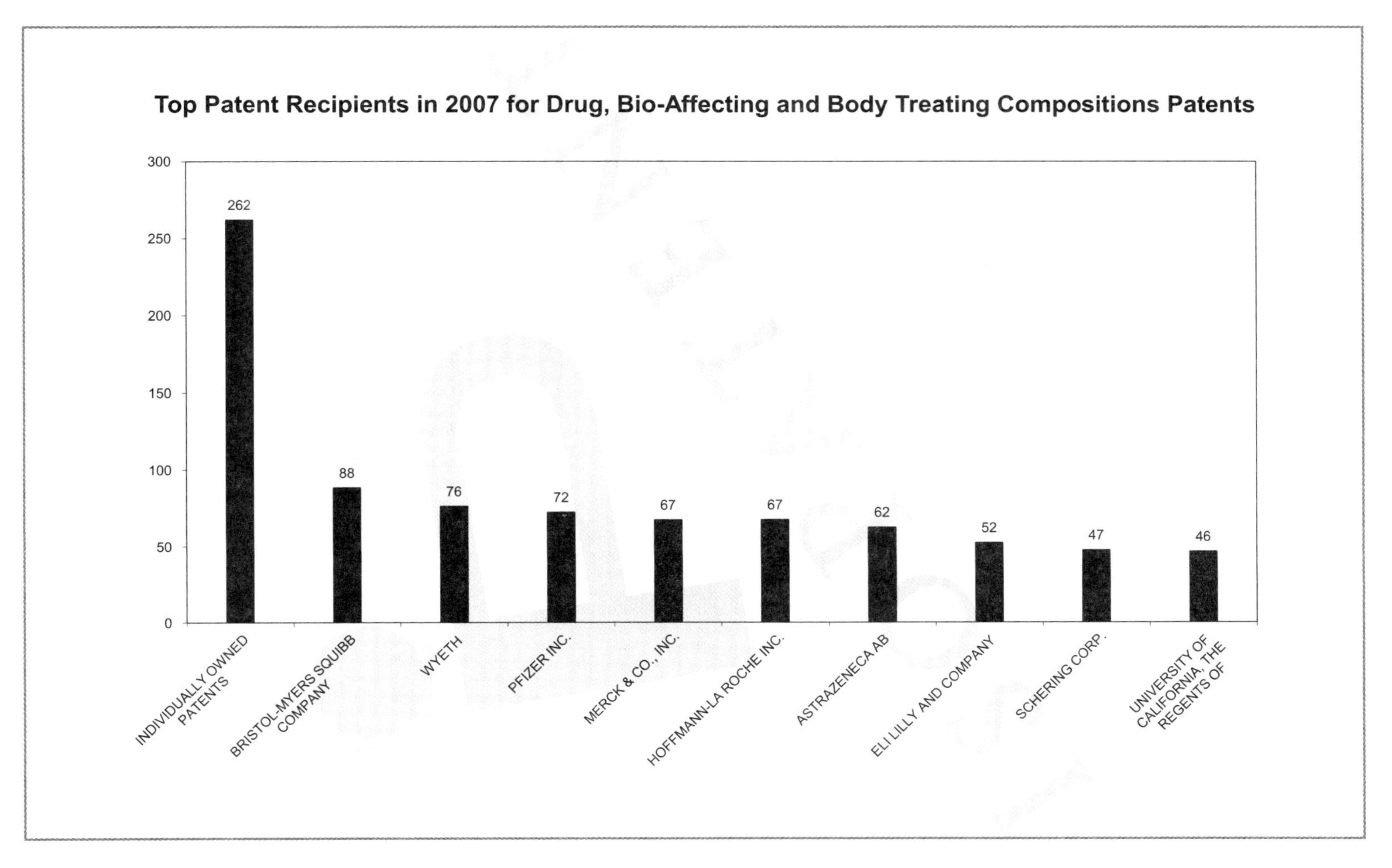
Top Patent Recipients in 2007 for Drug, Bio-Affecting and Body Treating Compositions Patents
300
250
200
150
100
50
0
262
88
76
72
67
67
62
52
47
46
INDIVIDUALLY OWNED PATENTS
BRISTOL-MYERS SQUIBB COMPANY
WYETH
PFIZER INC.
MERCK & CO., INC.
HOFFMANN-LA ROCHE INC.
ASTRAZENECA AB
ELI LILLY AND COMPANY
SCHERING CORP.
UNIVERSITY OF CALIFORNIA, THE REGENTS OF

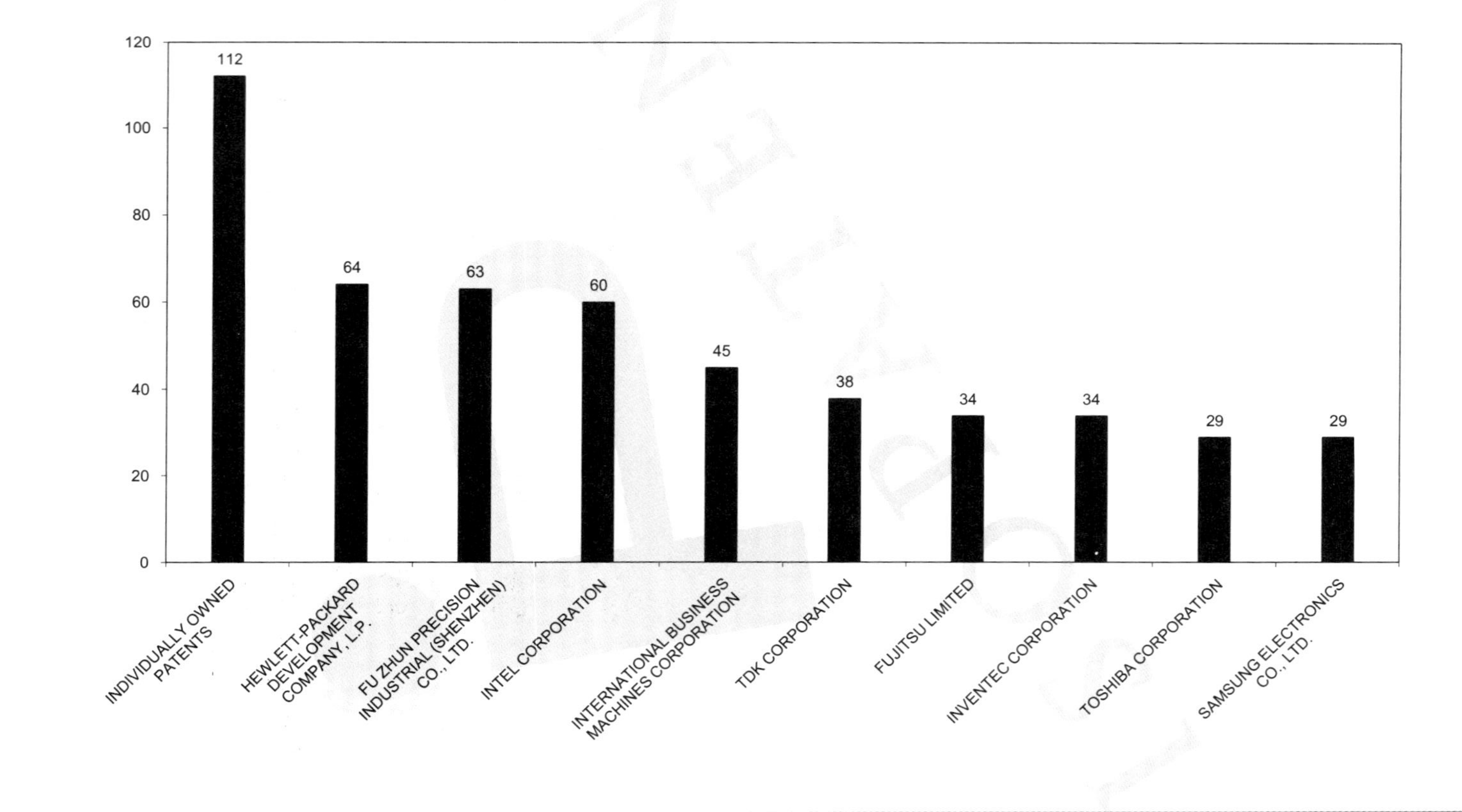
Top Patent Recipients in 2007 for Electricity--Electrical Systems and Devices Patents
120
100
80
60
40
20
0
112
64
63
60
45
38
34
34
29
29
INDIVIDUALLY OWNED PATENTS
HEWLETT-PACKARD DEVELOPMENT COMPANY, L.P.
FU ZHUN PRECISION INDUSTRIAL (SHENZHEN) CO., LTD.
INTEL CORPORATION
INTERNATIONAL BUSINESS MACHINES CORPORATION
TDK CORPORATION
FUJITSU LIMITED
INVENTEC CORPORATION
TOSHIBA CORPORATION
SAMSUNG ELECTRONICS CO., LTD.

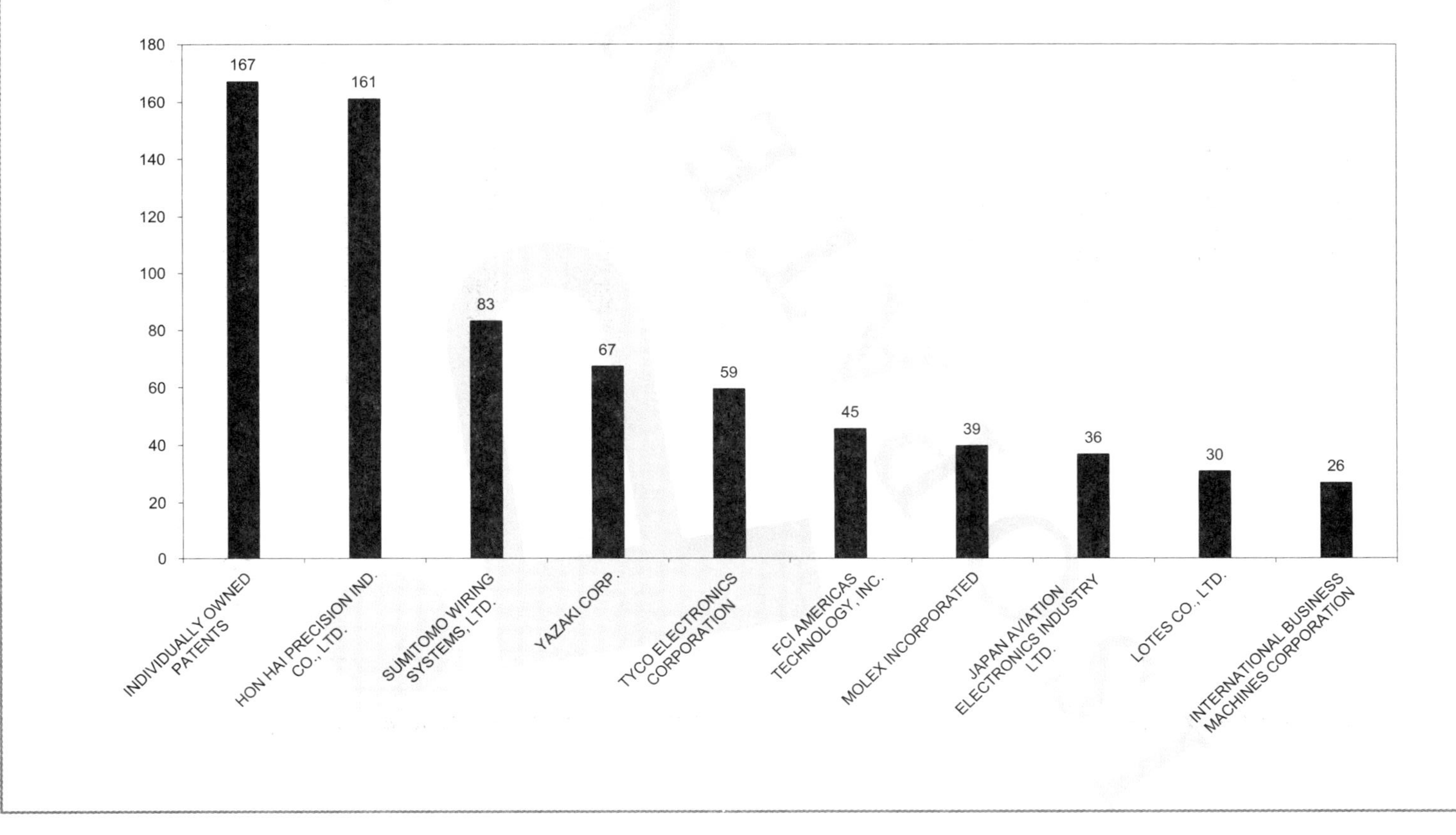

Top Patent Recipients in 2007 for Electrical Connectors Patents
180
160
140
120
100
80
60
40
20
0
167
161
83
67
59
45
39
36
30
26
INDIVIDUALLY OWNED PATENTS
HON HAI PRECISION IND. CO., LTD.
SUMITOMO WIRING SYSTEMS, LTD.
YAZAKI CORP.
TYCO ELECTRONICS CORPORATION
FCI AMERICAS TECHNOLOGY, INC.
MOLEX INCORPORATED
JAPAN AVIATION ELECTRONICS INDUSTRY LTD.
LOTES CO., LTD.
INTERNATIONAL BUSINESS MACHINES CORPORATION

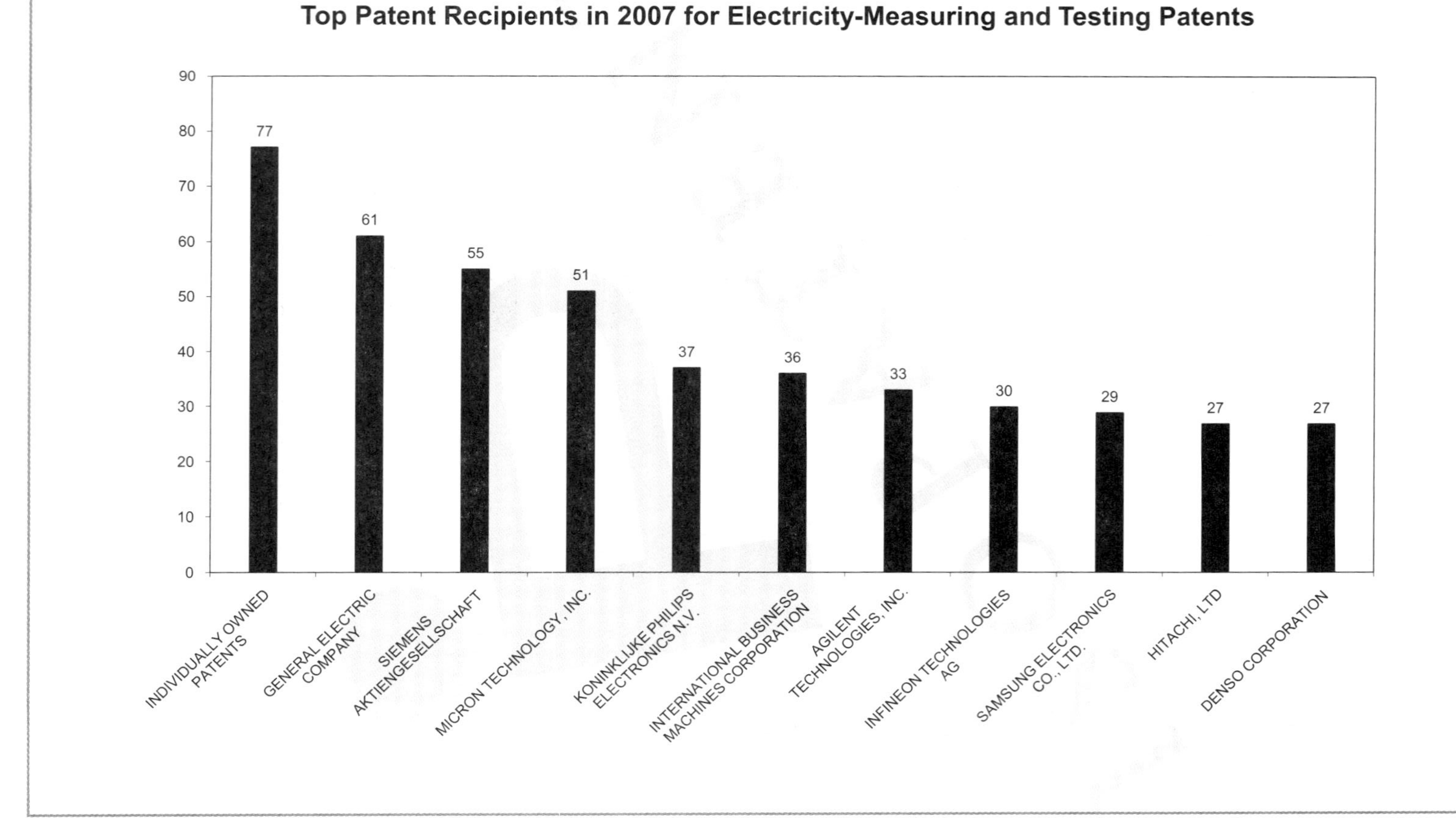
Top Patent Recipients in 2007 for Electricity-Measuring and Testing Patents
90
80
70
60
50
40
30
20
10
0
77
61
55
51
37
36
33
30
29
27
27
INDIVIDUALLY OWNED PATENTS
GENERAL ELECTRIC COMPANY
SIEMENS AKTIENGESELLSCHAFT
MICRON TECHNOLOGY, INC.
KONINKLIJKE PHILIPS ELECTRONICS N.V.
INTERNATIONAL BUSINESS MACHINES CORPORATION
AGILENT TECHNOLOGIES, INC.
INFINEON TECHNOLOGIES AG
SAMSUNG ELECTRONICS CO., LTD.
HITACHI, LTD
DENSO CORPORATION

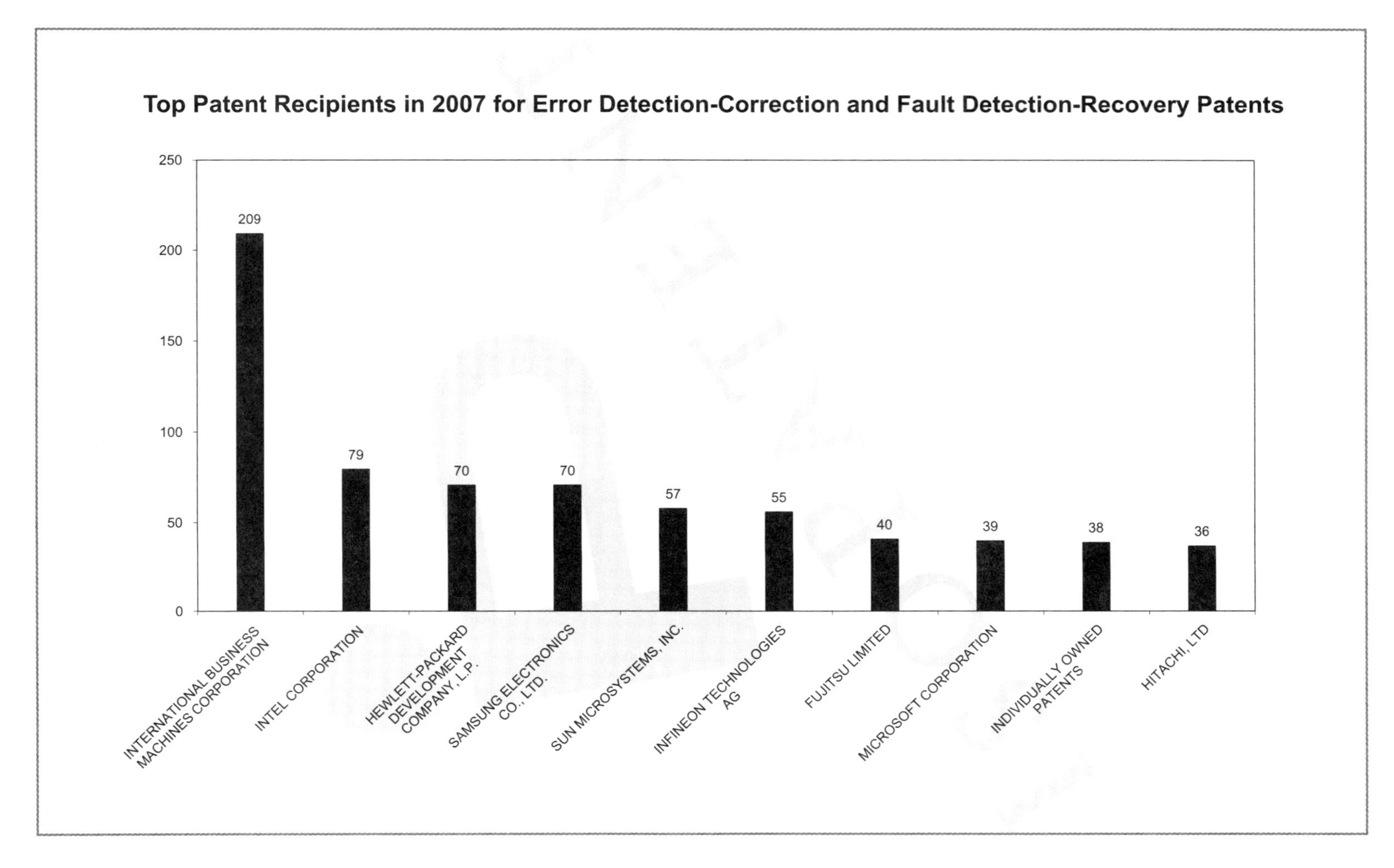

Top Patent Recipients in 2007 for Error Detection-Correction and Fault Detection-Recovery Patents
250
200
150
100
50
0
209
79
70
70
57
55
40
39
38
36
INTERNATIONAL BUSINESS MACHINES CORPORATION
INTEL CORPORATION
HEWLETT-PACKARD DEVELOPMENT COMPANY, L.P.
SAMSUNG ELECTRONICS CO., LTD.
SUN MICROSYSTEMS, INC.
INFINEON TECHNOLOGIES AG
FUJITSU LIMITED
MICROSOFT CORPORATION
INDIVIDUALLY OWNED PATENTS
HITACHI, LTD

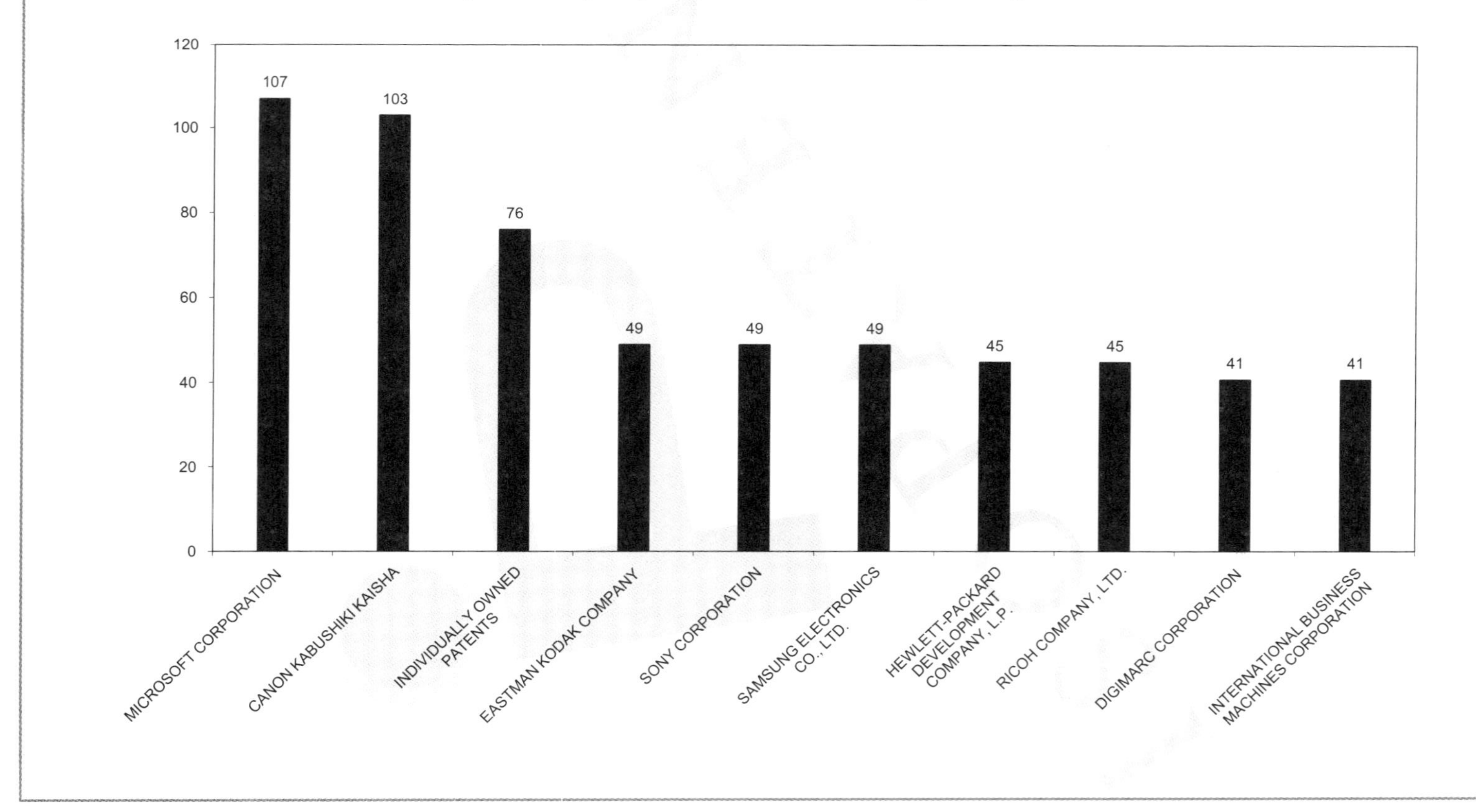

Top Patent Recipients in 2007 for Image Analysis Patents
120
100
80
60
40
20
0
107
103
76
49
49
49
45
45
41
41
MICROSOFT CORPORATION
CANON KABUSHIKI KAISHA
INDIVIDUALLY OWNED PATENTS
EASTMAN KODAK COMPANY
SONY CORPORATION
SAMSUNG ELECTRONICS CO., LTD.
HEWLETT-PACKARD DEVELOPMENT COMPANY, L.P.
RICOH COMPANY, LTD.
DIGIMARC CORPORATION
INTERNATIONAL BUSINESS MACHINES CORPORATION

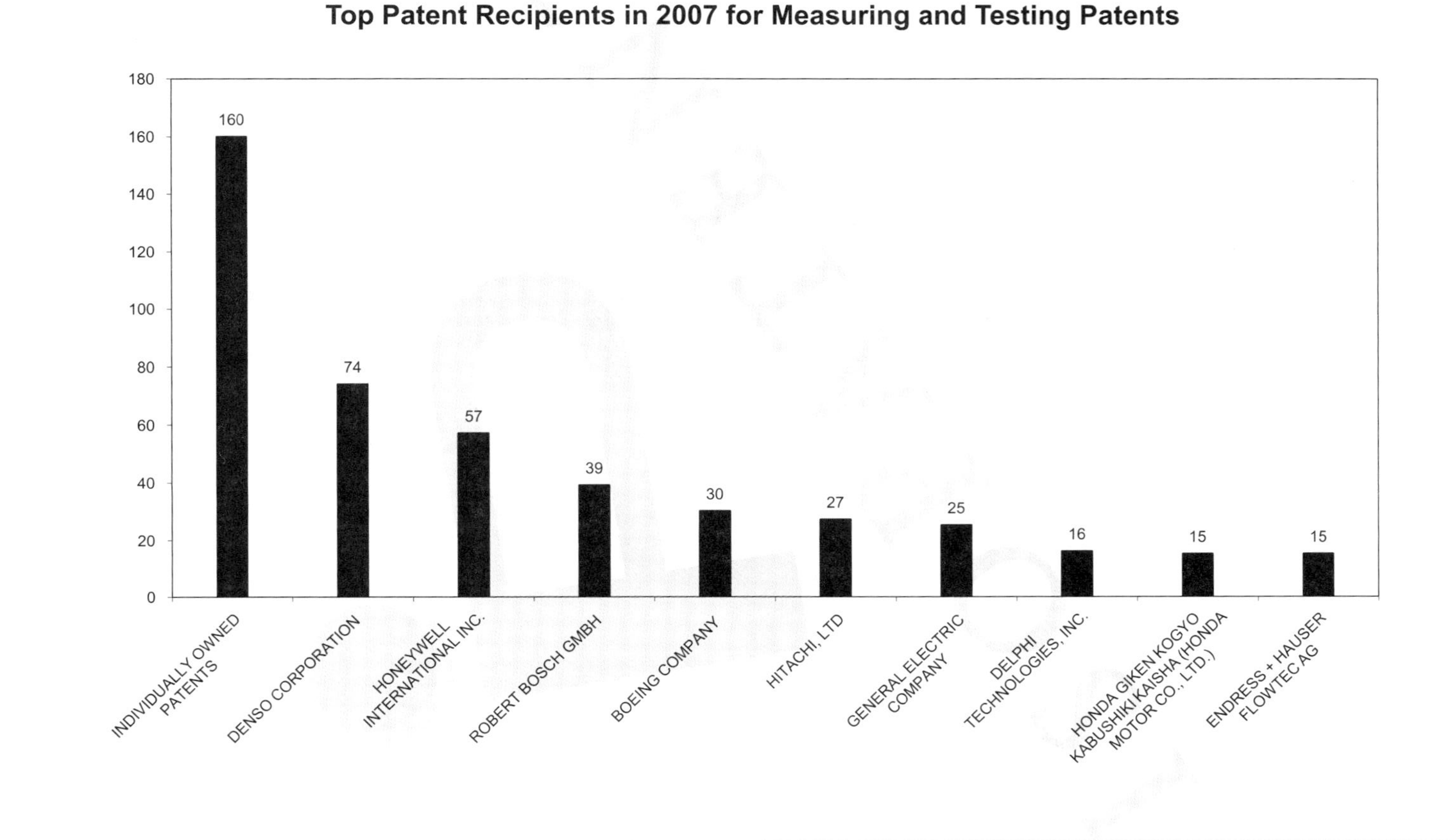
Top Patent Recipients in 2007 for Measuring and Testing Patents
180
160
140
120
100
80
60
40
20
0
160
74
57
39
30
27
25
16
15
15
INDIVIDUALLY OWNED PATENTS
DENSO CORPORATION
HONEYWELL INTERNATIONAL INC.
ROBERT BOSCH GMBH
BOEING COMPANY
HITACHI, LTD
GENERAL ELECTRIC COMPANY
DELPHI TECHNOLOGIES, INC.
HONDA GIKEN KOGYO KABUSHIKI KAISHA (HONDA MOTOR CO., LTD.)
ENDRESS + HAUSER FLOWTEC AG

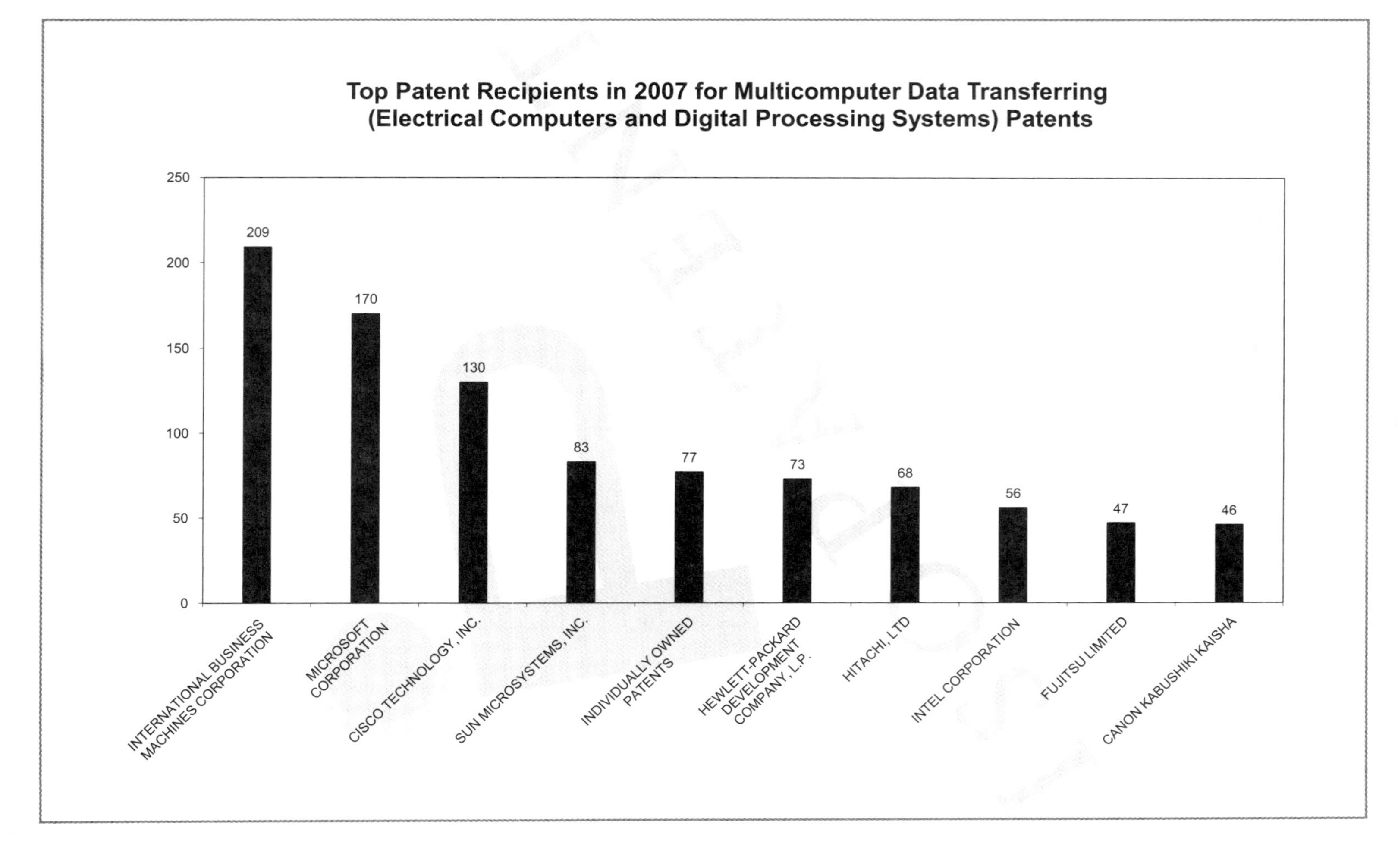
Top Patent Recipients in 2007 for Multicomputer Data Transferring
(Electrical Computers and Digital Processing Systems) Patents
250
200
150
100
50
0
209
170
130
83
77
73
68
56
47
46
INTERNATIONAL BUSINESS MACHINES CORPORATION
MICROSOFT CORPORATION
CISCO TECHNOLOGY, INC.
SUN MICROSYSTEMS, INC.
INDIVIDUALLY OWNED PATENTS
HEWLETT-PACKARD DEVELOPMENT COMPANY, L.P.
HITACHI, LTD
INTEL CORPORATION
FUJITSU LIMITED
CANON KABUSHIKI KAISHA

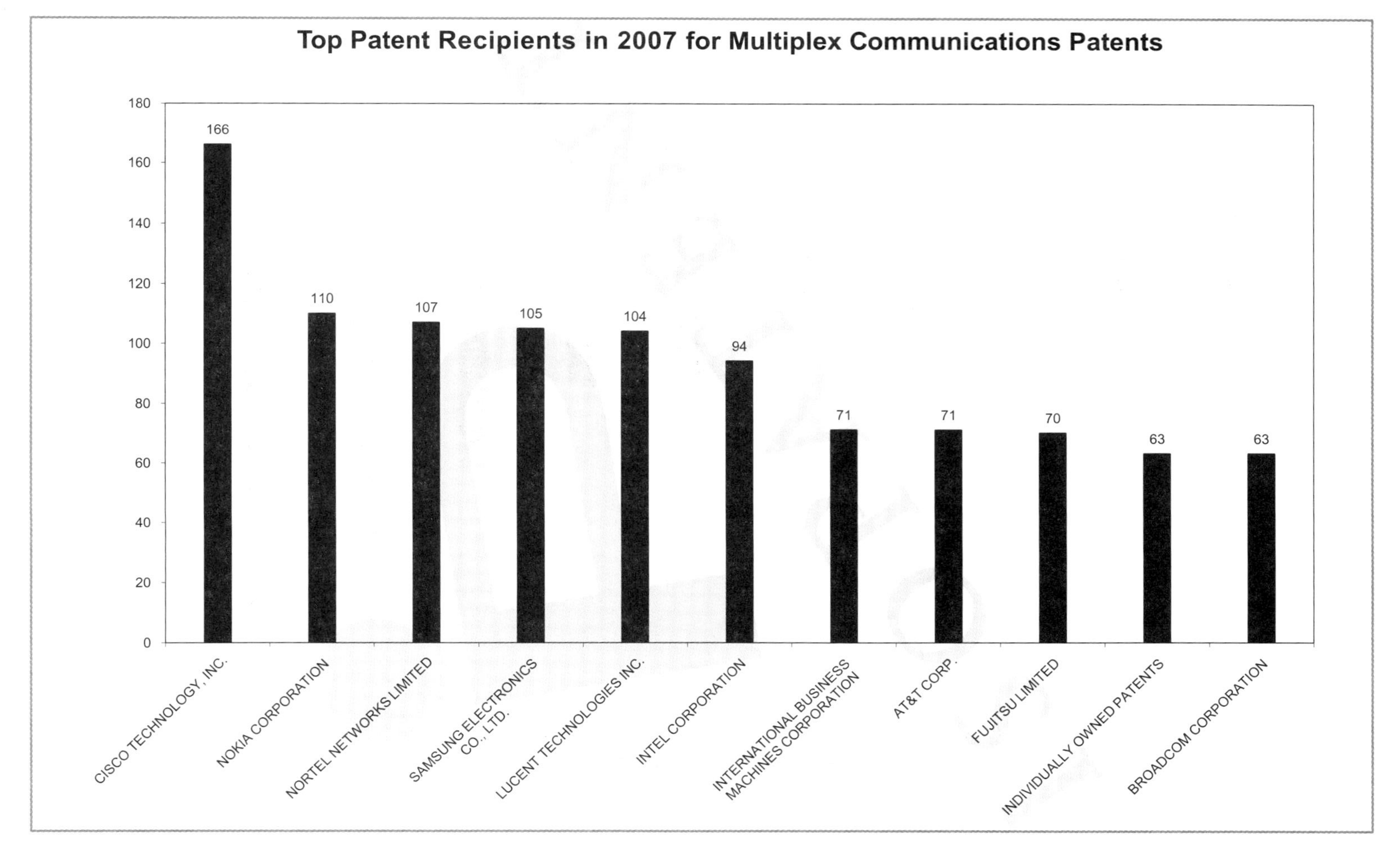
Top Patent Recipients in 2007 for Multiplex Communications Patents
180
160
140
120
100
80
60
40
20
0
166
110
107
105
104
94
71
71
70
63
63
CISCO TECHNOLOGY, INC.
NOKIA CORPORATION
NORTEL NETWORKS LIMITED
SAMSUNG ELECTRONICS CO., LTD.
LUCENT TECHNOLOGIES INC.
INTEL CORPORATION
INTERNATIONAL BUSINESS MACHINES CORPORATION
AT&T CORP.
FUJITSU LIMITED
INDIVIDUALLY OWNED PATENTS
BROADCOM CORPORATION

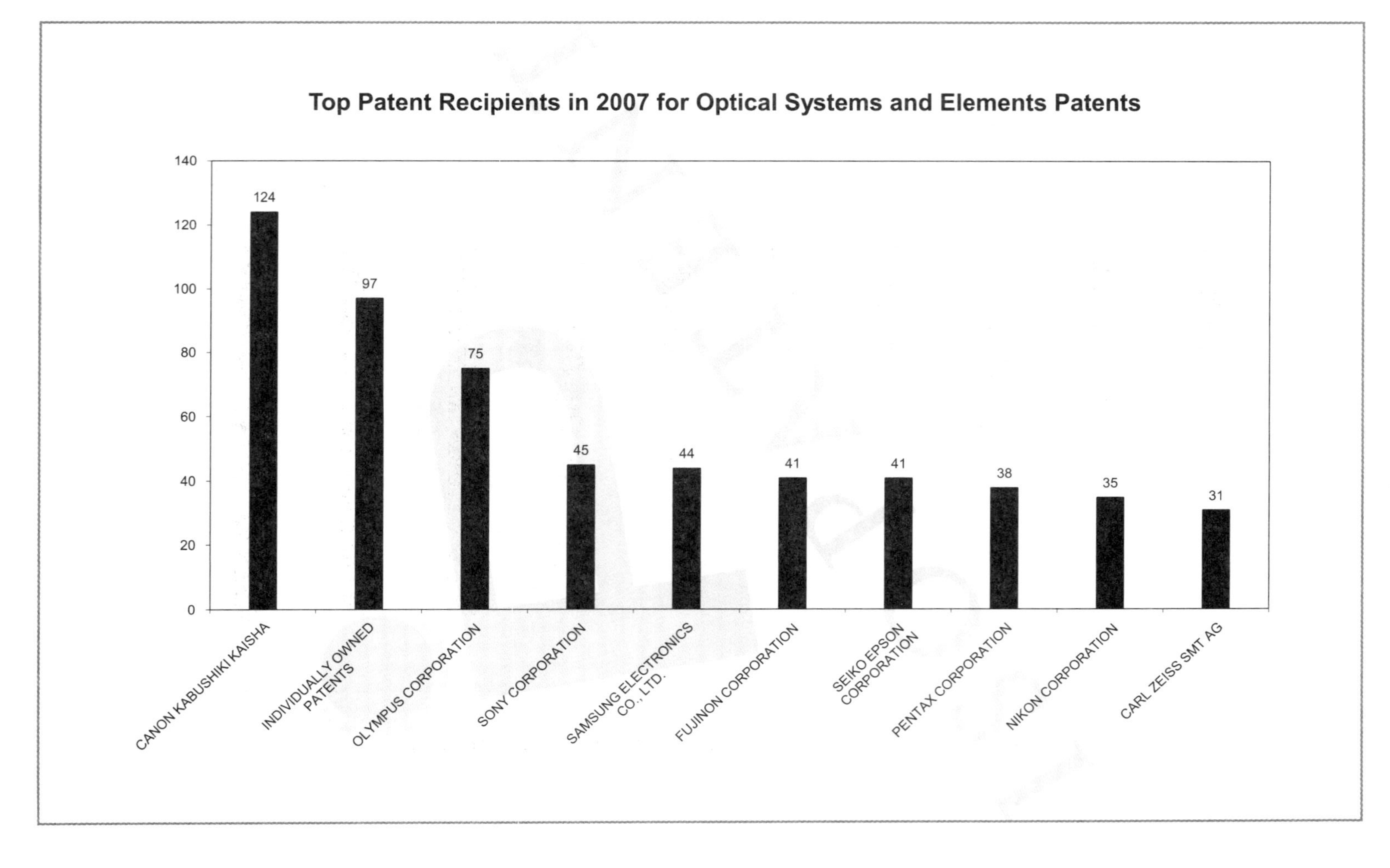

Top Patent Recipients in 2007 for Optical Systems and Elements Patents
140
120
100
80
60
40
20
0
124
97
75
45
44
41
41
38
35
31
CANON KABUSHIKI KAISHA
INDIVIDUALLY OWNED PATENTS
OLYMPUS CORPORATION
SONY CORPORATION
SAMSUNG ELECTRONICS CO., LTD.
FUJINON CORPORATION
SEIKO EPSON CORPORATION
PENTAX CORPORATION
NIKON CORPORATION
CARL ZEISS SMT AG

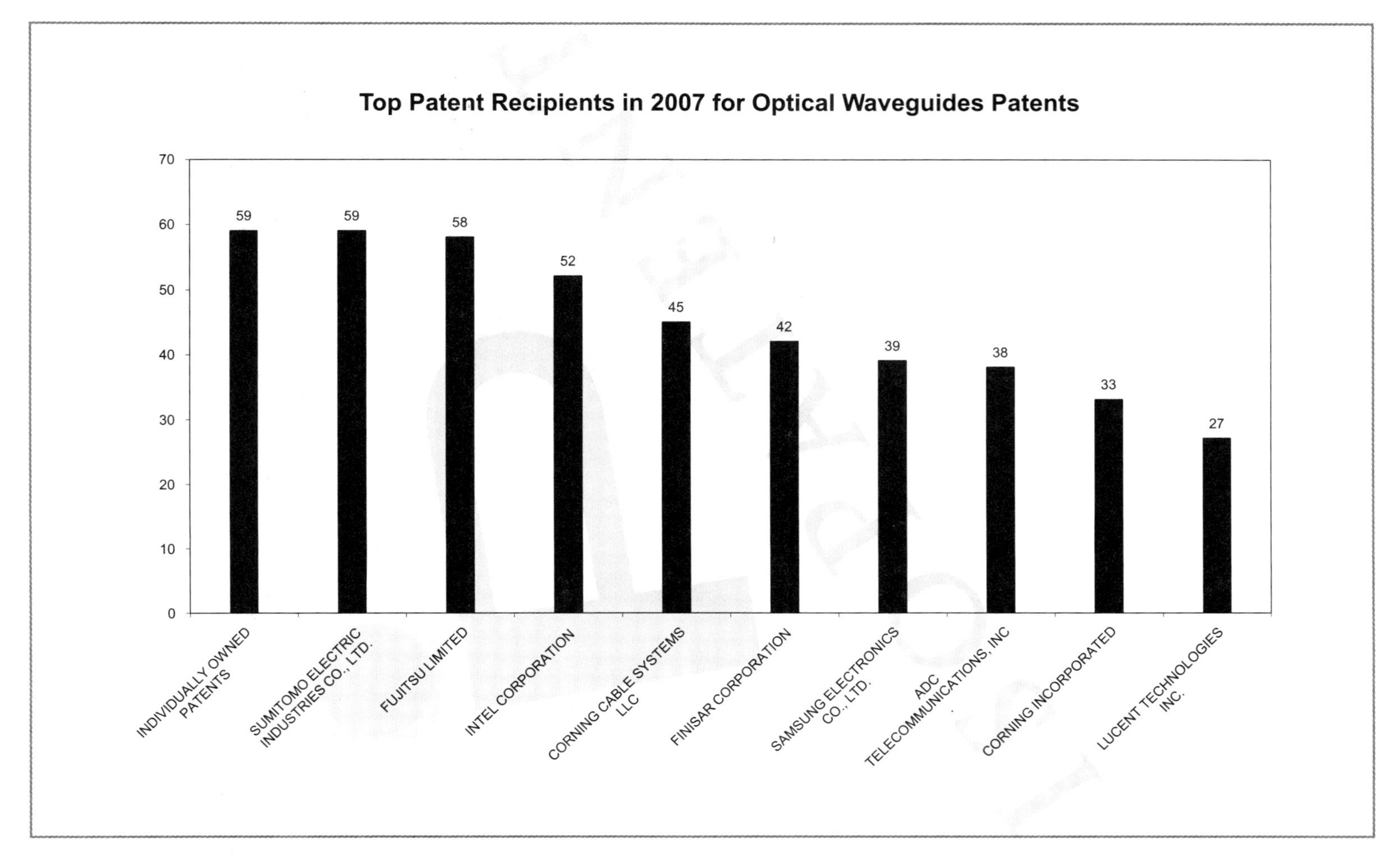
Top Patent Recipients in 2007 for Optical Waveguides Patents
70
60
50
40
30
20
10
0
59
59
58
52
45
42
39
38
33
27
INDIVIDUALLY OWNED PATENTS
SUMITOMO ELECTRIC INDUSTRIES CO., LTD.
FUJITSU LIMITED
INTEL CORPORATION
CORNING CABLE SYSTEMS LLC
FINISAR CORPORATION
SAMSUNG ELECTRONICS CO., LTD.
ADC TELECOMMUNICATIONS, INC
CORNING INCORPORATED
LUCENT TECHNOLOGIES INC.

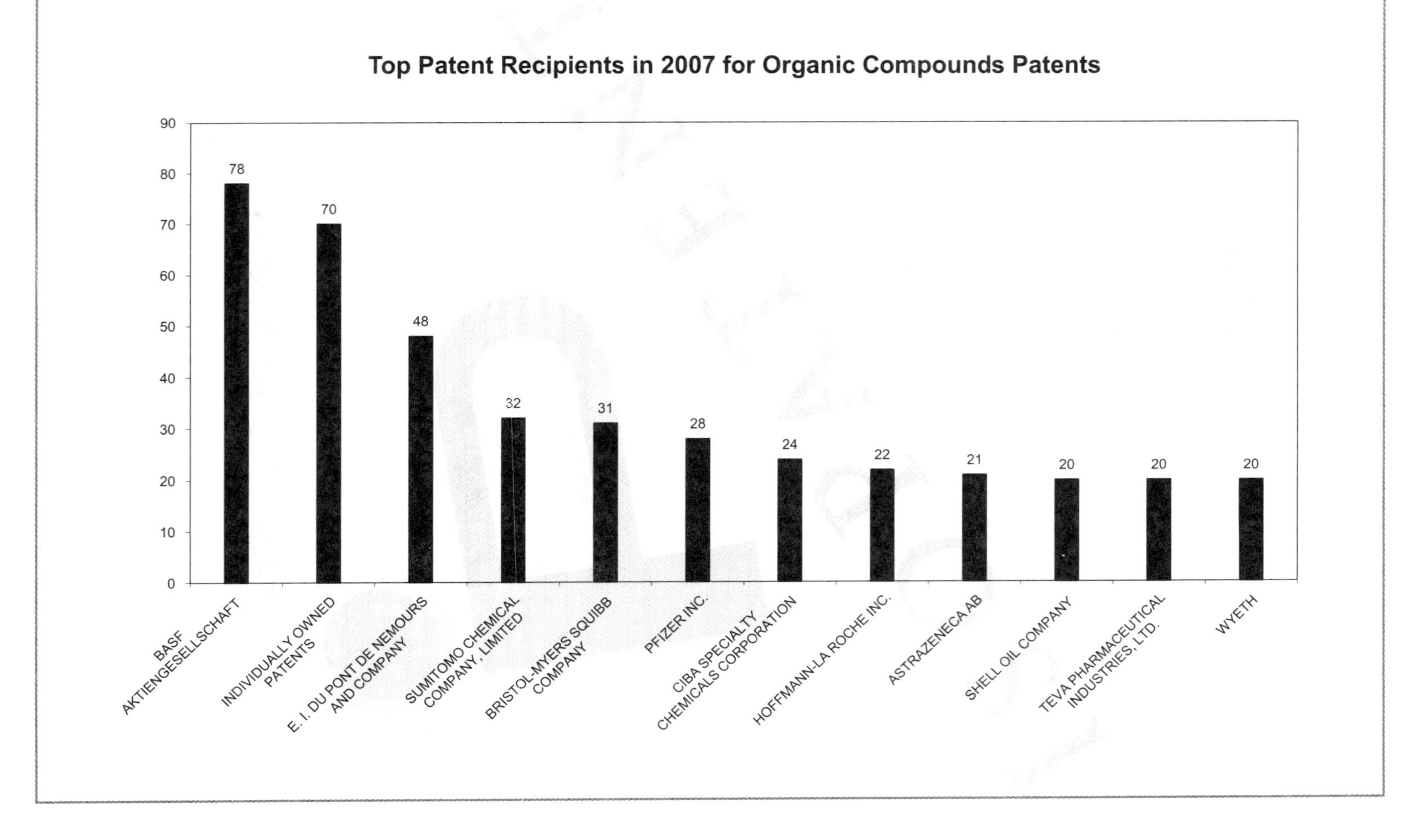
Top Patent Recipients in 2007 for Organic Compounds Patents
90
80
70
60
50
40
30
20
10
0
78
70
48
32
31
28
24
22
21
20
20
20
BASF AKTIENGESELLSCHAFT
INDIVIDUALLY OWNED PATENTS
E. I. DU PONT DE NEMOURS AND COMPANY
SUMITOMO CHEMICAL COMPANY, LIMITED
BRISTOL-MYERS SQUIBB COMPANY
PFIZER INC.
CIBA SPECIALTY CHEMICALS CORPORATION
HOFFMANN-LA ROCHE INC.
ASTRAZENECA AB
SHELL OIL COMPANY
TEVA PHARMACEUTICAL INDUSTRIES, LTD.
WYETH

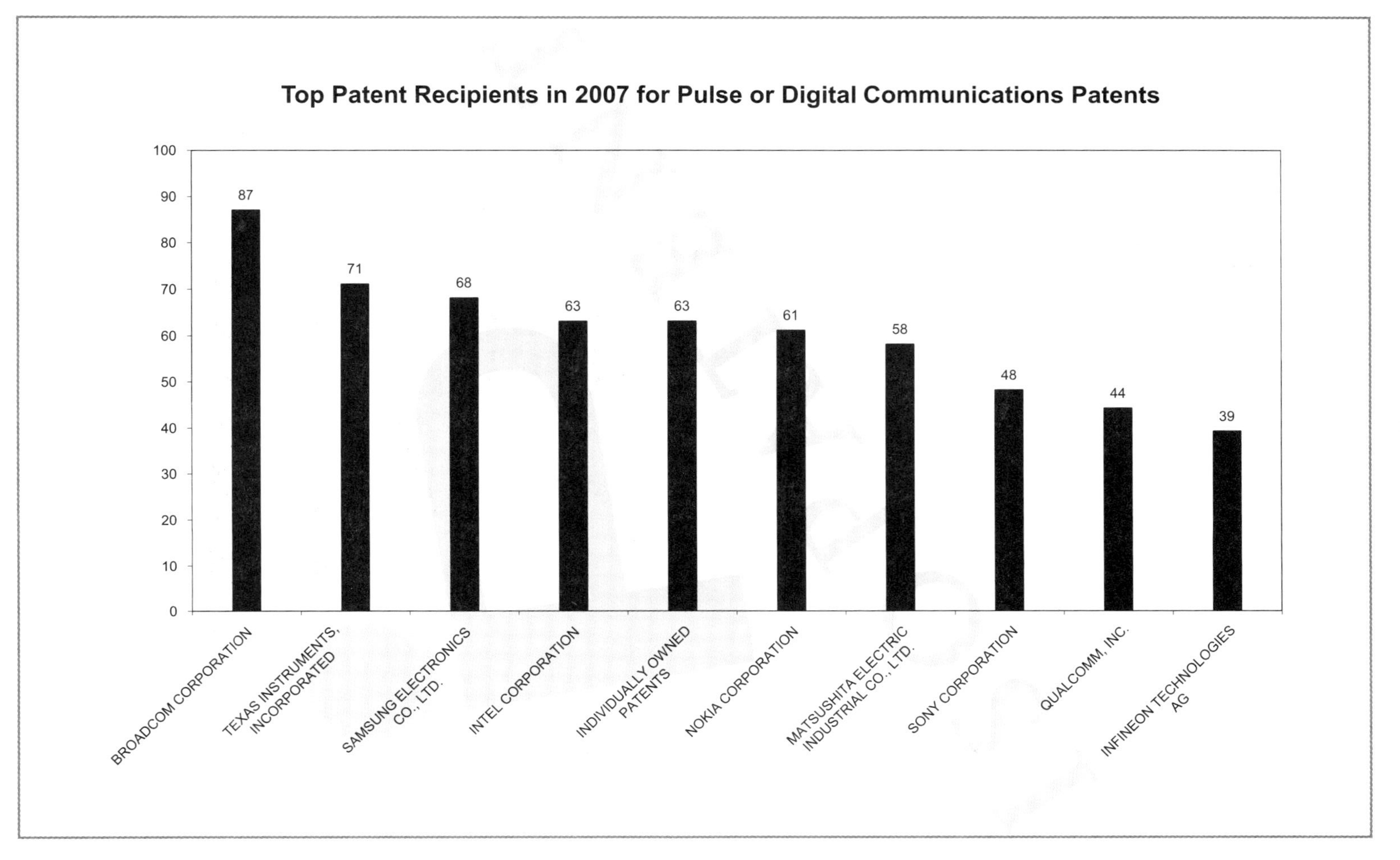

Top Patent Recipients in 2007 for Pulse or Digital Communications Patents
100
90
80
70
60
50
40
30
20
10
0
87
71
68
63
63
61
58
48
44
39
BROADCOM CORPORATION
TEXAS INSTRUMENTS, INCORPORATED
SAMSUNG ELECTRONICS CO., LTD.
INTEL CORPORATION
INDIVIDUALLY OWNED PATENTS
NOKIA CORPORATION
MATSUSHITA ELECTRIC INDUSTRIAL CO., LTD.
SONY CORPORATION
QUALCOMM, INC.
INFINEON TECHNOLOGIES AG

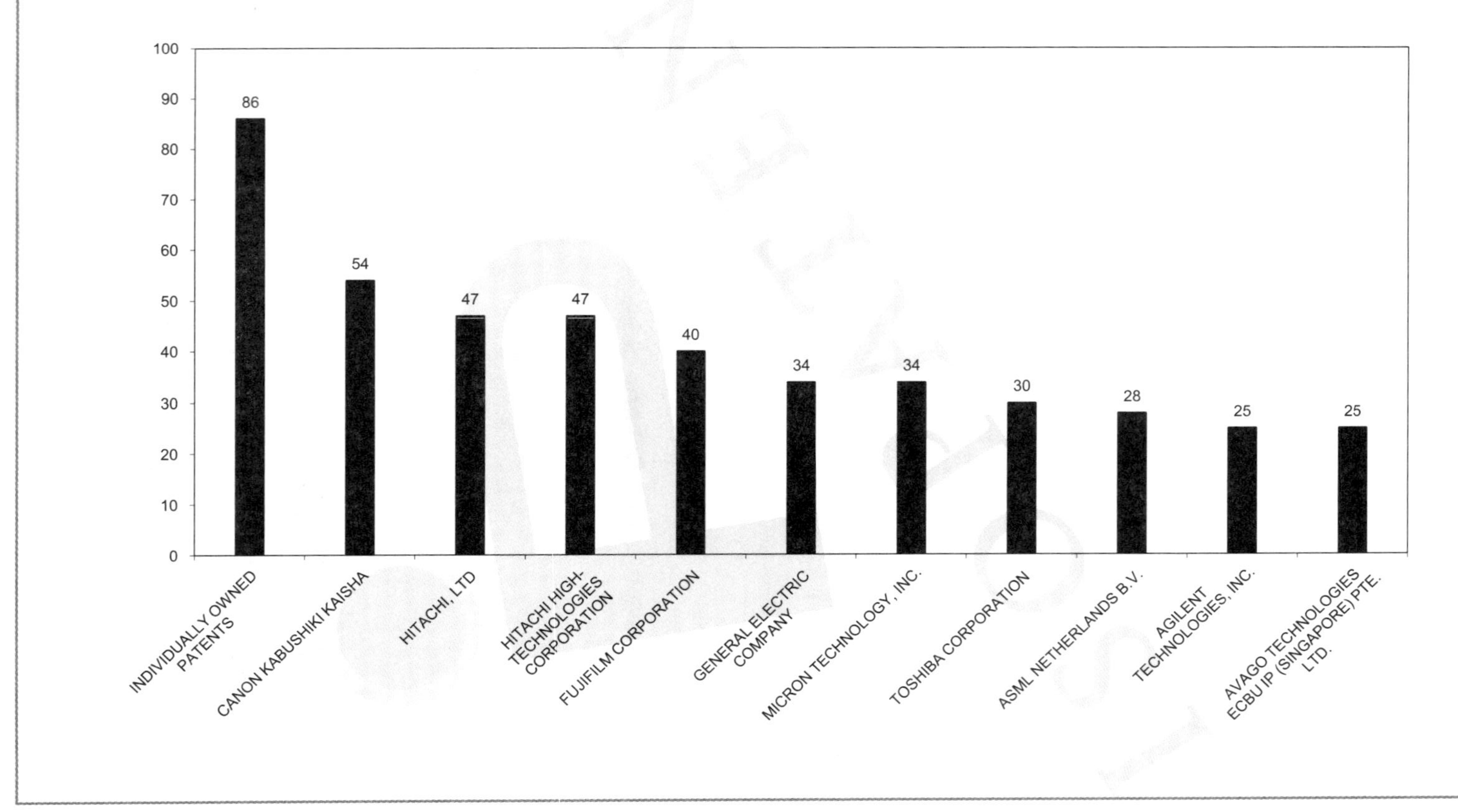
Top Patent Recipients in 2007 for Radiant Energy Patents
100
90
80
70
60
50
40
30
20
10
0
86
54
47
47
40
34
34
30
28
25
25
INDIVIDUALLY OWNED PATENTS
CANON KABUSHIKI KAISHA
HITACHI, LTD
HITACHI HIGH-TECHNOLOGIES CORPORATION
FUJIFILM CORPORATION
GENERAL ELECTRIC COMPANY
MICRON TECHNOLOGY, INC.
TOSHIBA CORPORATION
ASML NETHERLANDS B.V.
AGILENT TECHNOLOGIES, INC.
AVAGO TECHNOLOGIES ECBU IP (SINGAPORE) PTE. LTD.

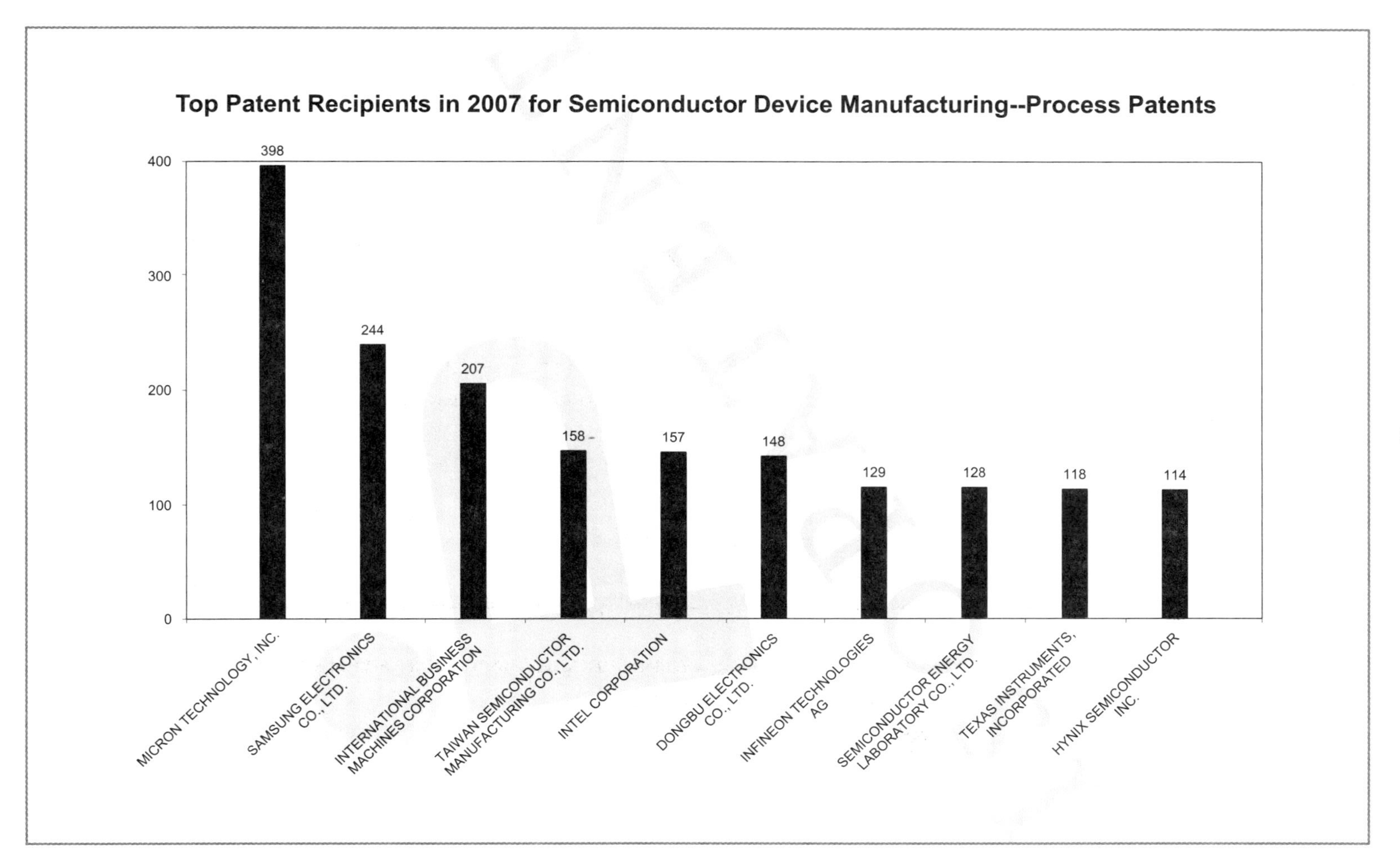
Top Patent Recipients in 2007 for Semiconductor Device Manufacturing--Process Patents
400
300
200
100
0
398
244
207
158
157
148
129
128
118
114
MICRON TECHNOLOGY, INC.
SAMSUNG ELECTRONICS CO., LTD.
INTERNATIONAL BUSINESS MACHINES CORPORATION
TAIWAN SEMICONDUCTOR MANUFACTURING CO., LTD.
INTEL CORPORATION
DONGBU ELECTRONICS CO., LTD.
INFINEON TECHNOLOGIES AG
SEMICONDUCTOR ENERGY LABORATORY CO., LTD.
TEXAS INSTRUMENTS, INCORPORATED
HYNIX SEMICONDUCTOR INC.

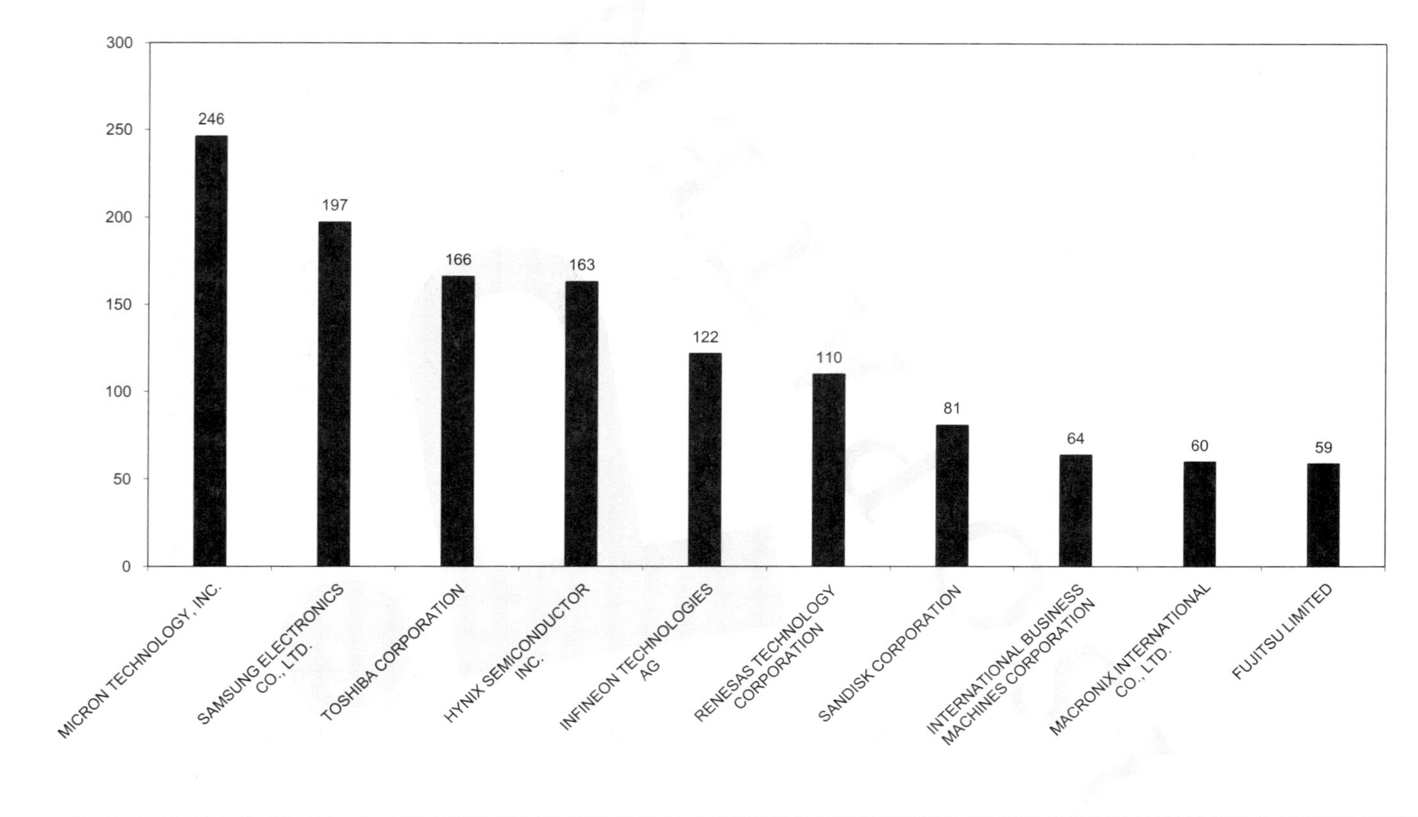

Top Patent Recipients in 2007 for Static Information Storage and Retrieval Patents
300
250
200
150
100
50
0
246
197
166
163
122
110
81
64
60
59
MICRON TECHNOLOGY, INC.
SAMSUNG ELECTRONICS CO., LTD.
TOSHIBA CORPORATION
HYNIX SEMICONDUCTOR INC.
INFINEON TECHNOLOGIES AG
RENESAS TECHNOLOGY CORPORATION
SANDISK CORPORATION
INTERNATIONAL BUSINESS MACHINES CORPORATION
MACRONIX INTERNATIONAL CO., LTD.
FUJITSU LIMITED

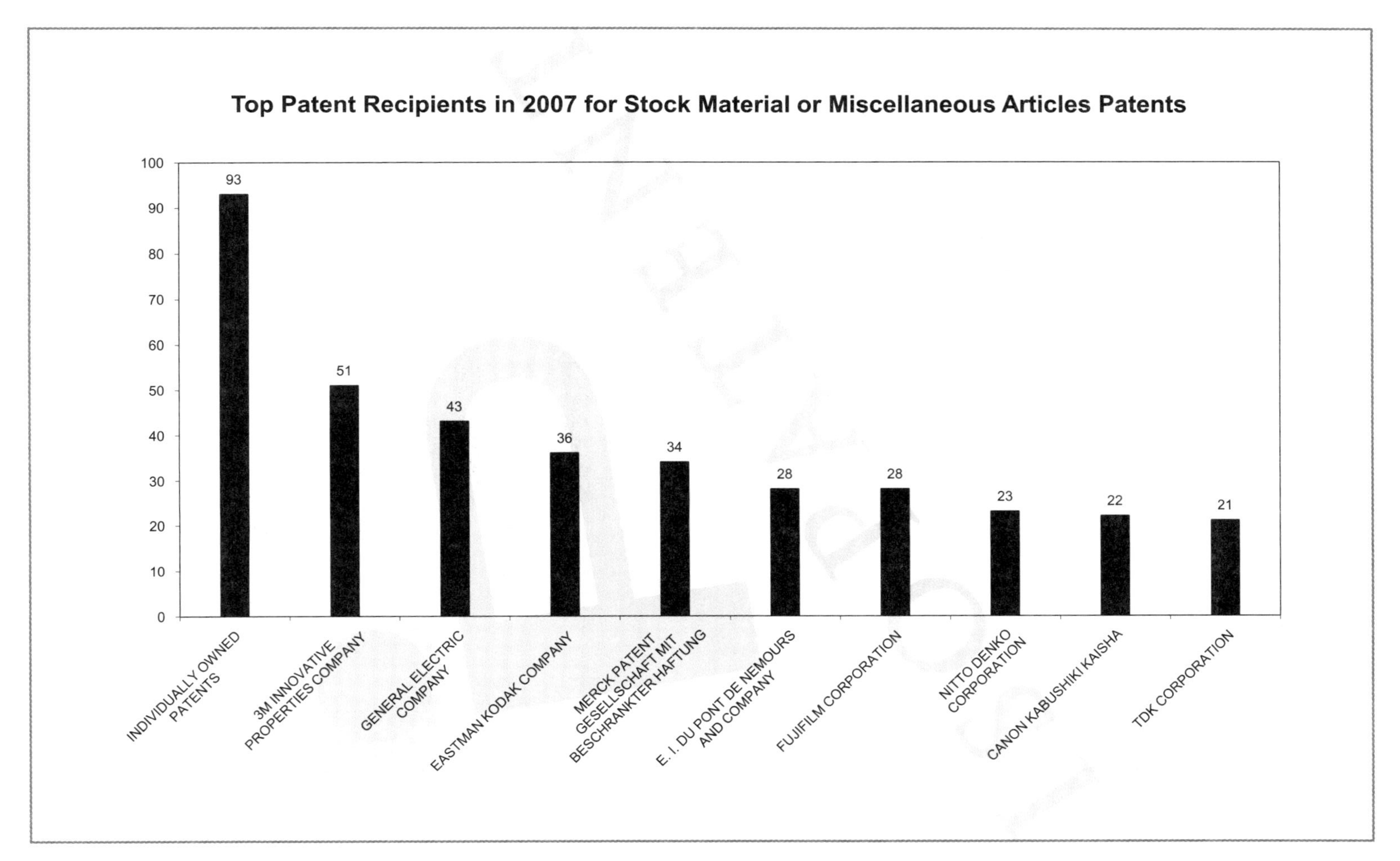
Top Patent Recipients in 2007 for Stock Material or Miscellaneous Articles Patents
100
90
80
70
60
50
40
30
20
10
0
93
51
43
36
34
28
28
23
22
21
INDIVIDUALLY OWNED PATENTS
3M INNOVATIVE PROPERTIES COMPANY
GENERAL ELECTRIC COMPANY
EASTMAN KODAK COMPANY
MERCK PATENT GESELLSCHAFT MIT BESCHRANKTER HAFTUNG
E. I. DU PONT DE NEMOURS AND COMPANY
FUJIFILM CORPORATION
NITTO DENKO CORPORATION
CANON KABUSHIKI KAISHA
TDK CORPORATION

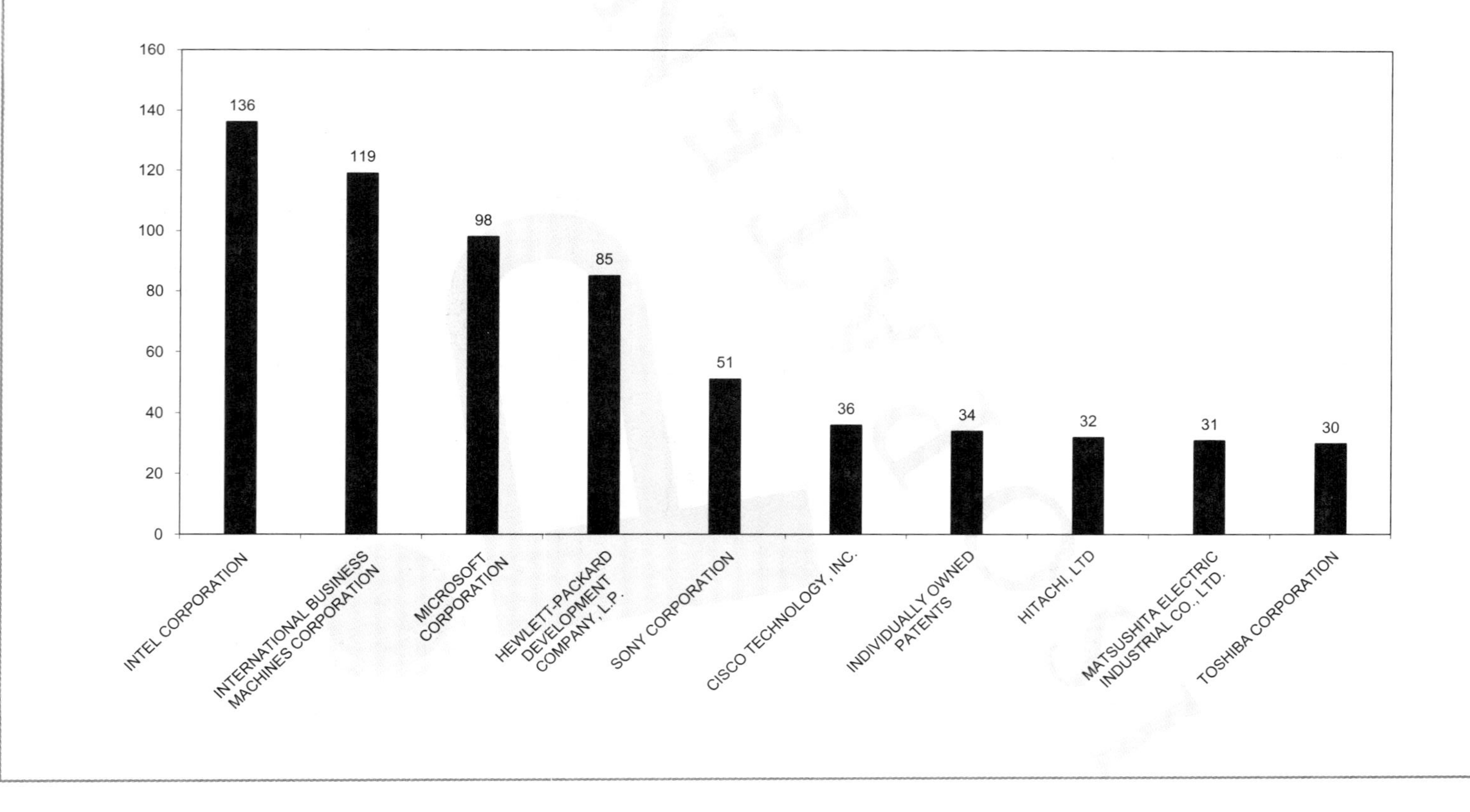
Top Patent Recipients in 2007 for Support--Electrical Computers and Digital Processing Sytems Patents
160
140
120
100
80
60
40
20
0
136
119
98
85
51
36
34
32
31
30
INTEL CORPORATION
INTERNATIONAL BUSINESS MACHINES CORPORATION
MICROSOFT CORPORATION
HEWLETT-PACKARD DEVELOPMENT COMPANY, L.P.
SONY CORPORATION
CISCO TECHNOLOGY, INC.
INDIVIDUALLY OWNED PATENTS
HITACHI, LTD
MATSUSHITA ELECTRIC INDUSTRIAL CO., LTD.
TOSHIBA CORPORATION

Top Patent Recipients in 2007 for Synthetic Resins or Natural Rubbers Patents

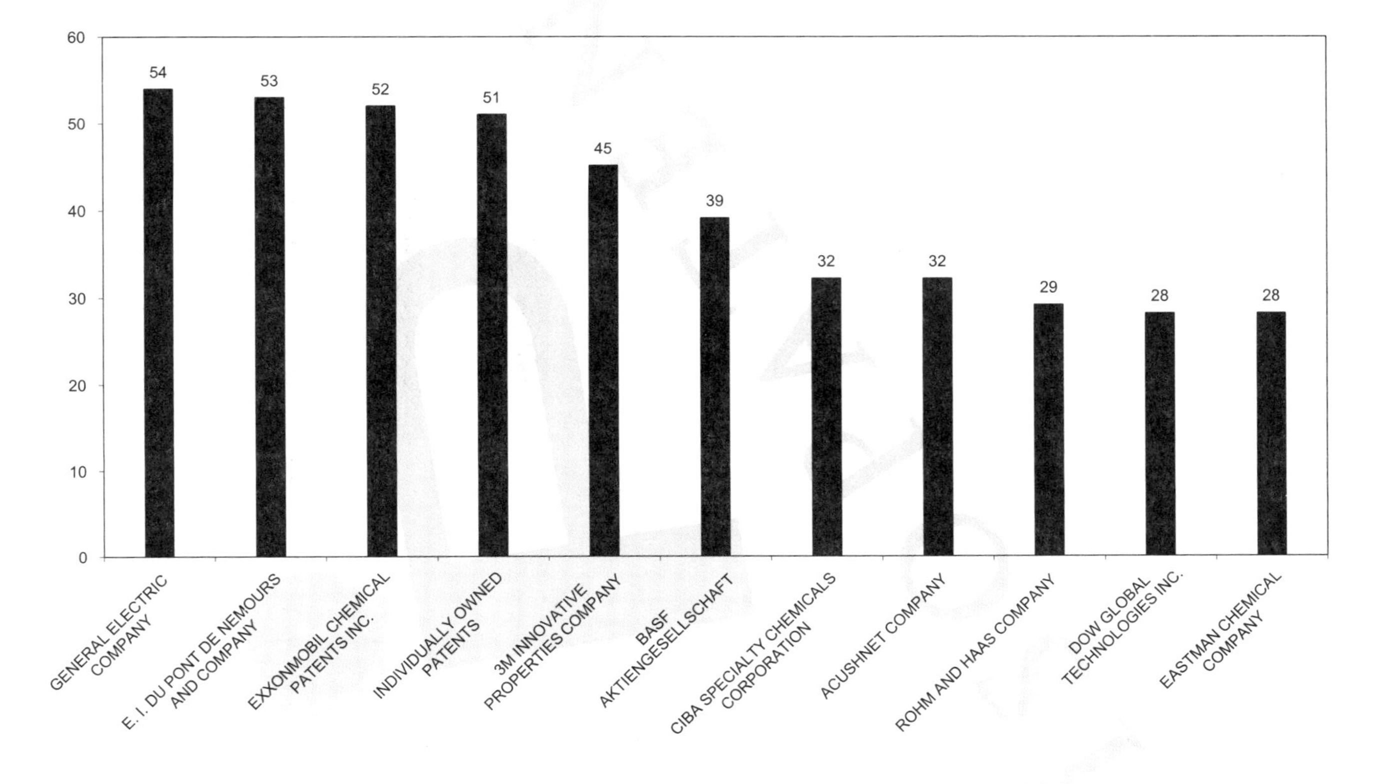
60
50
40
30
20
10
0
54
53
52
51
45
39
32
32
29
28
28
GENERAL ELECTRIC COMPANY
E. I. DU PONT DE NEMOURS AND COMPANY
EXXONMOBIL CHEMICAL PATENTS INC.
INDIVIDUALLY OWNED PATENTS
3M INNOVATIVE PROPERTIES COMPANY
BASF AKTIENGESELLSCHAFT
CIBA SPECIALTY CHEMICALS CORPORATION
ACUSHNET COMPANY
ROHM AND HAAS COMPANY
DOW GLOBAL TECHNOLOGIES INC.
EASTMAN CHEMICAL COMPANY

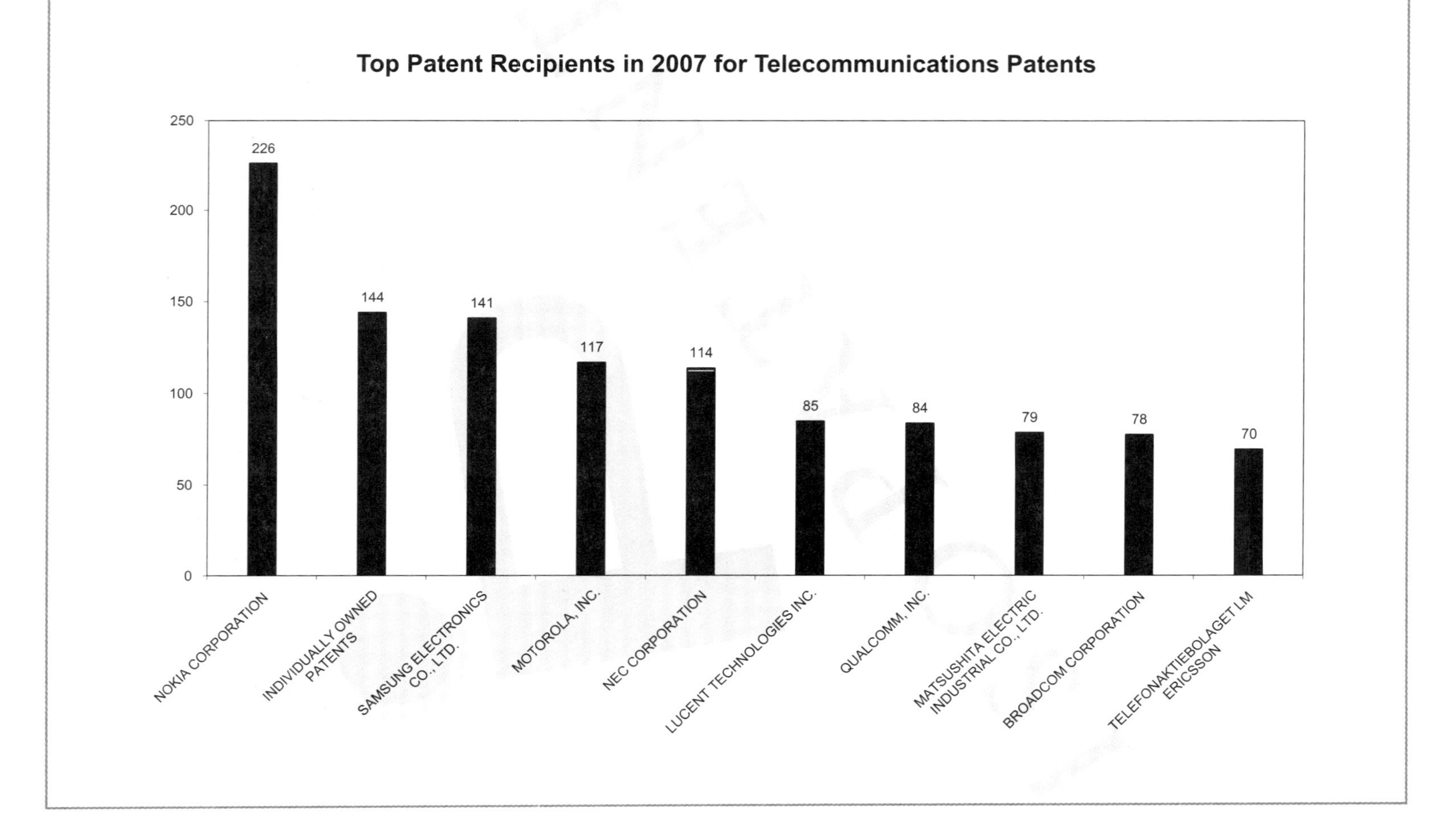
Top Patent Recipients in 2007 for Telecommunications Patents
250
200
150
100
50
0
226
144
141
117
114
85
84
79
78
70
NOKIA CORPORATION
INDIVIDUALLY OWNED PATENTS
SAMSUNG ELECTRONICS CO., LTD.
MOTOROLA, INC.
NEC CORPORATION
LUCENT TECHNOLOGIES INC.
QUALCOMM, INC.
MATSUSHITA ELECTRIC INDUSTRIAL CO., LTD.
BROADCOM CORPORATION
TELEFONAKTIEBOLAGET LM ERICSSON

Chapter 15 Demographics and Patent Information for Each of the Fifty United States[71]

Alabama

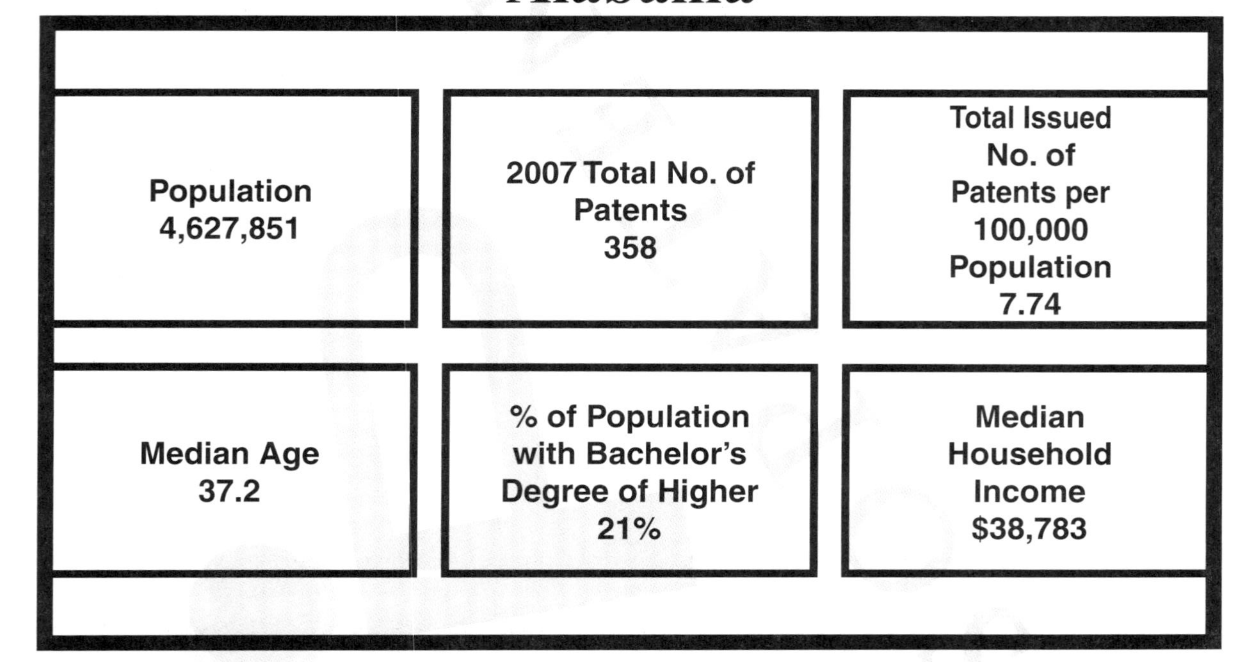
Population
4,627,851
2007 Total No. of
Patents
358
Total Issued
No. of
Patents per
100,000
Population
7.74
Median Age
37.2
% of Population
with Bachelor's
Degree of Higher
21%
Median
Household
Income
$38,783

Alaska

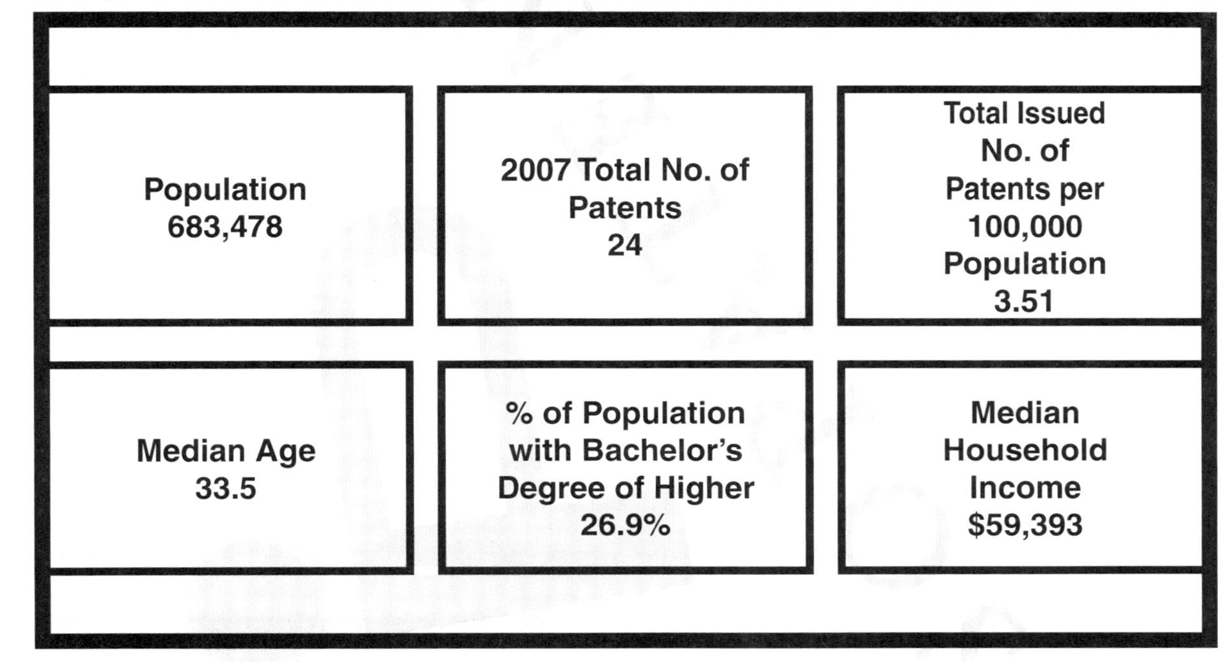

Population
683,478
2007 Total No. of
Patents
24
Total Issued
No. of
Patents per
100,000
Population
3.51
Median Age
33.5
% of Population
with Bachelor's
Degree of Higher
26.9%
Median
Household
Income
$59,393

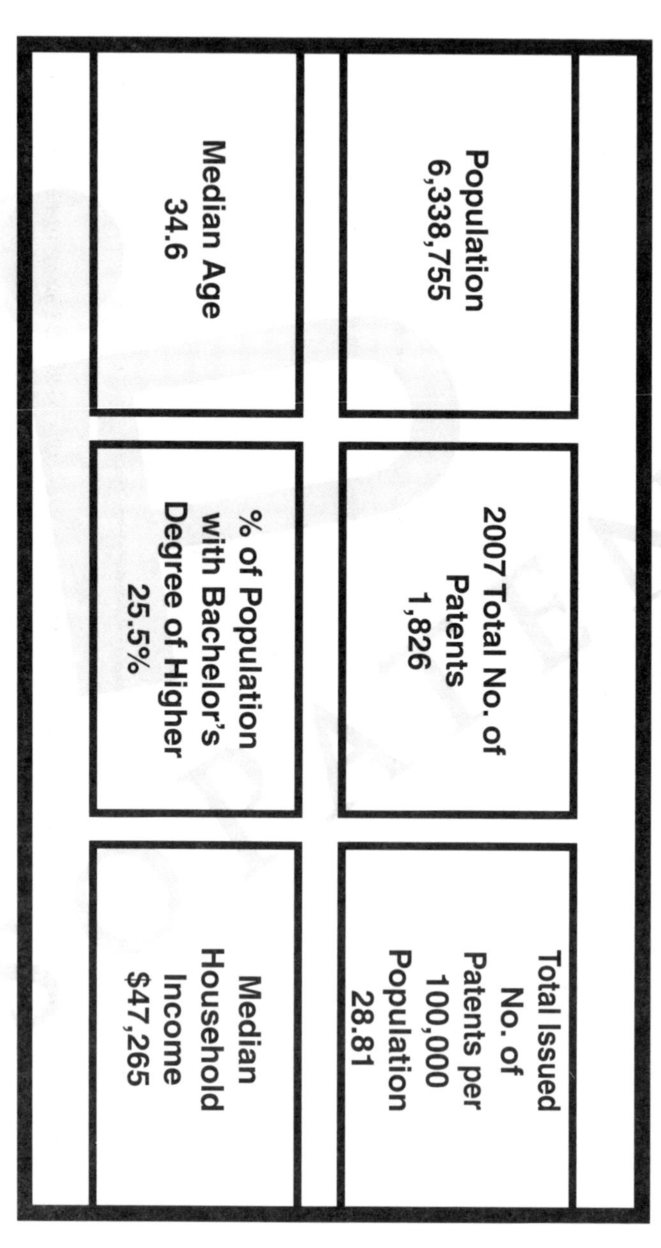
Arizona
Population
6,338,755
2007 Total No. of Patents
1,826
Total Issued No. of Patents per 100,000 Population
28.81
Median Age
34.6
% of Population with Bachelor's Degree of Higher
25.5%
Median Household Income
$47,265

Arkansas

Population 2,834,797	2007 Total No. of Patents 169	Total Issued No. of Patents per 100,000 Population 5.96
Median Age 37.1	% of Population with Bachelor's Degree of Higher 18.2%	Median Household Income $36,599

California

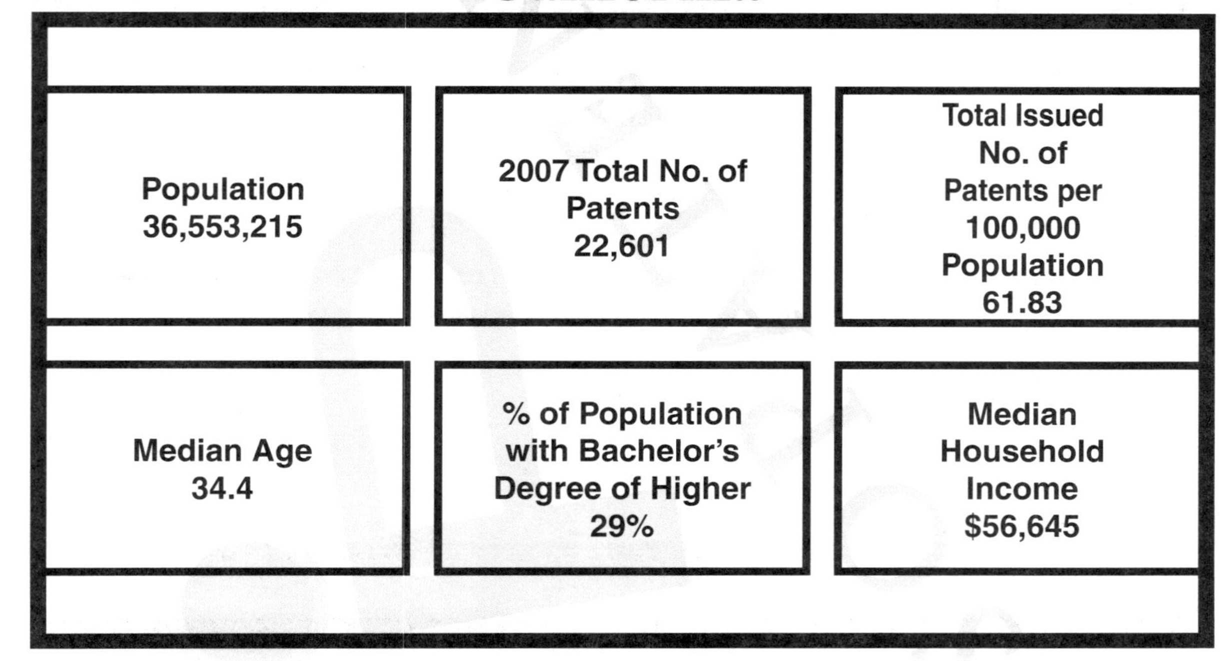
Population
36,553,215
2007 Total No. of
Patents
22,601
Total Issued
No. of
Patents per
100,000
Population
61.83
Median Age
34.4
% of Population
with Bachelor's
Degree of Higher
29%
Median
Household
Income
$56,645

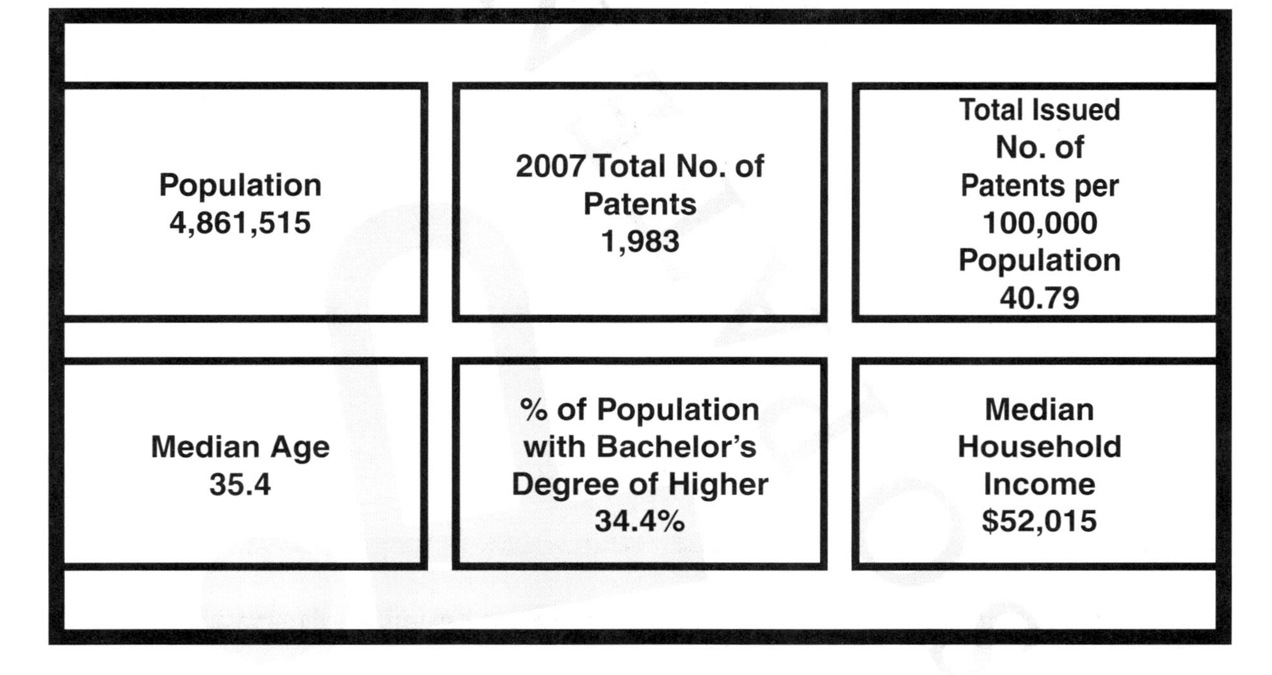
Colorado
Population
4,861,515
2007 Total No. of
Patents
1,983
Total Issued
No. of
Patents per
100,000
Population
40.79
Median Age
35.4
% of Population
with Bachelor's
Degree of Higher
34.4%
Median
Household
Income
$52,015

Connecticut

Population 3,502,309	2007 Total No. of Patents 1,611	Total Issued No. of Patents per 100,000 Population 46.00
Median Age 39.1	% of Population with Bachelor's Degree of Higher 33.7%	Median Household Income $63,422

Delaware

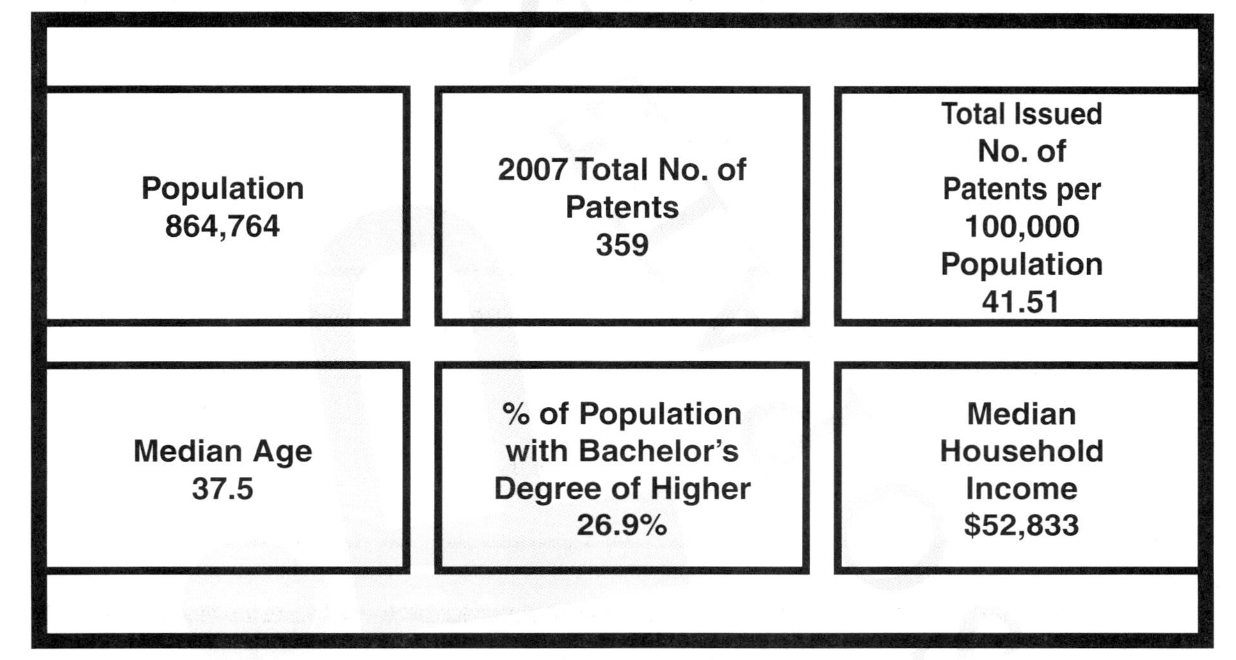
Population
864,764
2007 Total No. of
Patents
359
Total Issued
No. of
Patents per
100,000
Population
41.51
Median Age
37.5
% of Population
with Bachelor's
Degree of Higher
26.9%
Median
Household
Income
$52,833

Florida

Population 18,251,243	2007 Total No. of Patents 3,092	Total Issued No. of Patents per 100,000 Population 16.94
Median Age 39.8	% of Population with Bachelor's Degree of Higher 25.3%	Median Household Income $45,495

Georgia

Population 9,544,750	2007 Total No. of Patents 1,580	Total Issued No. of Patents per 100,000 Population 16.55
Median Age 34.6	% of Population with Bachelor's Degree of Higher 26.5%	Median Household Income $46,832

Hawaii

Population 1,283,388	2007 Total No. of Patents 82	Total Issued No. of Patents per 100,000 Population 6.39
Median Age 37.2	% of Population with Bachelor's Degree of Higher 29.7%	Median Household Income $61,160

Idaho

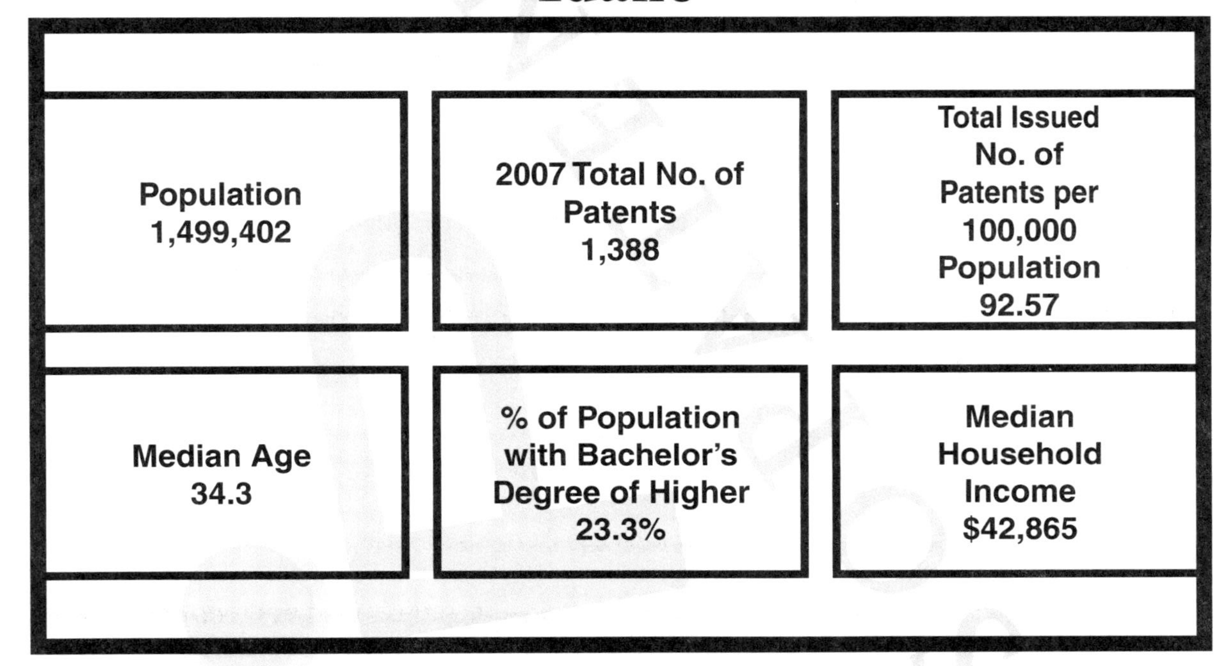
Population
1,499,402
2007 Total No. of
Patents
1,388
Total Issued
No. of
Patents per
100,000
Population
92.57
Median Age
34.3
% of Population
with Bachelor's
Degree of Higher
23.3%
Median
Household
Income
$42,865

Illinois

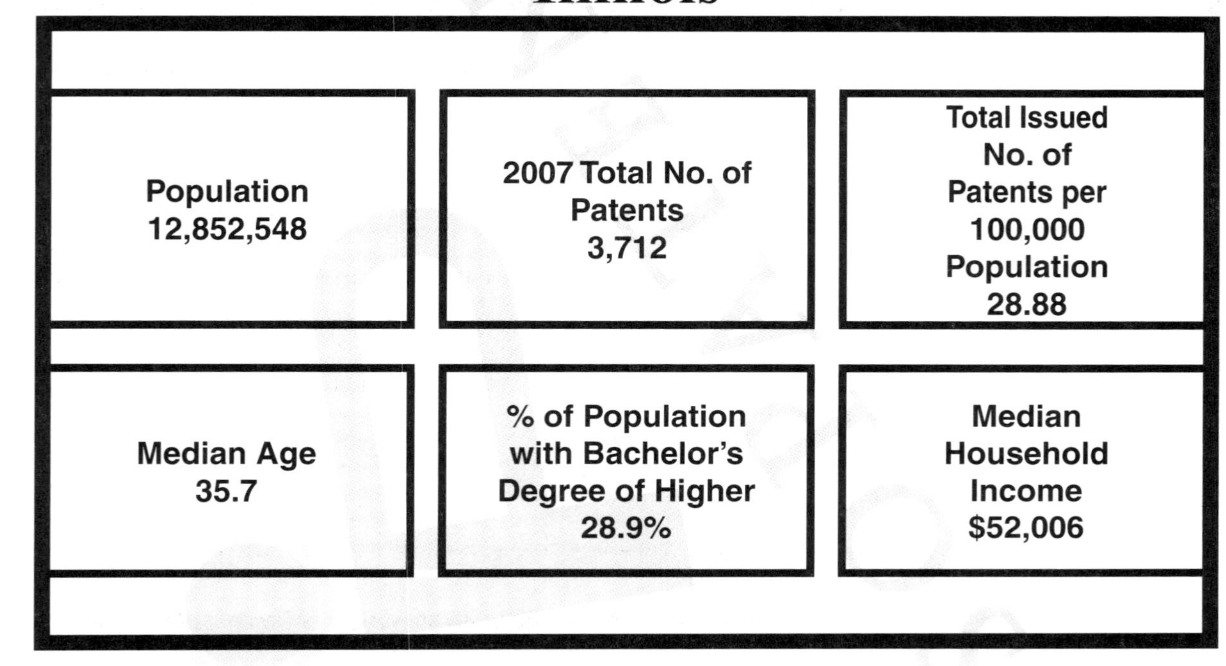
Population
12,852,548
2007 Total No. of
Patents
3,712
Total Issued
No. of
Patents per
100,000
Population
28.88
Median Age
35.7
% of Population
with Bachelor's
Degree of Higher
28.9%
Median
Household
Income
$52,006

Indiana

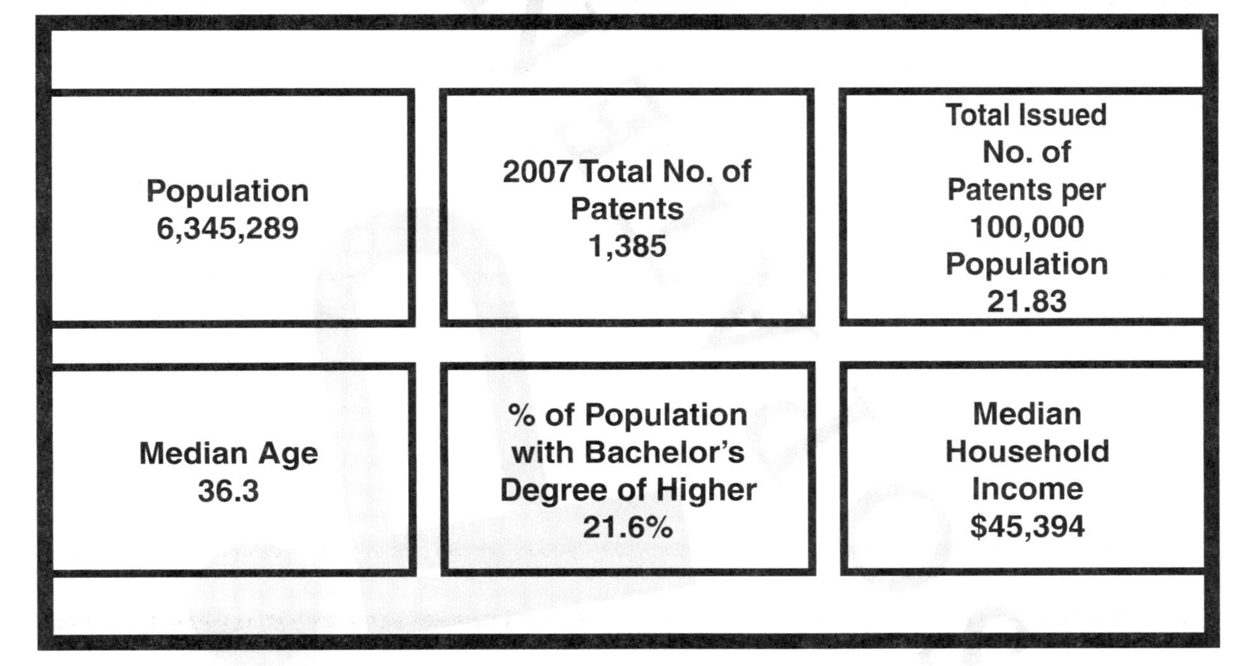
Population
6,345,289
2007 Total No. of
Patents
1,385
Total Issued
No. of
Patents per
100,000
Population
21.83
Median Age
36.3
% of Population
with Bachelor's
Degree of Higher
21.6%
Median
Household
Income
$45,394

Iowa

Population 2,988,046	2007 Total No. of Patents 664	Total Issued No. of Patents per 100,000 Population 22.22
Median Age 37.8	% of Population with Bachelor's Degree of Higher 24%	Median Household Income $44,491

Kansas

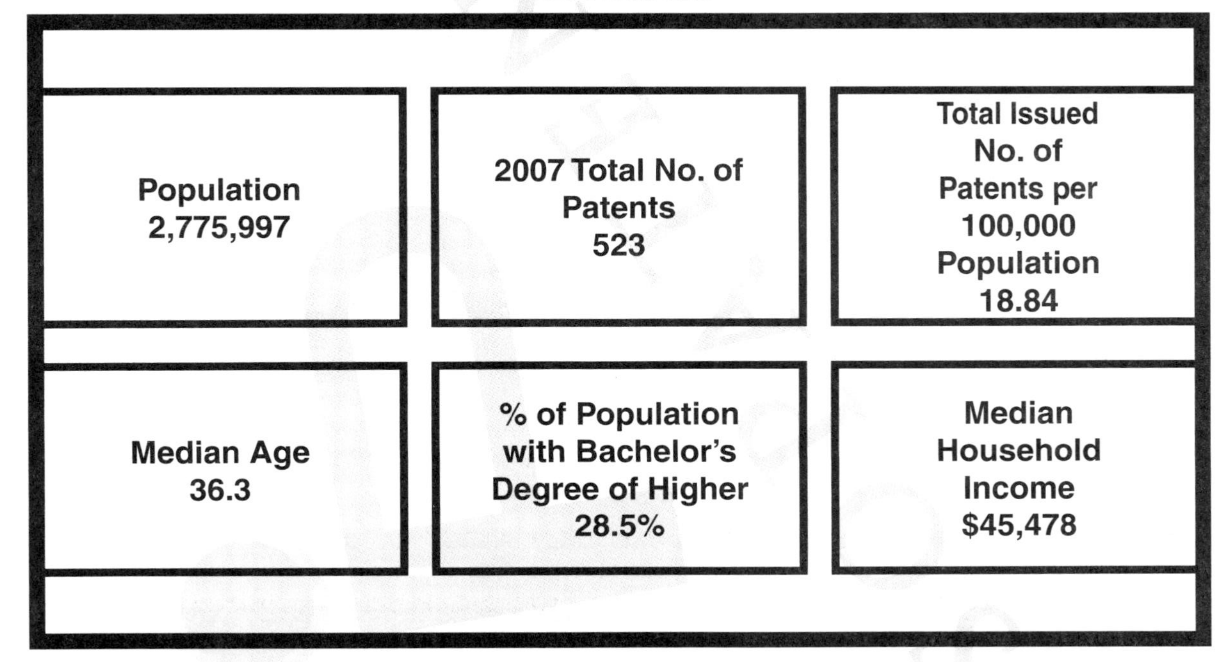
Population
2,775,997
2007 Total No. of
Patents
523
Total Issued
No. of
Patents per
100,000
Population
18.84
Median Age
36.3
% of Population
with Bachelor's
Degree of Higher
28.5%
Median
Household
Income
$45,478

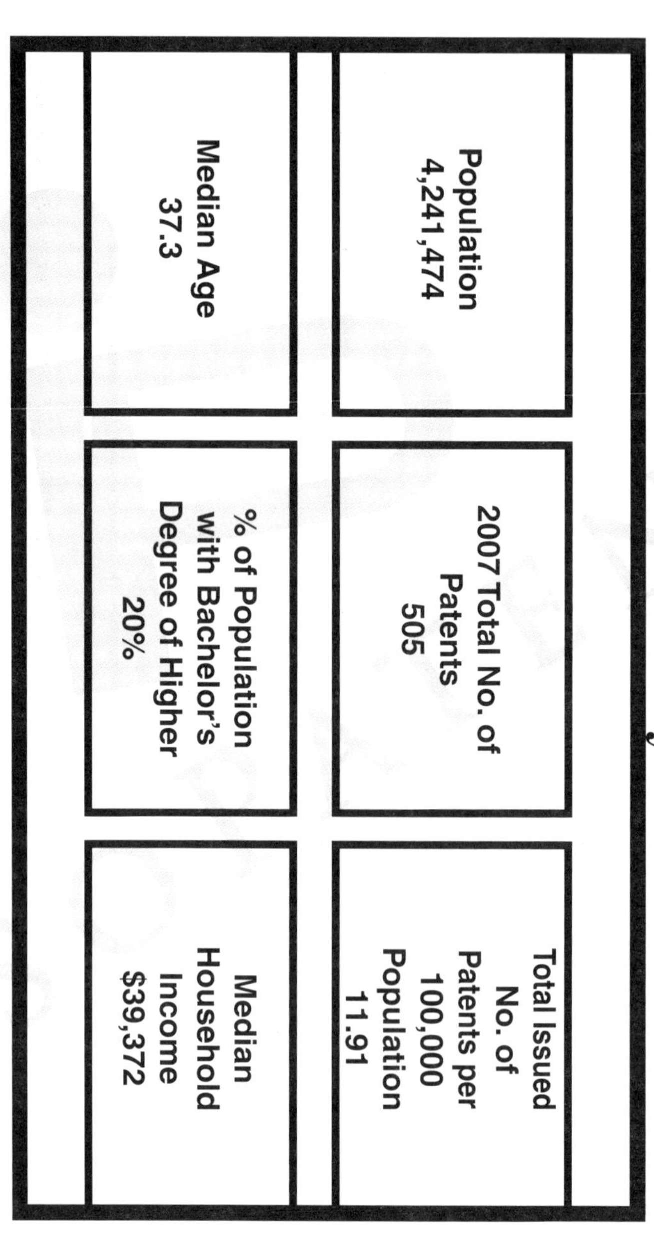
Kentucky
Population
4,241,474
2007 Total No. of Patents
505
Total Issued No. of Patents per 100,000 Population
11.91
Median Age
37.3
% of Population with Bachelor's Degree of Higher
20%
Median Household Income
$39,372

Louisiana

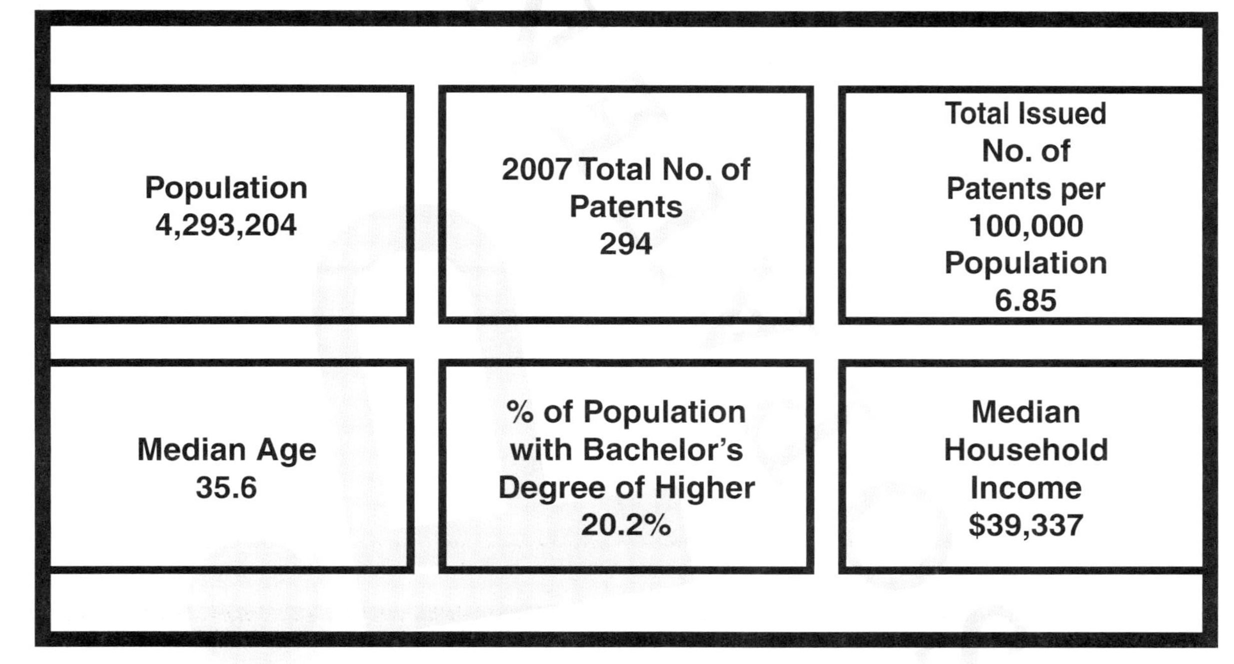

Population
4,293,204
2007 Total No. of
Patents
294
Total Issued
No. of
Patents per
100,000
Population
6.85
Median Age
35.6
% of Population
with Bachelor's
Degree of Higher
20.2%
Median
Household
Income
$39,337

Maine

Population
1,317,207

2007 Total No. of Patents
126

Total Issued No. of Patents per 100,000 Population
9.57

Median Age
41

% of Population with Bachelor's Degree of Higher
25.8%

Median Household Income
$43,439

Maryland

Population 5,618,344	2007 Total No. of Patents 1,409	Total Issued No. of Patents per 100,000 Population 25.08
Median Age 37.3	% of Population with Bachelor's Degree of Higher 35.1%	Median Household Income $65,144

Massachusetts

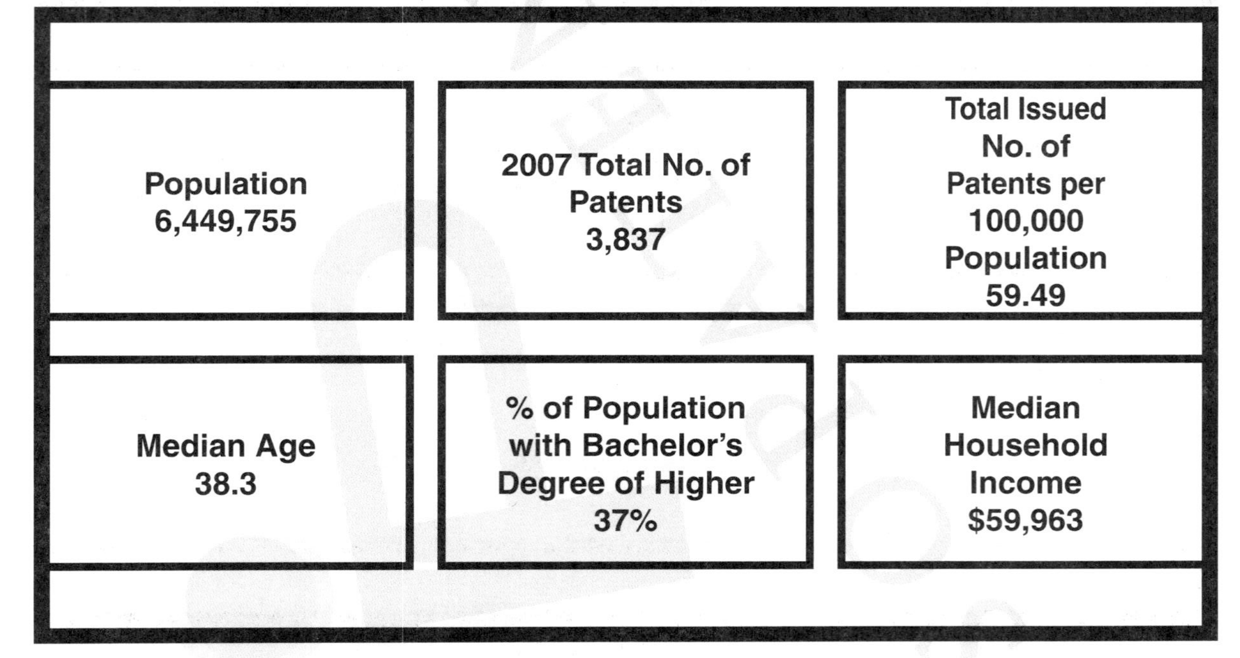

Michigan

Population 10,071,822	2007 Total No. of Patents 3,695	Total Issued No. of Patents per 100,000 Population 36.69
Median Age 37.3	% of Population with Bachelor's Degree of Higher 24.5%	Median Household Income $47,182

Minnesota

Population 5,197,621	2007 Total No. of Patents 2,920	Total Issued No. of Patents per 100,000 Population 56.18
Median Age 36.8	% of Population with Bachelor's Degree of Higher 30.4%	Median Household Income $54,023

Mississippi

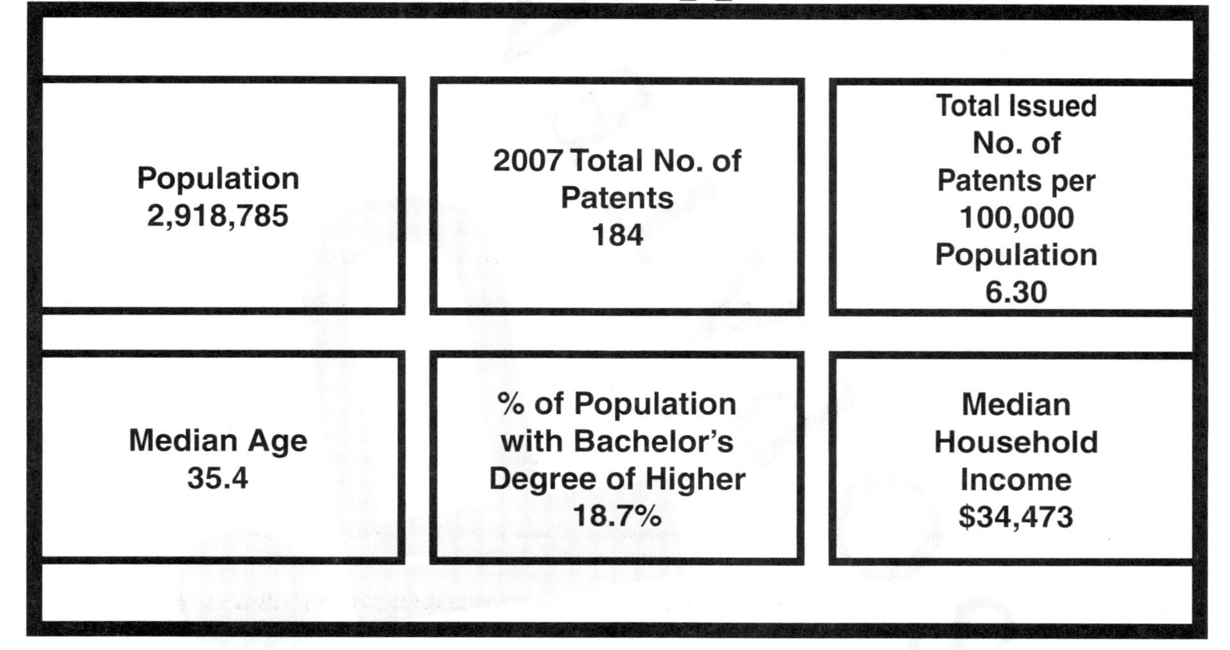

Missouri

Population 5,878,415	2007 Total No. of Patents 861	Total Issued No. of Patents per 100,000 Population 14.65
Median Age 37.2	% of Population with Bachelor's Degree of Higher 24.3%	Median Household Income $42,841

Montana

Population 957,861	2007 Total No. of Patents 131	Total Issued No. of Patents per 100,000 Population 13.68
Median Age 39.5	% of Population with Bachelor's Degree of Higher 27.4%	Median Household Income $40,627

Nebraska

Population 1,774,571	2007 Total No. of Patents 254	Total Issued No. of Patents per 100,000 Population 14.31
Median Age 36	% of Population with Bachelor's Degree of Higher 26.9%	Median Household Income $45,474

Nevada

Population 2,565,382	2007 Total No. of Patents 468	Total Issued No. of Patents per 100,000 Population 18.24
Median Age 35.6	% of Population with Bachelor's Degree of Higher 20.8%	Median Household Income $52,998

New Hampshire

Population 1,315,828	2007 Total No. of Patents 618	Total Issued No. of Patents per 100,000 Population 46.97
Median Age 39.3	% of Population with Bachelor's Degree of Higher 31.9%	Median Household Income $59,683

New Jersey

Population 8,685,920	2007 Total No. of Patents 3,154	Total Issued No. of Patents per 100,000 Population 36.31
Median Age 38.2	% of Population with Bachelor's Degree of Higher 33.4%	Median Household Income $64,470

New Mexico

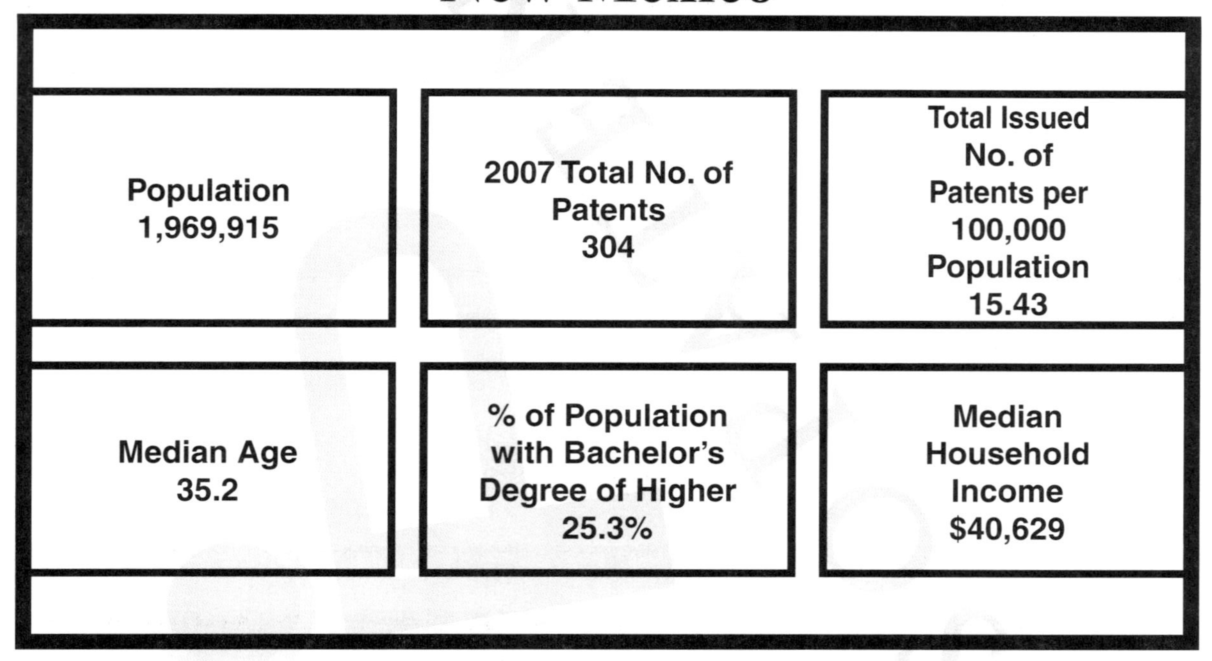

Population
1,969,915
2007 Total No. of
Patents
304
Total Issued
No. of
Patents per
100,000
Population
15.43
Median Age
35.2
% of Population
with Bachelor's
Degree of Higher
25.3%
Median
Household
Income
$40,629

New York

Population 19,297,729	2007 Total No. of Patents 6,025	Total Issued No. of Patents per 100,000 Population 31.22
Median Age 37.4	% of Population with Bachelor's Degree of Higher 31.2%	Median Household Income $51,384

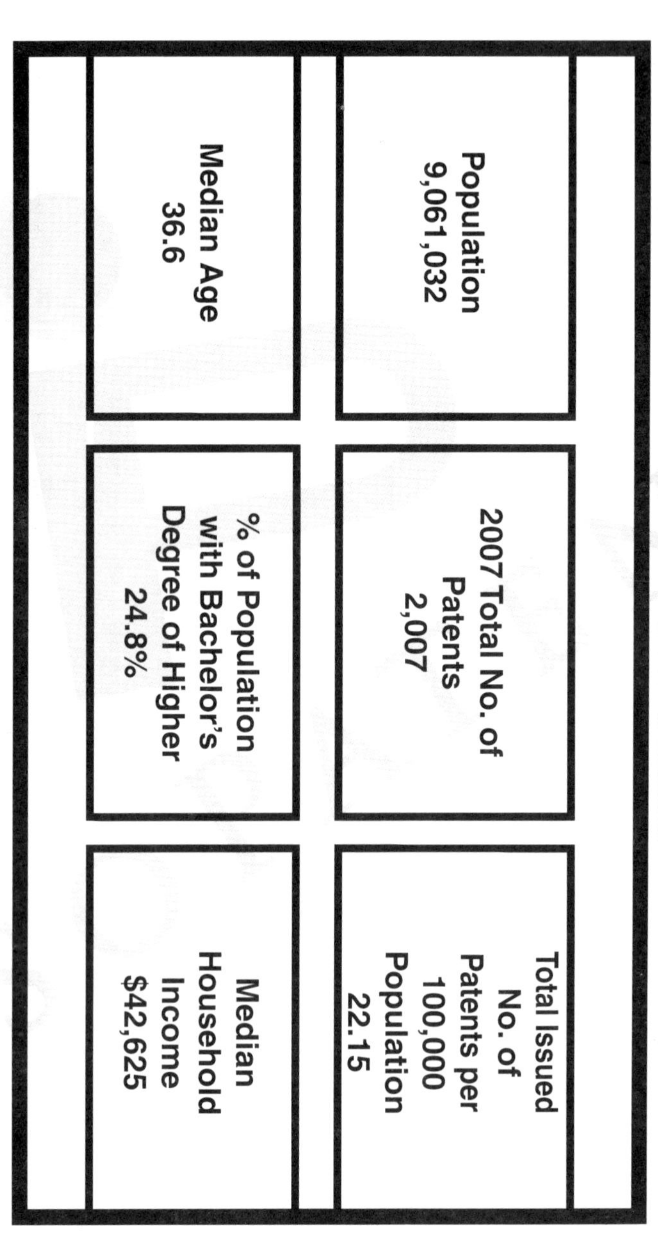
North Carolina
Population
9,061,032
2007 Total No. of Patents
2,007
Total Issued No. of Patents per 100,000 Population
22.15
Median Age
36.6
% of Population with Bachelor's Degree of Higher
24.8%
Median Household Income
$42,625

North Dakota

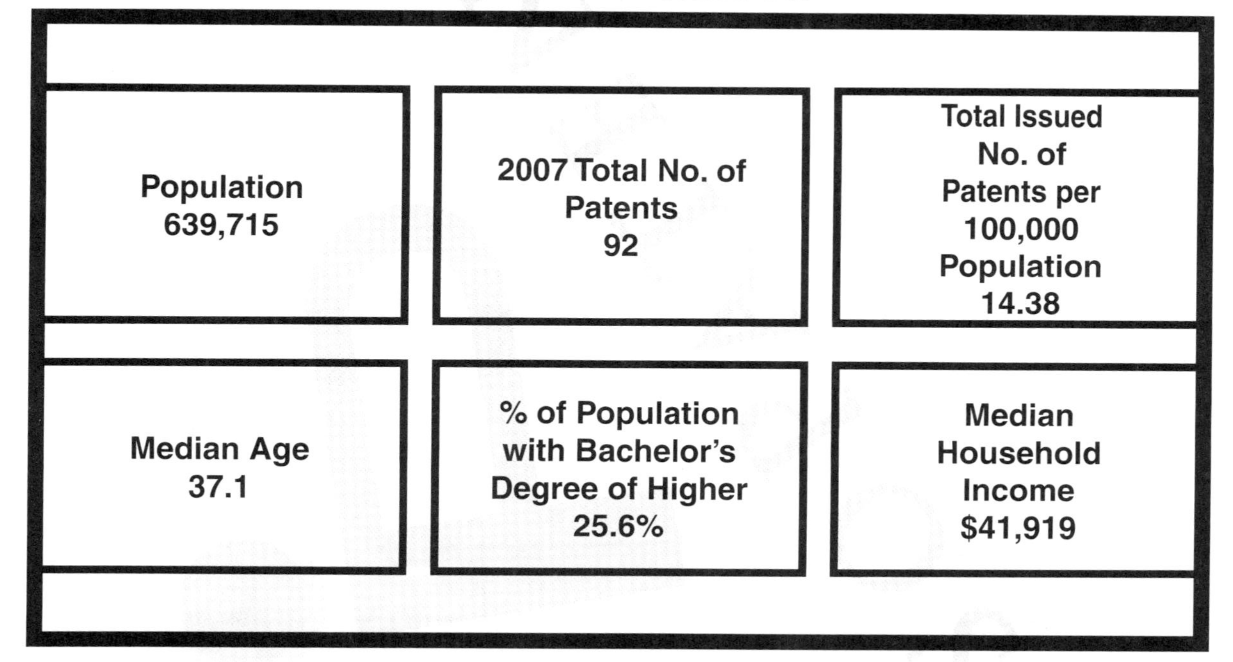

Ohio

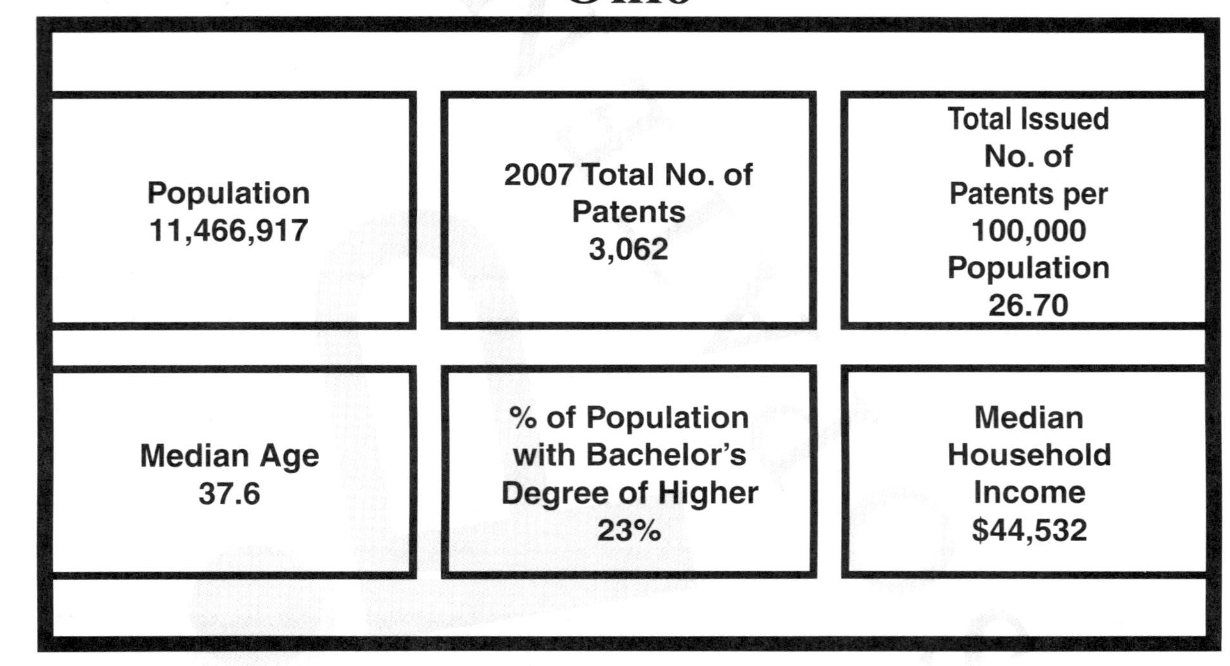
Population
11,466,917
2007 Total No. of
Patents
3,062
Total Issued
No. of
Patents per
100,000
Population
26.70
Median Age
37.6
% of Population
with Bachelor's
Degree of Higher
23%
Median
Household
Income
$44,532

Oklahoma

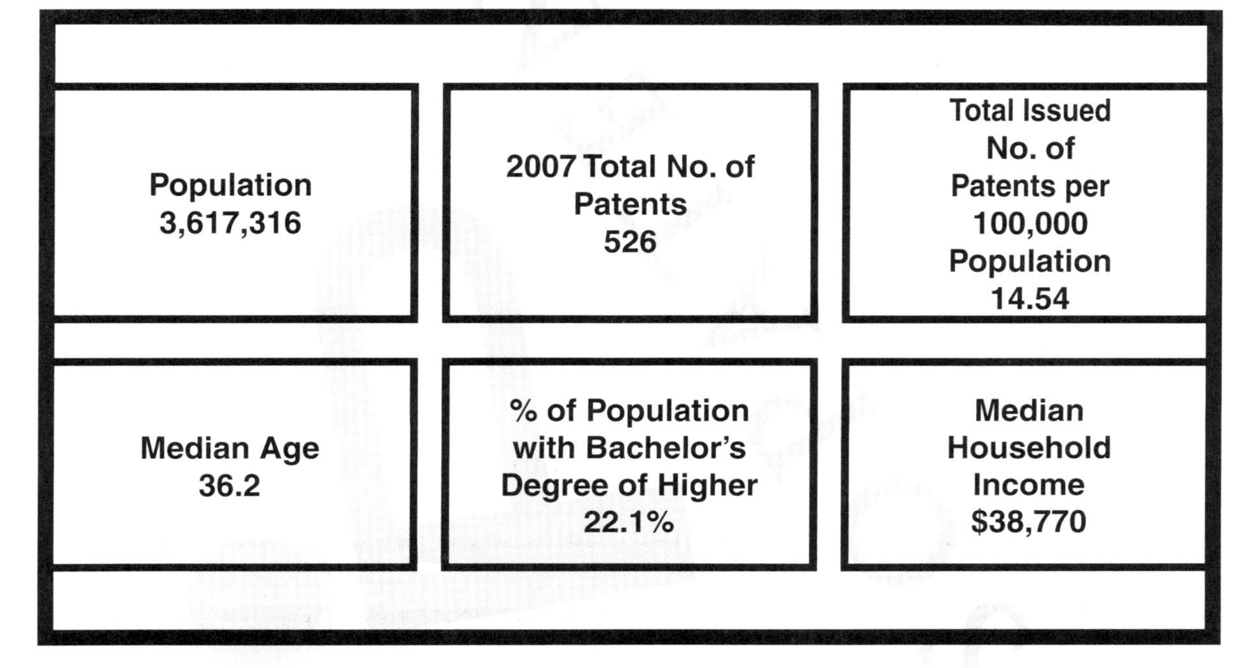
Population
3,617,316
2007 Total No. of
Patents
526
Total Issued
No. of
Patents per
100,000
Population
14.54
Median Age
36.2
% of Population
with Bachelor's
Degree of Higher
22.1%
Median
Household
Income
$38,770

Oregon

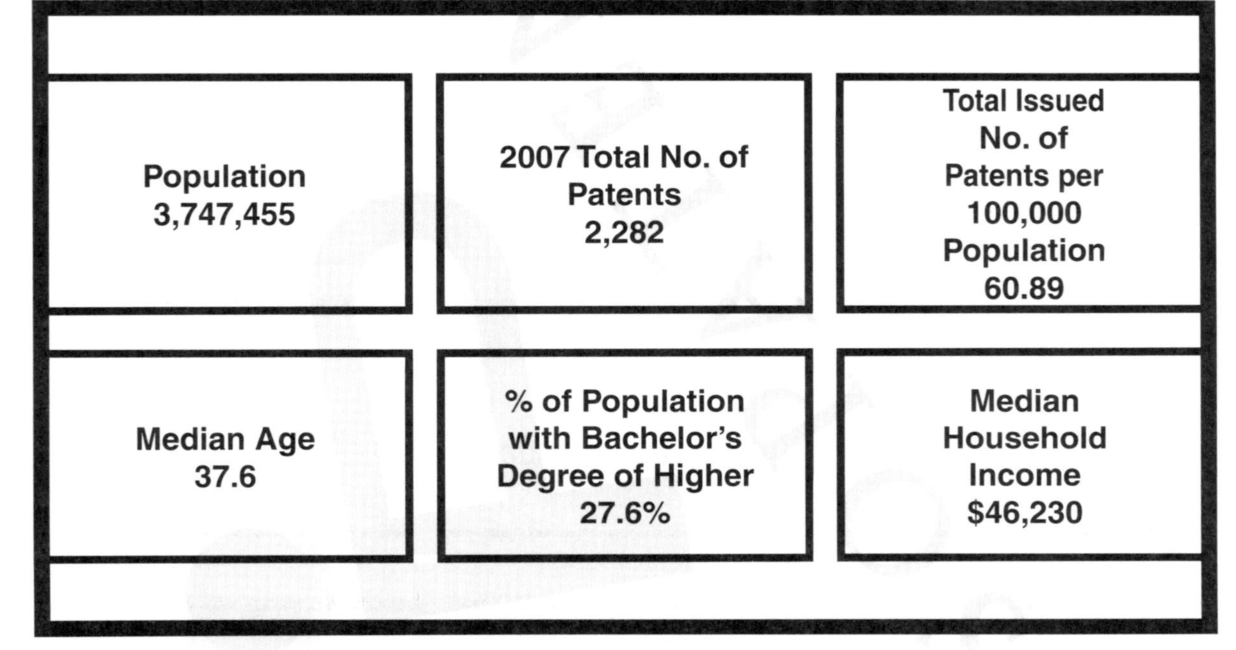
Population
3,747,455
2007 Total No. of
Patents
2,282
Total Issued
No. of
Patents per
100,000
Population
60.89
Median Age
37.6
% of Population
with Bachelor's
Degree of Higher
27.6%
Median
Household
Income
$46,230

Pennsylvania

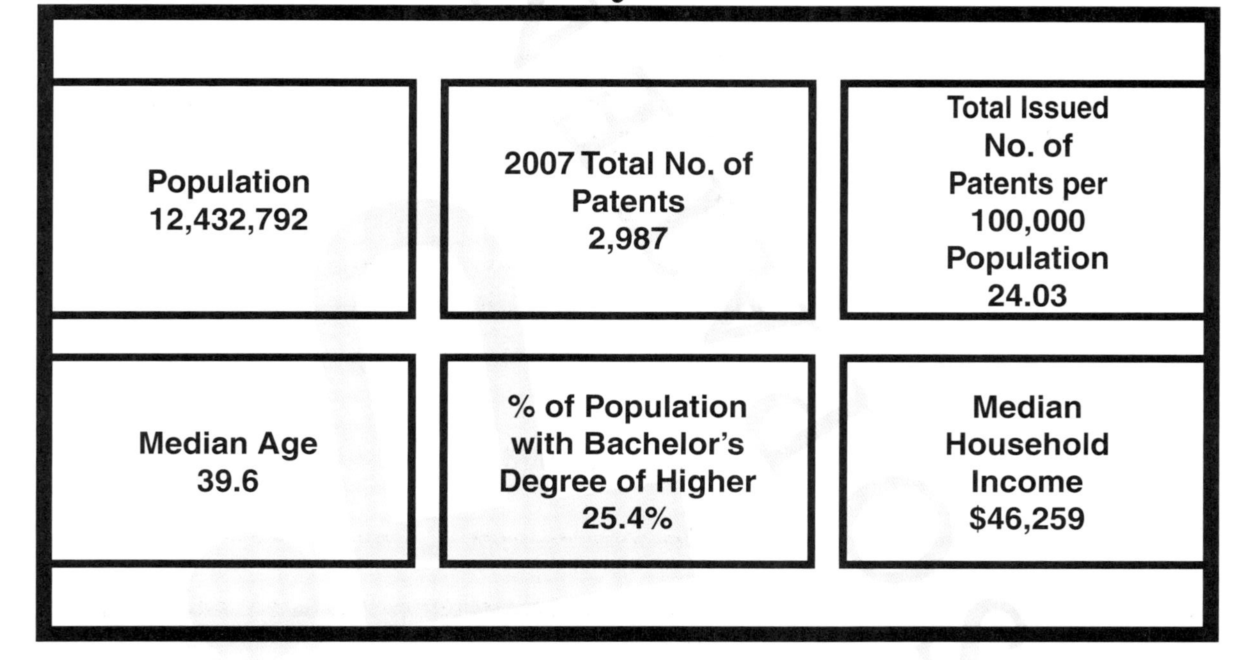

Rhode Island

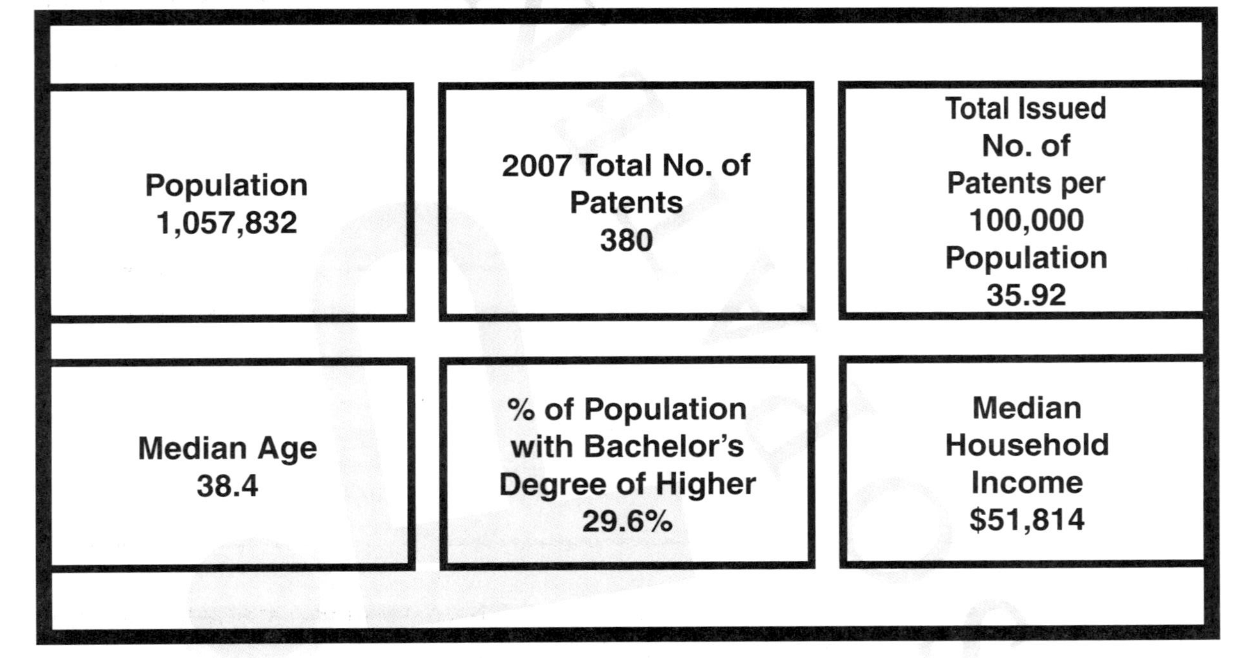

South Carolina

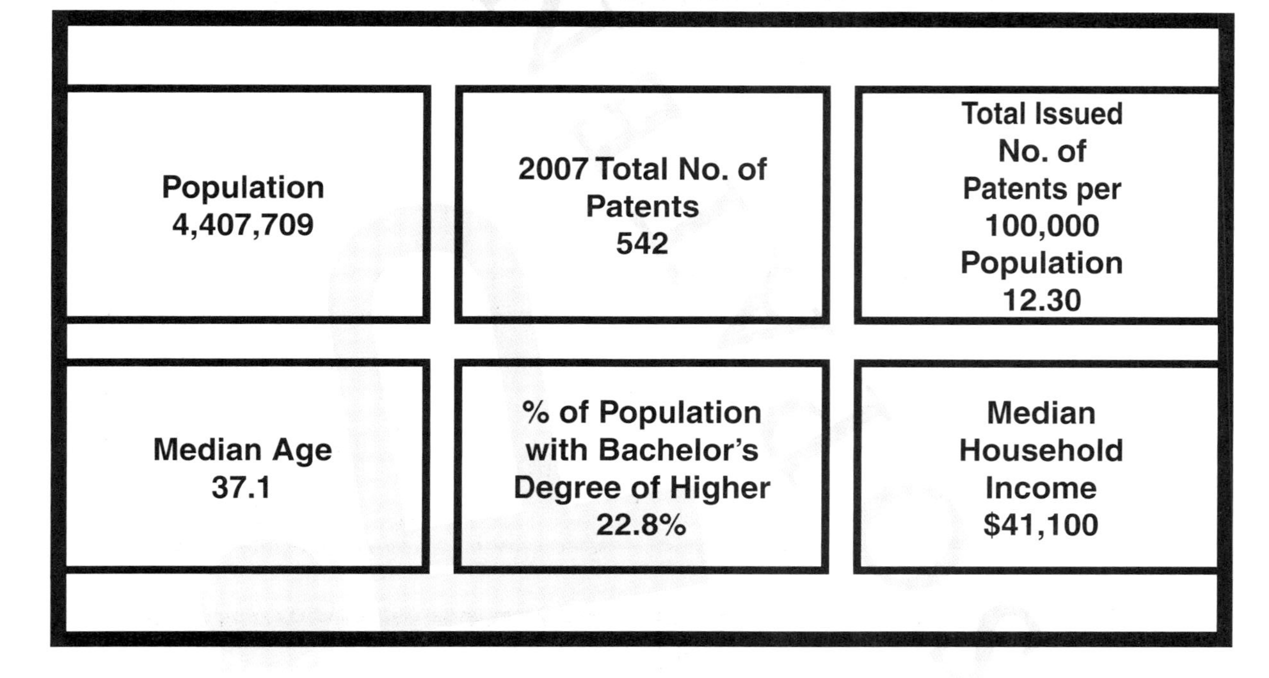

South Dakota

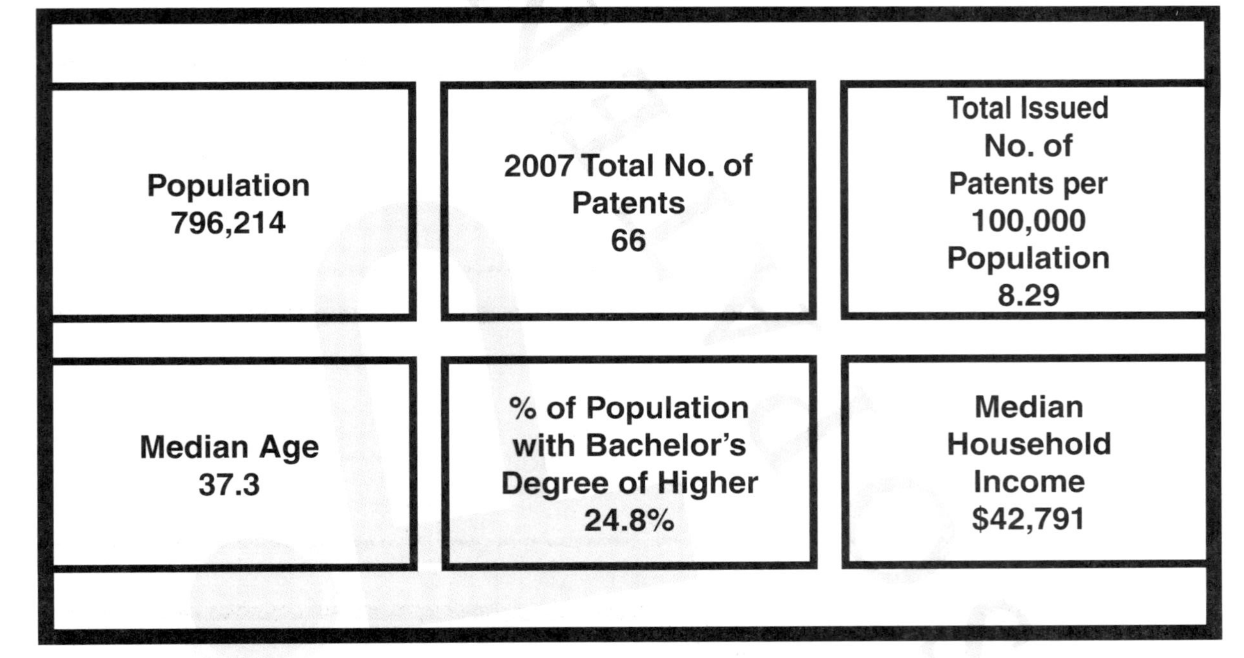

Tennessee

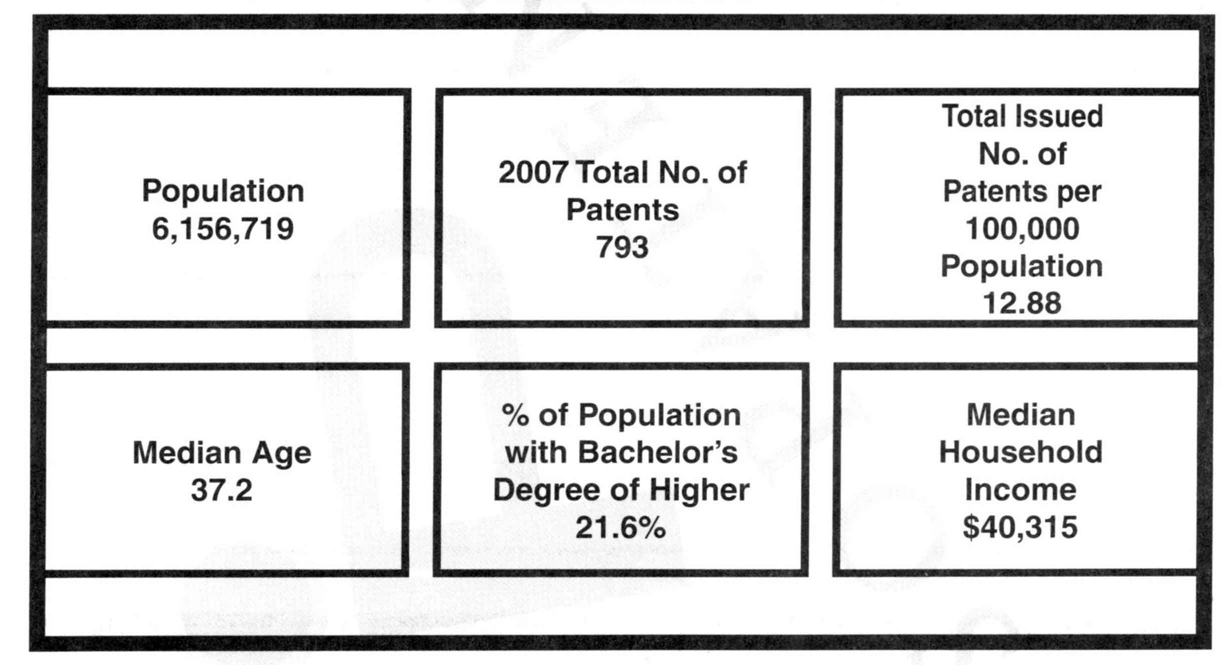
Population
6,156,719
2007 Total No. of
Patents
793
Total Issued
No. of
Patents per
100,000
Population
12.88
Median Age
37.2
% of Population
with Bachelor's
Degree of Higher
21.6%
Median
Household
Income
$40,315

Texas

Population 23,904,380	2007 Total No. of Patents 6,228	Total Issued No. of Patents per 100,000 Population 26.05
Median Age 33.1	% of Population with Bachelor's Degree of Higher 24.7%	Median Household Income $44,922

Utah

Population 2,645,330	2007 Total No. of Patents 766	Total Issued No. of Patents per 100,000 Population 28.96
Median Age 28.4	% of Population with Bachelor's Degree of Higher 28.6%	Median Household Income $51,309

Vermont

Population 621,254	2007 Total No. of Patents 546	Total Issued No. of Patents per 100,000 Population 87.89
Median Age 40.6	% of Population with Bachelor's Degree of Higher 32.5%	Median Household Income $47,665

Virginia

Population 7,712,091	2007 Total No. of Patents 1,165	Total Issued No. of Patents per 100,000 Population 15.11
Median Age 36.9	% of Population with Bachelor's Degree of Higher 32.7%	Median Household Income $56,277

Washington

Population 6,468,424	2007 Total No. of Patents 3,894	Total Issued No. of Patents per 100,000 Population 60.20
Median Age 36.7	% of Population with Bachelor's Degree of Higher 30.5%	Median Household Income $52,583

West Virginia

Population 1,812,035	2007 Total No. of Patents 116	Total Issued No. of Patents per 100,000 Population 6.40
Median Age 40.7	% of Population with Bachelor's Degree of Higher 16.6%	Median Household Income $35,059

Wisconsin

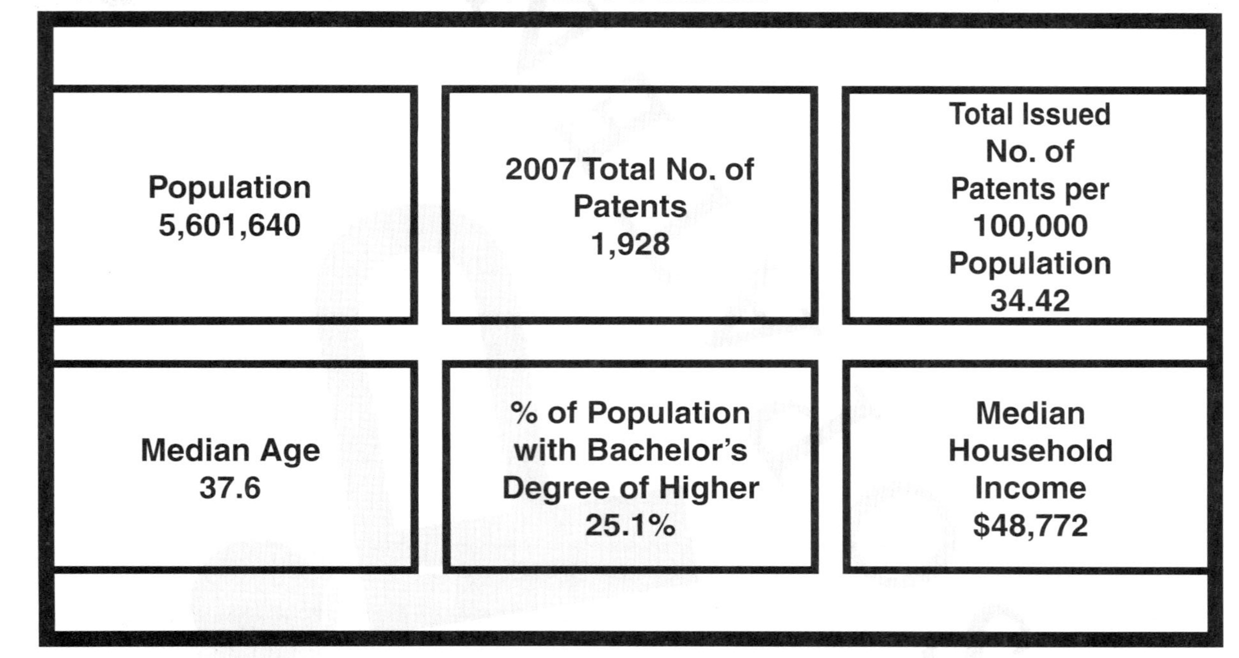
Population
5,601,640
2007 Total No. of
Patents
1,928
Total Issued
No. of
Patents per
100,000
Population
34.42
Median Age
37.6
% of Population
with Bachelor's
Degree of Higher
25.1%
Median
Household
Income
$48,772

Wyoming

Population 522,830	2007 Total No. of Patents 62	Total Issued No. of Patents per 100,000 Population 11.86
Median Age 37.5	% of Population with Bachelor's Degree of Higher 22.8%	Median Household Income $47,423

United States

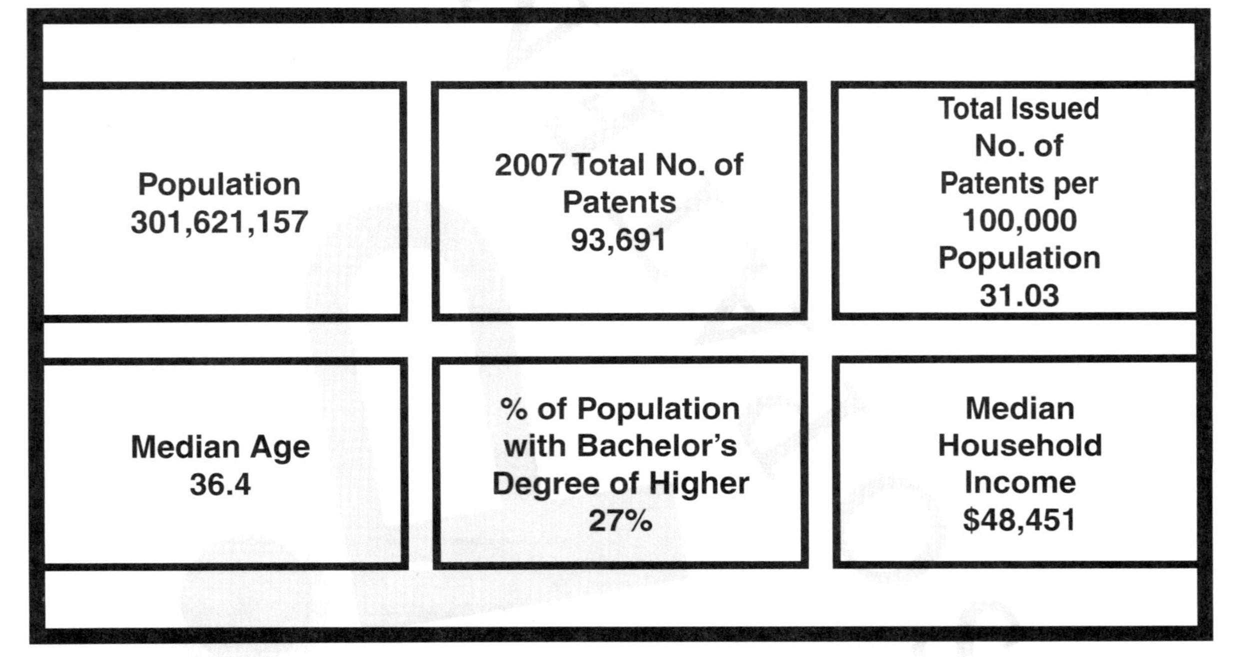

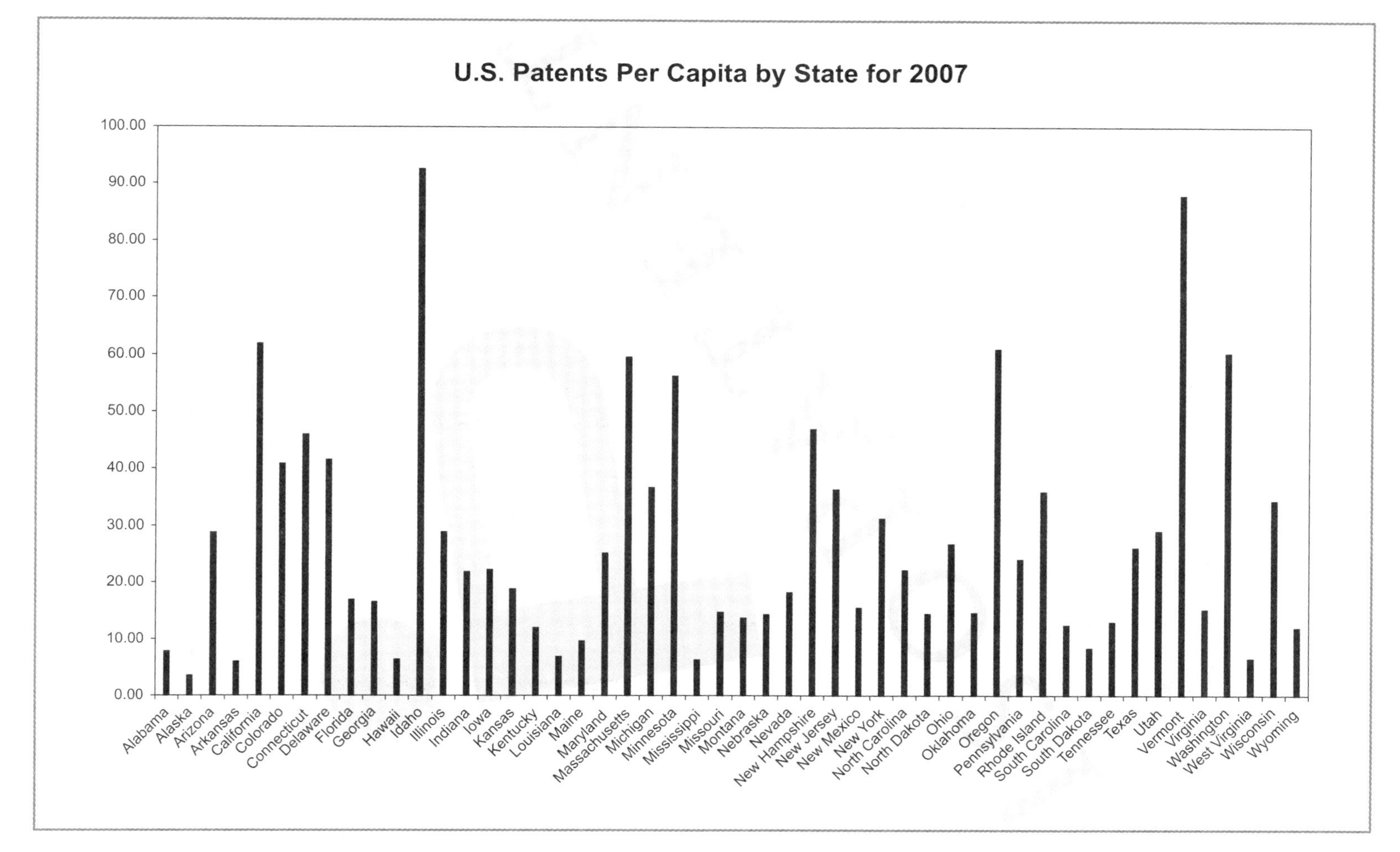
U.S. Patents Per Capita by State for 2007
100.00
90.00
80.00
70.00
60.00
50.00
40.00
30.00
20.00
10.00
0.00
Alabama
Alaska
Arizona
Arkansas
California
Colorado
Connecticut
Delaware
Florida
Georgia
Hawaii
Idaho
Illinois
Indiana
Iowa
Kansas
Kentucky
Louisiana
Maine
Maryland
Massachusetts
Michigan
Minnesota
Mississippi
Missouri
Montana
Nebraska
Nevada
New Hampshire
New Jersey
New Mexico
New York
North Carolina
North Dakota
Ohio
Oklahoma
Oregon
Pennsylvania
Rhode Island
South Carolina
South Dakota
Tennessee
Texas
Utah
Vermont
Virginia
Washington
West Virginia
Wisconsin
Wyoming

Summary of 2007 State Demographic and Patent Data

State	Population	Median House Hold Income	*% with Bachelors Degree	% with Graduate or Doctorate Degree	% with Bachelor's Degree or Higher	Median Age	2007 Number of Patents Issued	2007 Patents per 100K Population
Alabama	4,627,851	$38,783	13.30%	7.70%	21.00%	37.2	358	7.74
Alaska	683,478	$59,393	17.40%	9.50%	26.90%	33.5	24	3.51
Arizona	6,338,755	$47,265	16.30%	9.20%	25.50%	34.6	1,826	28.81
Arkansas	2,834,797	$36,599	12.00%	6.20%	18.20%	37.1	169	5.96
California	36,553,215	$56,645	18.60%	10.40%	29.00%	34.4	22,601	61.83
Colorado	4,861,515	$52,015	22.00%	12.40%	34.40%	35.4	1,983	40.79
Connecticut	3,502,309	$63,422	19.30%	14.40%	33.70%	39.1	1,611	46.00
Delaware	864,764	$52,833	16.40%	10.50%	26.90%	37.5	359	41.51
Florida	18,251,243	$45,495	16.40%	8.90%	25.30%	39.8	3,092	16.94
Georgia	9,544,750	$46,832	17.30%	9.20%	26.50%	34.6	1,580	16.55
Hawaii	1,283,388	$61,160	19.90%	9.80%	29.70%	37.2	82	6.39
Idaho	1,499,402	$42,865	16.20%	7.10%	23.30%	34.3	1,388	92.57
Illinois	12,852,548	$52,006	18.10%	10.80%	28.90%	35.7	3,712	28.88
Indiana	6,345,289	$45,394	13.60%	8.00%	21.60%	36.3	1,385	21.83
Iowa	2,988,046	$44,491	16.60%	7.40%	24.00%	37.8	664	22.22
Kansas	2,775,997	$45,478	18.70%	9.80%	28.50%	36.3	523	18.84
Kentucky	4,241,474	$39,372	11.80%	8.20%	20.00%	37.3	505	11.91
Louisiana	4,293,204	$39,337	13.40%	6.80%	20.20%	35.6	294	6.85
Maine	1,317,207	$43,439	16.90%	8.90%	25.80%	41	126	9.57
Maryland	5,618,344	$65,144	19.40%	15.70%	35.10%	37.3	1,409	25.08
Massachusetts	6,449,755	$59,963	21.40%	15.60%	37.00%	38.3	837	59.49
Michigan	10,071,822	$47,182	15.30%	9.20%	24.50%	37.3	3,695	36.69
Minnesota	5,197,621	$54,023	20.80%	9.60%	30.40%	36.8	2,920	56.18
Mississippi	2,918,785	$34,473	12.60%	6.10%	18.70%	35.4	184	6.30
Missouri	5,878,415	$42,841	15.60%	8.70%	24.30%	37.2	861	14.65
Montana	957,861	$40,627	19.00%	8.40%	27.40%	39.5	131	13.68
Nebraska	1,774,571	$45,474	18.50%	8.40%	26.90%	36	254	14.31
Nevada	2,565,382	$52,998	13.60%	7.20%	20.80%	35.6	468	18.24
New Hampshire	1,315,828	$59,683	20.70%	11.20%	31.90%	39.3	618	46.97
New Jersey	8,685,920	$64,470	21.00%	12.40%	33.40%	38.2	3,154	36.31
New Mexico	1,969,915	$40,629	14.40%	10.90%	25.30%	35.2	304	15.43
New York	19,297,729	$51,384	17.90%	13.30%	31.20%	37.4	6,025	31.22
North Carolina	9,061,032	$42,625	16.50%	8.30%	24.80%	36.6	2,007	22.15
North Dakota	639,715	$41,919	19.10%	6.50%	25.60%	37.1	92	14.38
Ohio	11,466,917	$44,532	14.70%	8.30%	23.00%	37.6	3,062	26.70
Oklahoma	3,617,316	$38,770	14.90%	7.20%	22.10%	36.2	526	14.54
Oregon	3,747,455	$46,230	17.60%	10.00%	27.60%	37.6	2,282	60.89
Pennsylvania	12,432,792	$46,259	15.80%	9.60%	25.40%	39.6	2,987	24.03
Rhode Island	1,057,832	$51,814	18.30%	11.30%	29.60%	38.4	380	35.92
South Carolina	4,407,709	$41,100	14.90%	7.90%	22.80%	37.1	542	12.30
South Dakota	796,214	$42,791	17.60%	7.20%	24.80%	37.3	66	8.29
Tennessee	6,156,719	$40,315	14.10%	7.50%	21.60%	37.2	793	12.88
Texas	23,904,380	$44,922	16.70%	8.00%	24.70%	33.1	6,228	26.05
Utah	2,645,330	$51,309	19.20%	9.40%	28.60%	28.4	766	28.96
Vermont	621,254	$47,665	19.70%	12.80%	32.50%	40.6	546	87.89
Virginia	7,712,091	$56,277	19.50%	13.20%	32.70%	36.9	1,165	15.11
Washington	6,468,424	$52,583	19.80%	10.70%	30.50%	36.7	3,894	60.20
West Virginia	1,812,035	$35,059	10.00%	6.60%	16.60%	40.7	116	6.40
Wisconsin	5,601,640	$48,772	16.70%	8.40%	25.10%	37.6	1,928	34.42
Wyoming	522,830	$47,423	15.40%	7.40%	22.80%	37.5	62	11.86
Total US[72]	301,621,157	$48,451	17.10%	9.90%	27.00%	36.4	93,584	31.03

Chapter 16
Top Technology Categories for Each of the Fifty United States

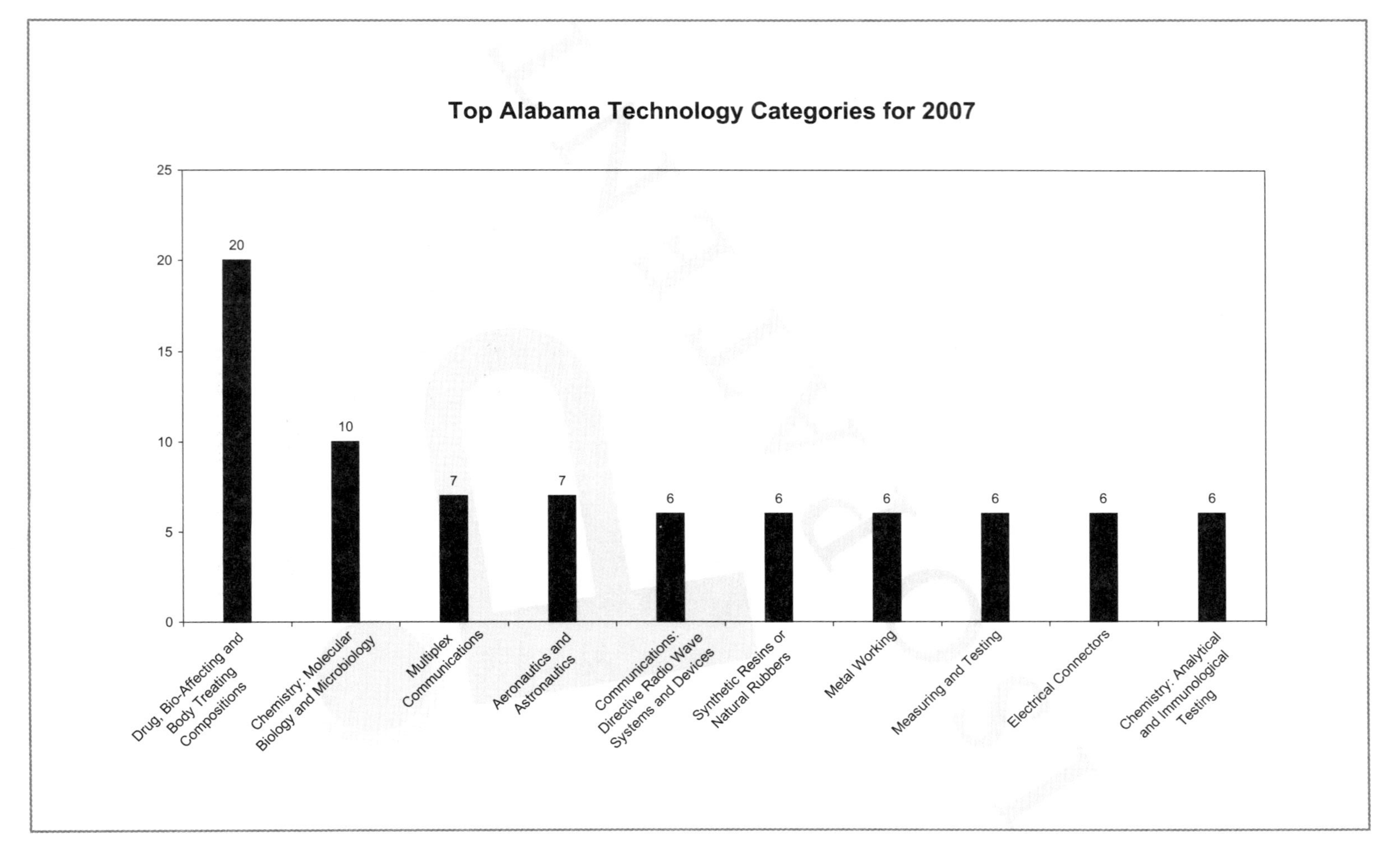
Top Alabama Technology Categories for 2007
25
20
15
10
5
0
20
10
7
7
6
6
6
6
6
6
Drug, Bio-Affecting and Body Treating Compositions
Chemistry: Molecular Biology and Microbiology
Multiplex Communications
Aeronautics and Astronautics
Communications: Directive Radio Wave Systems and Devices
Synthetic Resins or Natural Rubbers
Metal Working
Measuring and Testing
Electrical Connectors
Chemistry: Analytical and Immunological Testing

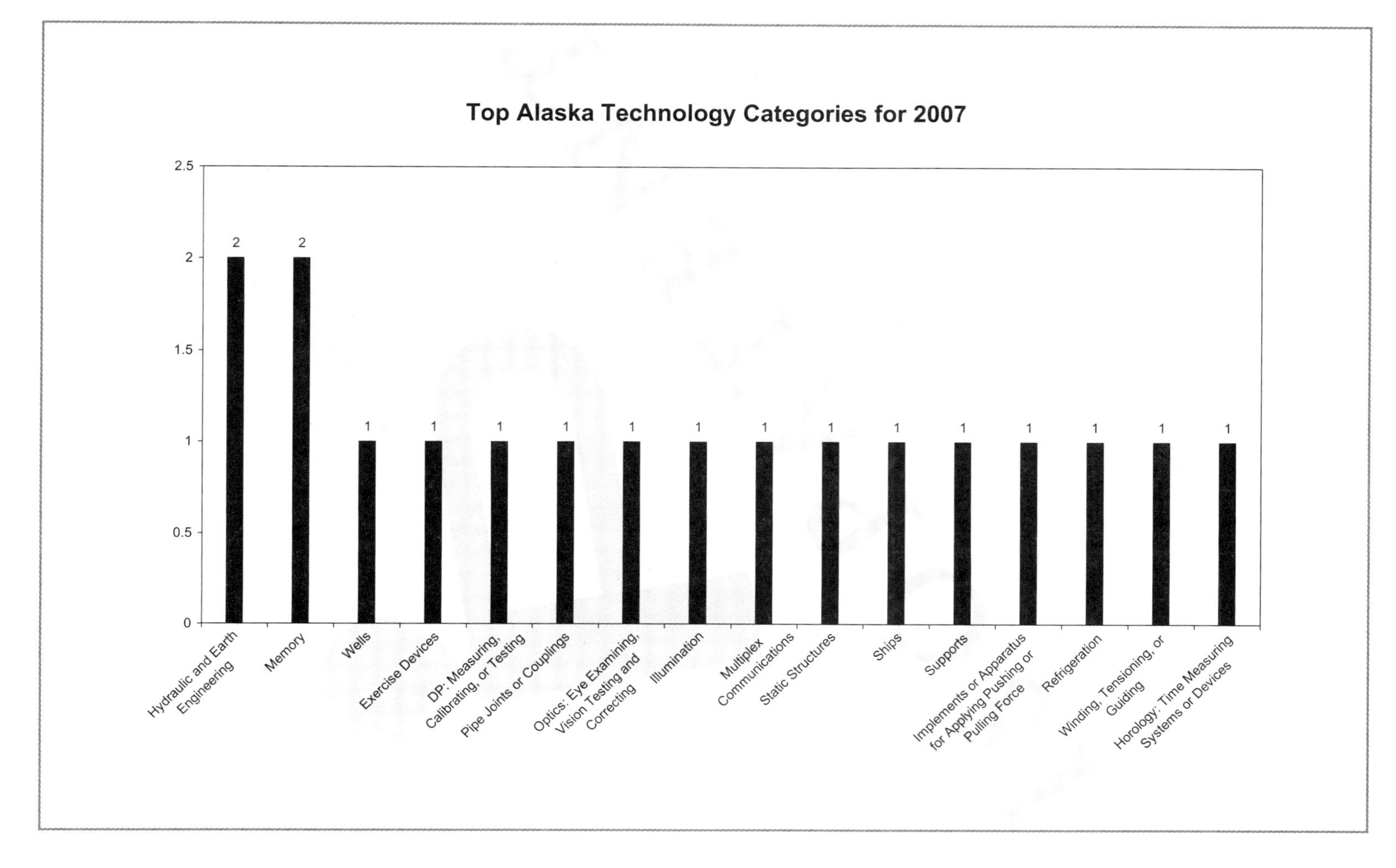
Top Alaska Technology Categories for 2007
2.5
2
1.5
1
0.5
0
2
2
1
1
1
1
1
1
1
1
1
1
1
1
1
1
Hydraulic and Earth Engineering
Memory
Wells
Exercise Devices
DP: Measuring, Calibrating, or Testing
Pipe Joints or Couplings
Optics: Eye Examining, Vision Testing and Correcting
Illumination
Multiplex Communications
Static Structures
Ships
Supports
Implements or Apparatus for Applying Pushing or Pulling Force
Refrigeration
Winding, Tensioning, or Guiding
Horology: Time Measuring Systems or Devices

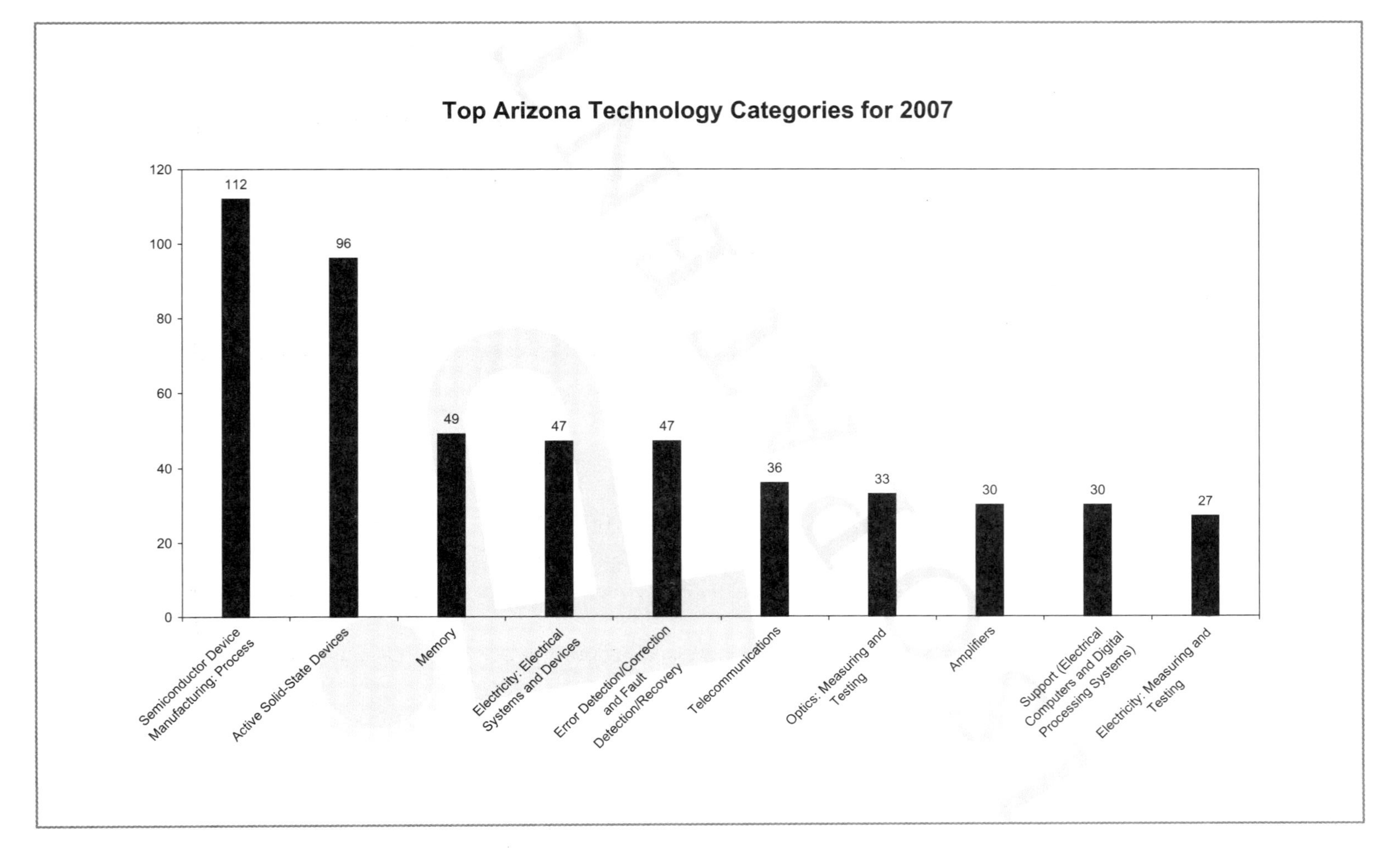

Top Arizona Technology Categories for 2007
0
20
40
60
80
100
120
112
96
49
47
47
36
33
30
30
27
Semiconductor Device Manufacturing: Process
Active Solid-State Devices
Memory
Electricity: Electrical Systems and Devices
Error Detection/Correction and Fault Detection/Recovery
Telecommunications
Optics: Measuring and Testing
Amplifiers
Support (Electrical Computers and Digital Processing Systems)
Electricity: Measuring and Testing

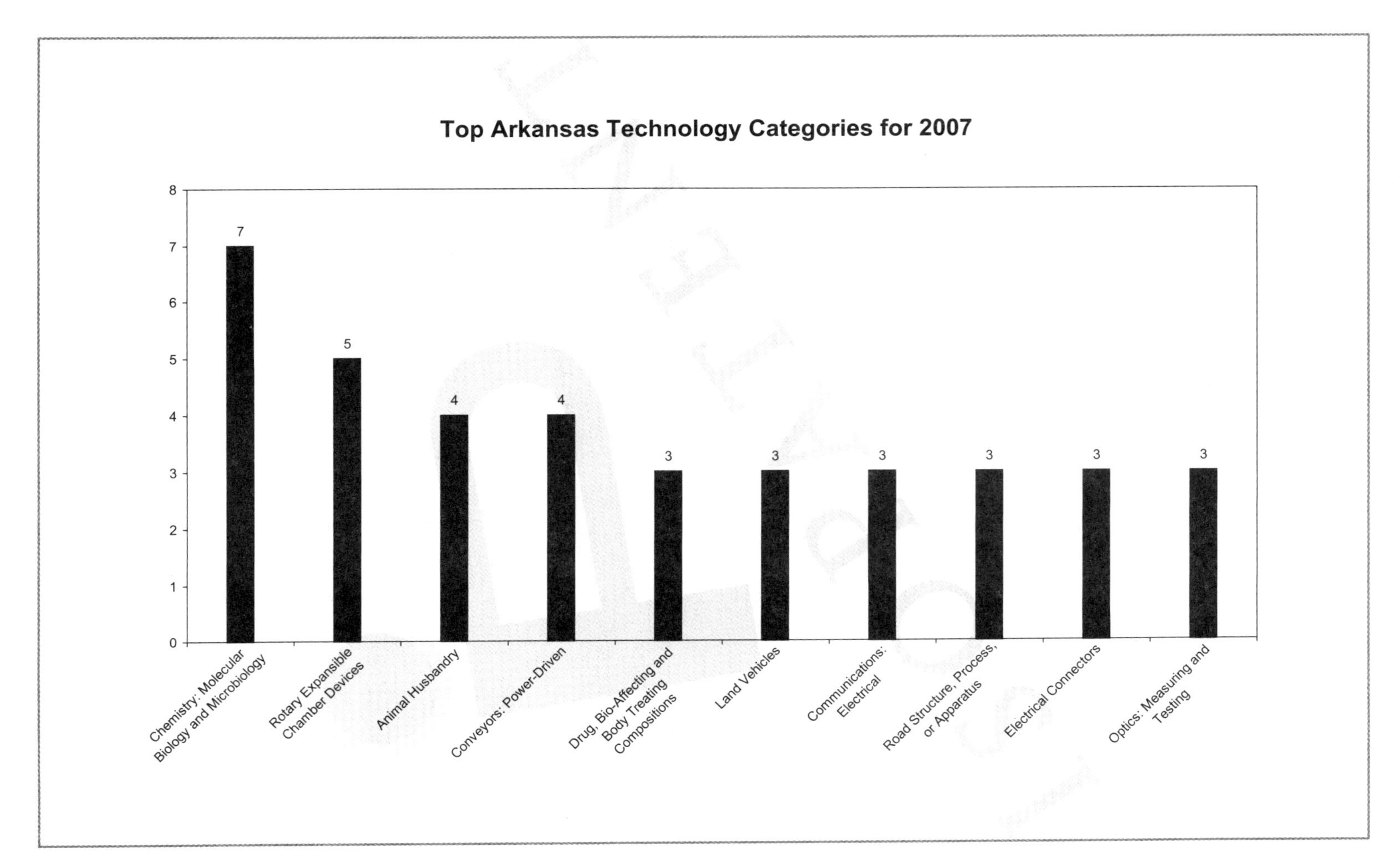

Top Arkansas Technology Categories for 2007
8
7
6
5
4
3
2
1
0
7
5
4
4
3
3
3
3
3
3
Chemistry: Molecular Biology and Microbiology
Rotary Expansible Chamber Devices
Animal Husbandry
Conveyors: Power-Driven
Drug, Bio-Affecting and Body Treating Compositions
Land Vehicles
Communications: Electrical
Road Structure, Process, or Apparatus
Electrical Connectors
Optics: Measuring and Testing

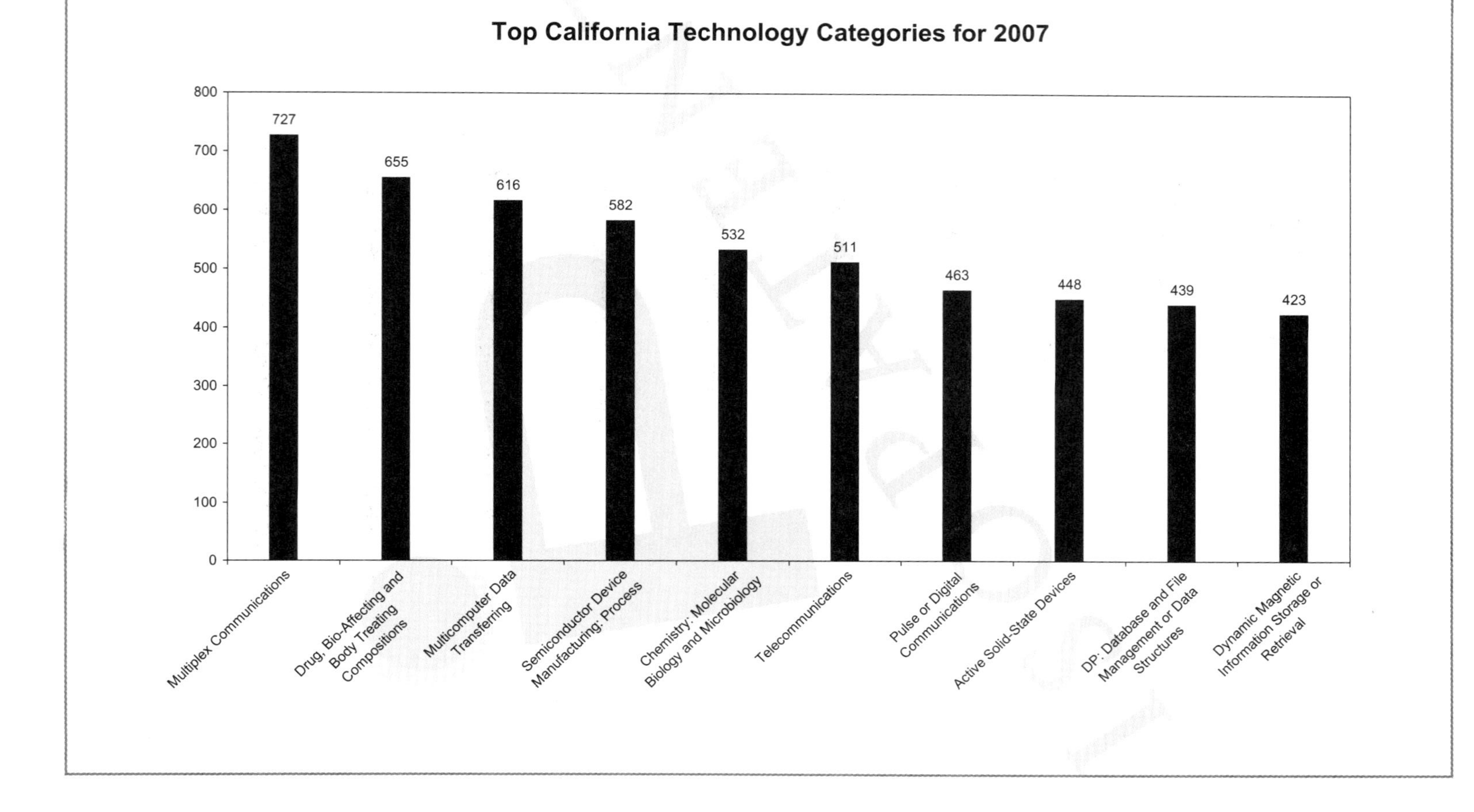

Top California Technology Categories for 2007
800
700
600
500
400
300
200
100
0
727
655
616
582
532
511
463
448
439
423
Multiplex Communications
Drug, Bio-Affecting and Body Treating Compositions
Multicomputer Data Transferring
Semiconductor Device Manufacturing: Process
Chemistry: Molecular Biology and Microbiology
Telecommunications
Pulse or Digital Communications
Active Solid-State Devices
DP: Database and File Management or Data Structures
Dynamic Magnetic Information Storage or Retrieval

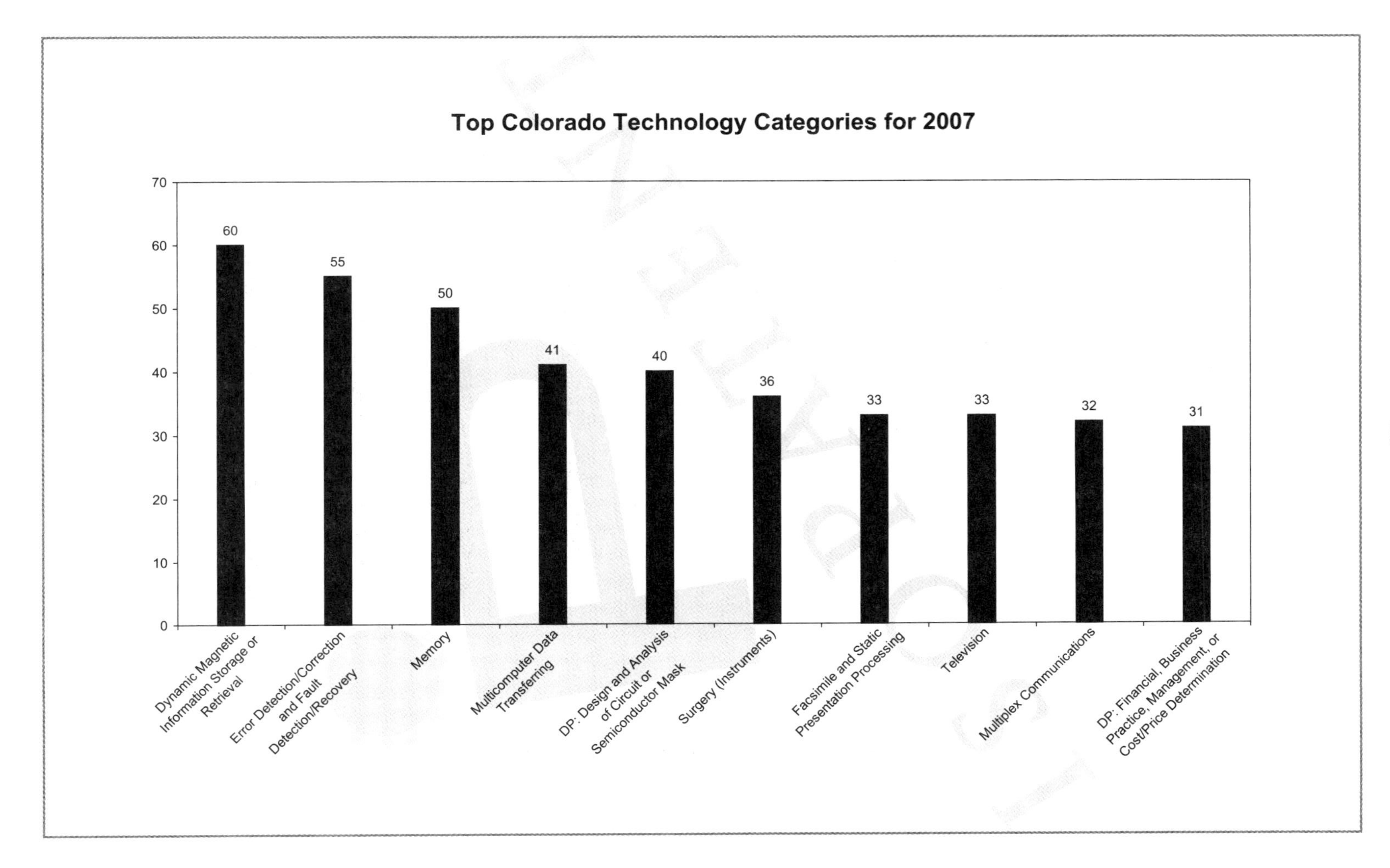
Top Colorado Technology Categories for 2007
70
60
50
40
30
20
10
0
60
55
50
41
40
36
33
33
32
31
Dynamic Magnetic Information Storage or Retrieval
Error Detection/Correction and Fault Detection/Recovery
Memory
Multicomputer Data Transferring
DP: Design and Analysis of Circuit or Semiconductor Mask
Surgery (Instruments)
Facsimile and Static Presentation Processing
Television
Multiplex Communications
DP: Financial, Business Practice, Management, or Cost/Price Determination

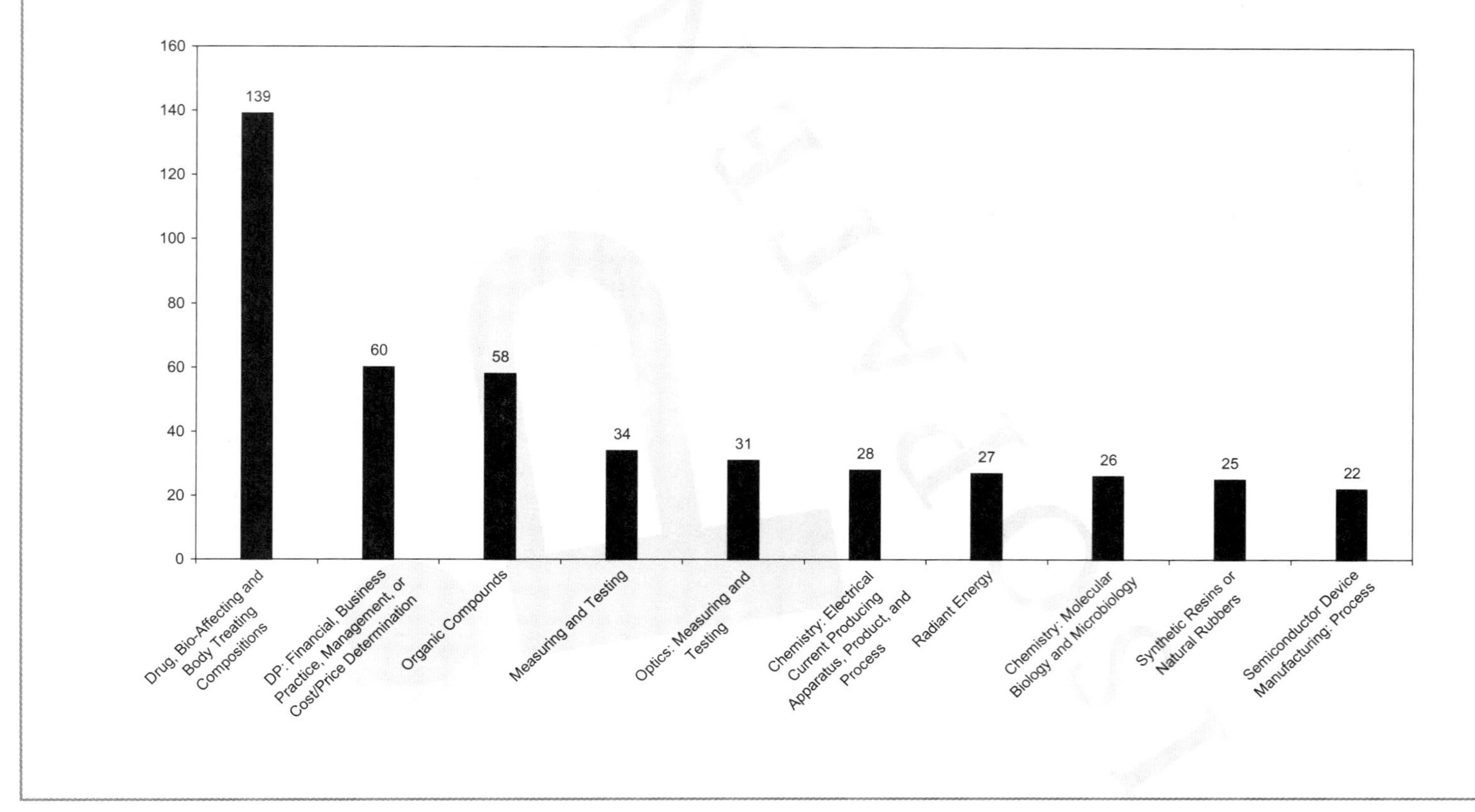

Top Connecticut Technology Categories for 2007
160
140
120
100
80
60
40
20
0
139
60
58
34
31
28
27
26
25
22
Drug, Bio-Affecting and Body Treating Compositions
DP: Financial, Business Practice, Management, or Cost/Price Determination
Organic Compounds
Measuring and Testing
Optics: Measuring and Testing
Chemistry: Electrical Current Producing Apparatus, Product, and Process
Radiant Energy
Chemistry: Molecular Biology and Microbiology
Synthetic Resins or Natural Rubbers
Semiconductor Device Manufacturing: Process

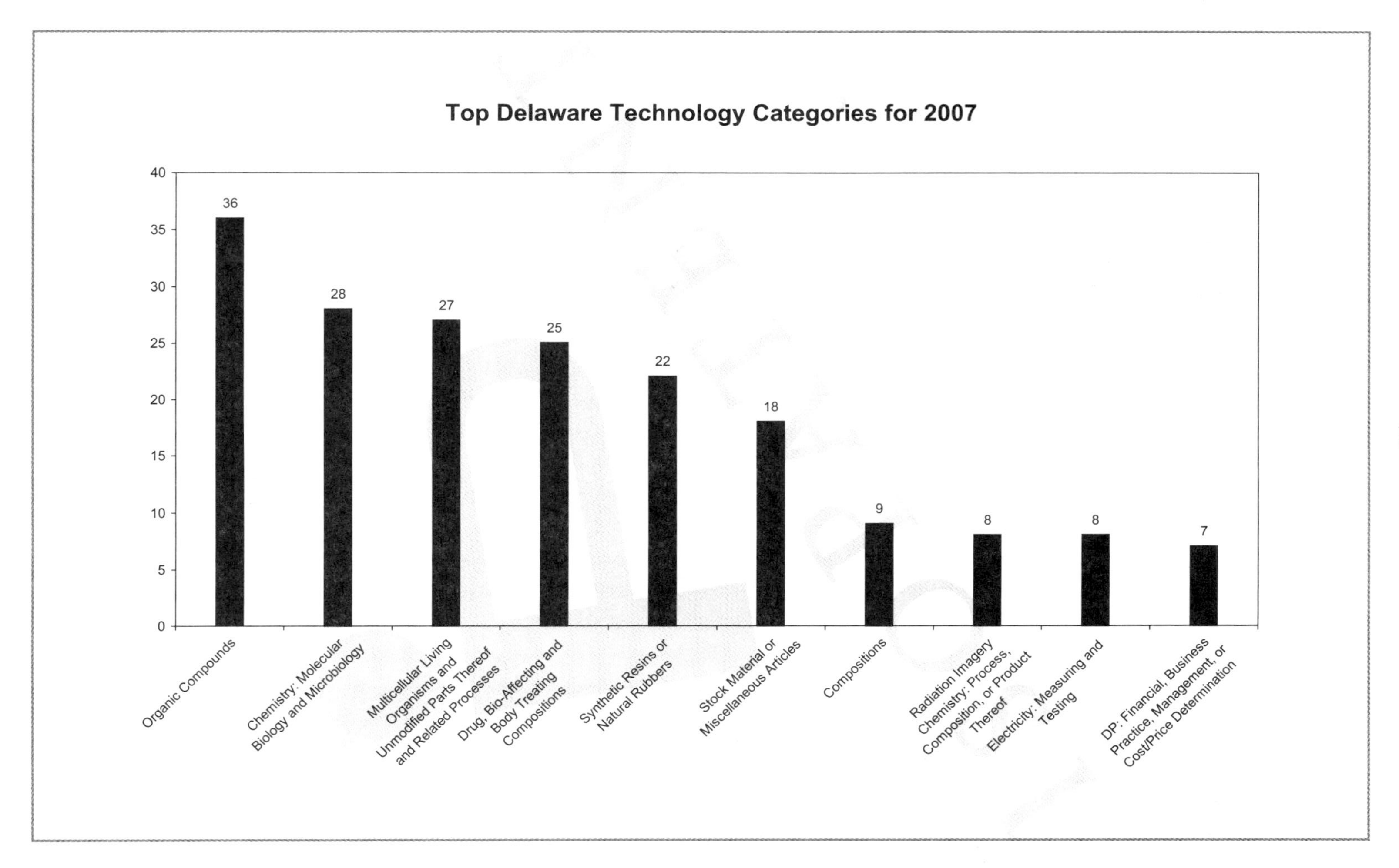
Top Delaware Technology Categories for 2007
40
35
30
25
20
15
10
5
0
36
28
27
25
22
18
9
8
8
7
Organic Compounds
Chemistry: Molecular Biology and Microbiology
Multicellular Living Organisms and Unmodified Parts Thereof and Related Processes
Drug, Bio-Affecting and Body Treating Compositions
Synthetic Resins or Natural Rubbers
Stock Material or Miscellaneous Articles
Compositions
Radiation Imagery Chemistry: Process, Composition, or Product Thereof
Electricity: Measuring and Testing
DP: Financial, Business Practice, Management, or Cost/Price Determination

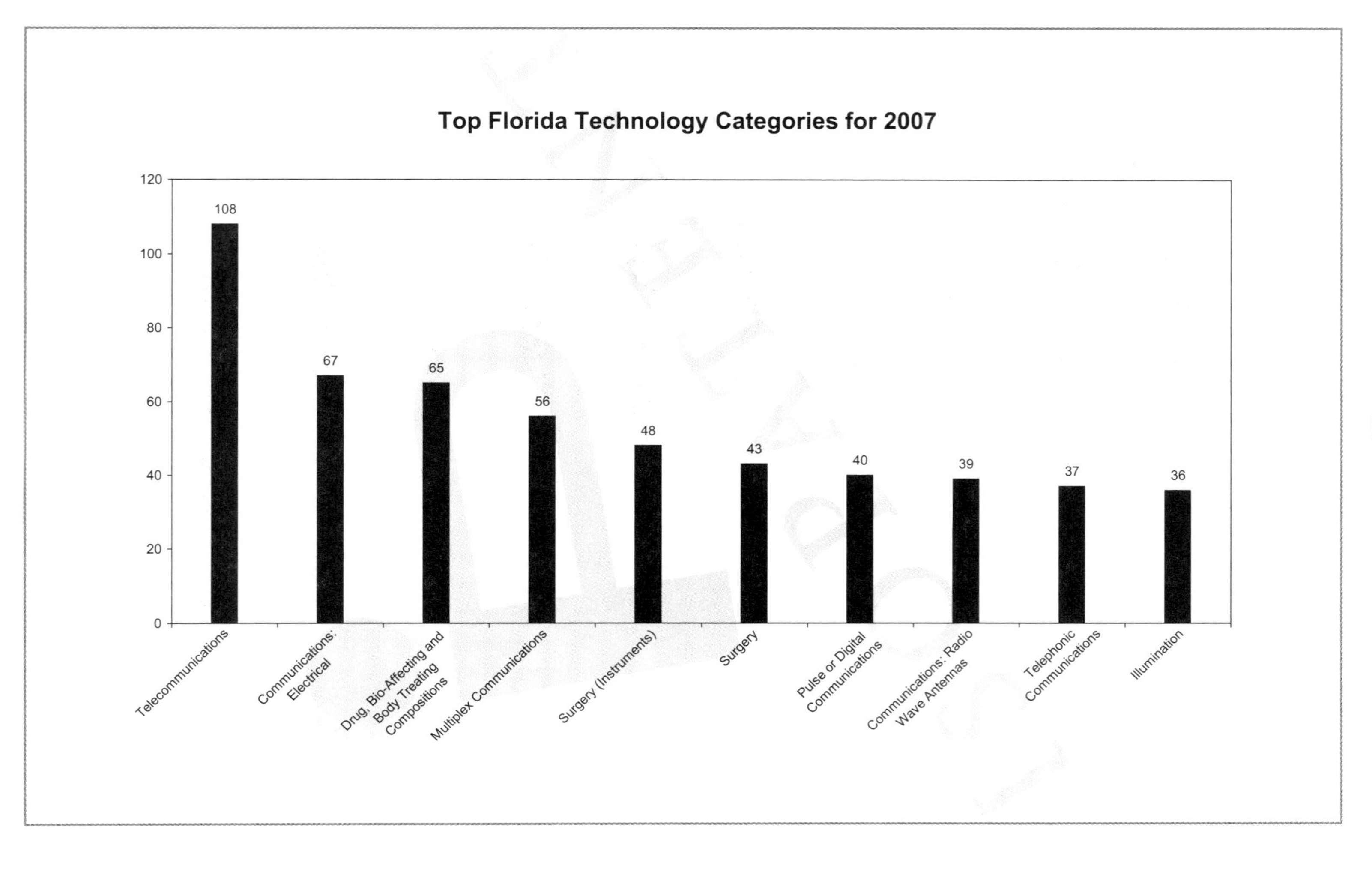
Top Florida Technology Categories for 2007
120
100
80
60
40
20
0
108
67
65
56
48
43
40
39
37
36
Telecommunications
Communications: Electrical
Drug, Bio-Affecting and Body Treating Compositions
Multiplex Communications
Surgery (Instruments)
Surgery
Pulse or Digital Communications
Communications: Radio Wave Antennas
Telephonic Communications
Illumination

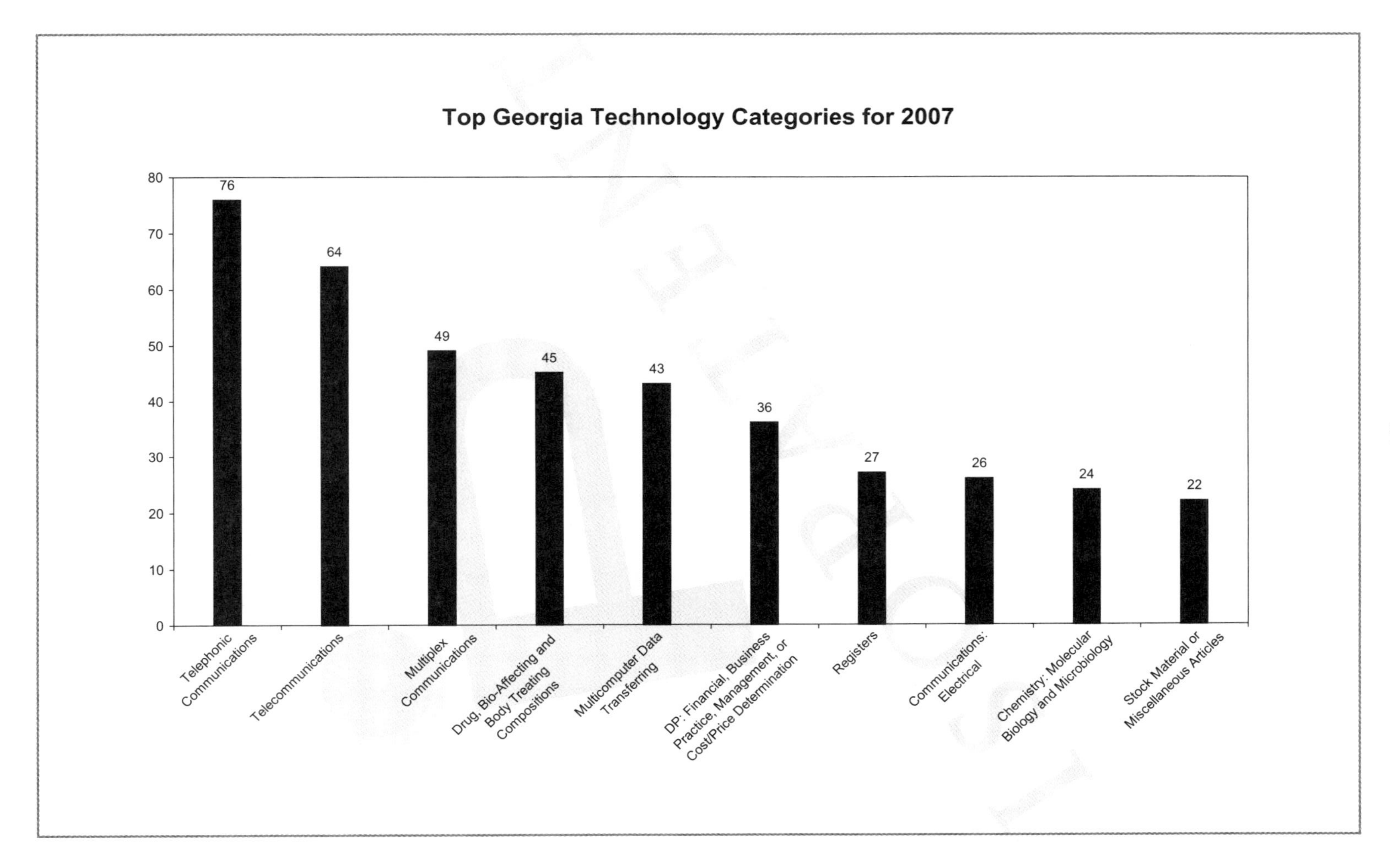
Top Georgia Technology Categories for 2007
80
70
60
50
40
30
20
10
0
76
64
49
45
43
36
27
26
24
22
Telephonic Communications
Telecommunications
Multiplex Communications
Drug, Bio-Affecting and Body Treating Compositions
Multicomputer Data Transferring
DP: Financial, Business Practice, Management, or Cost/Price Determination
Registers
Communications: Electrical
Chemistry: Molecular Biology and Microbiology
Stock Material or Miscellaneous Articles

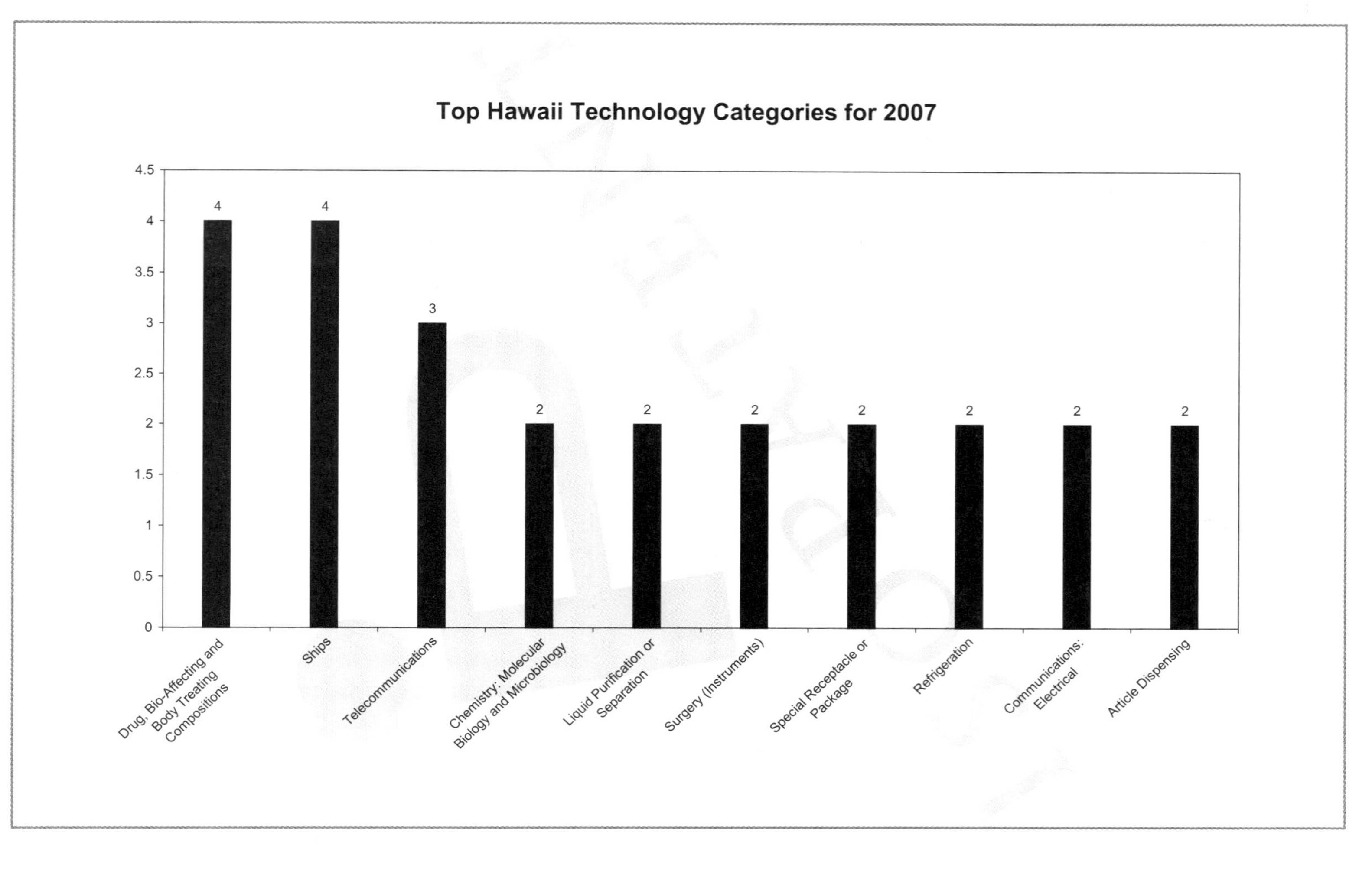

Top Hawaii Technology Categories for 2007
4.5
4
3.5
3
2.5
2
1.5
1
0.5
0
4
4
3
2
2
2
2
2
2
2
Drug, Bio-Affecting and Body Treating Compositions
Ships
Telecommunications
Chemistry: Molecular Biology and Microbiology
Liquid Purification or Separation
Surgery (Instruments)
Special Receptacle or Package
Refrigeration
Communications: Electrical
Article Dispensing

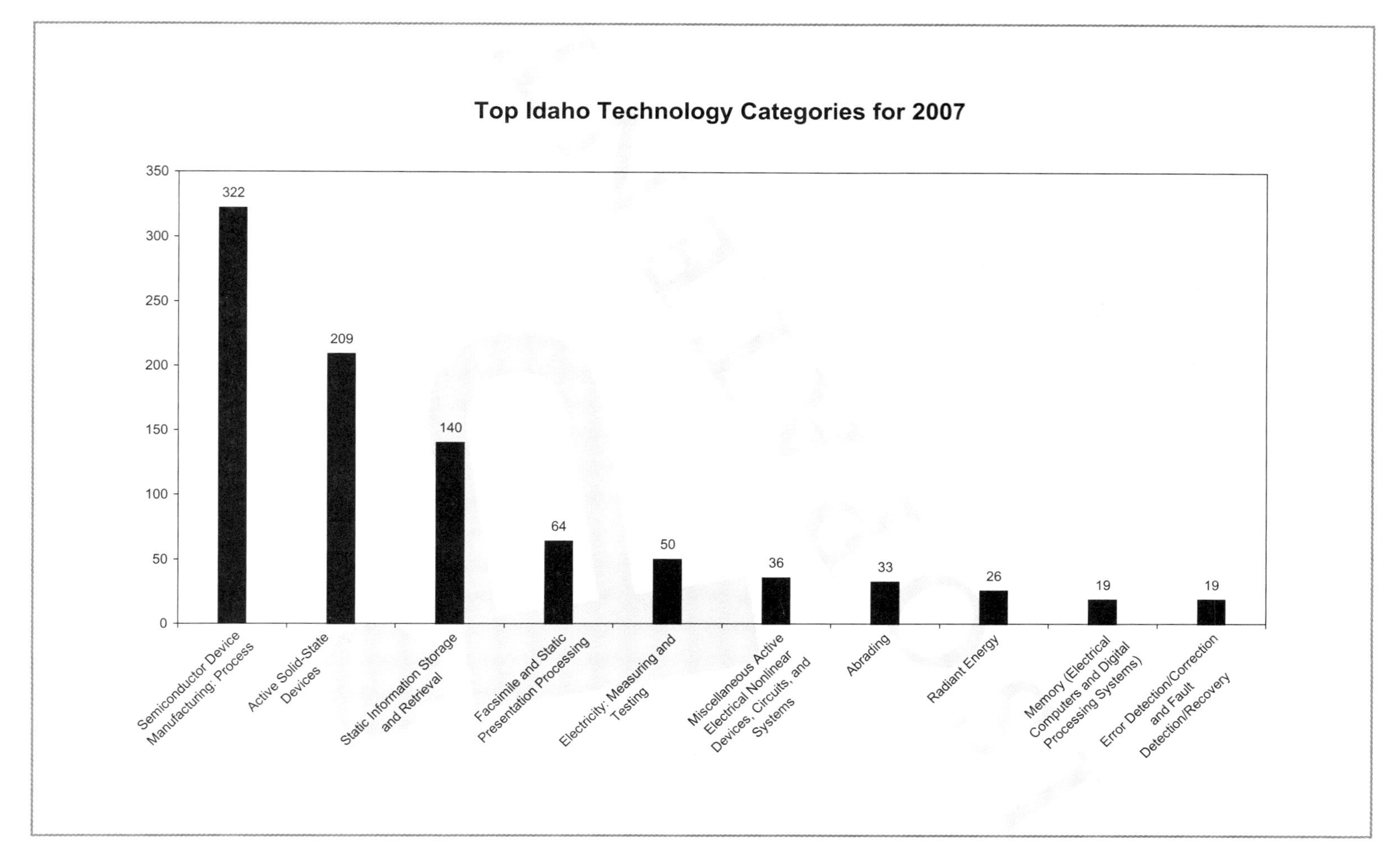
Top Idaho Technology Categories for 2007
350
300
250
200
150
100
50
0
322
209
140
64
50
36
33
26
19
19
Semiconductor Device Manufacturing: Process
Active Solid-State Devices
Static Information Storage and Retrieval
Facsimile and Static Presentation Processing
Electricity: Measuring and Testing
Miscellaneous Active Electrical Nonlinear Devices, Circuits, and Systems
Abrading
Radiant Energy
Memory (Electrical Computers and Digital Processing Systems)
Error Detection/Correction and Fault Detection/Recovery

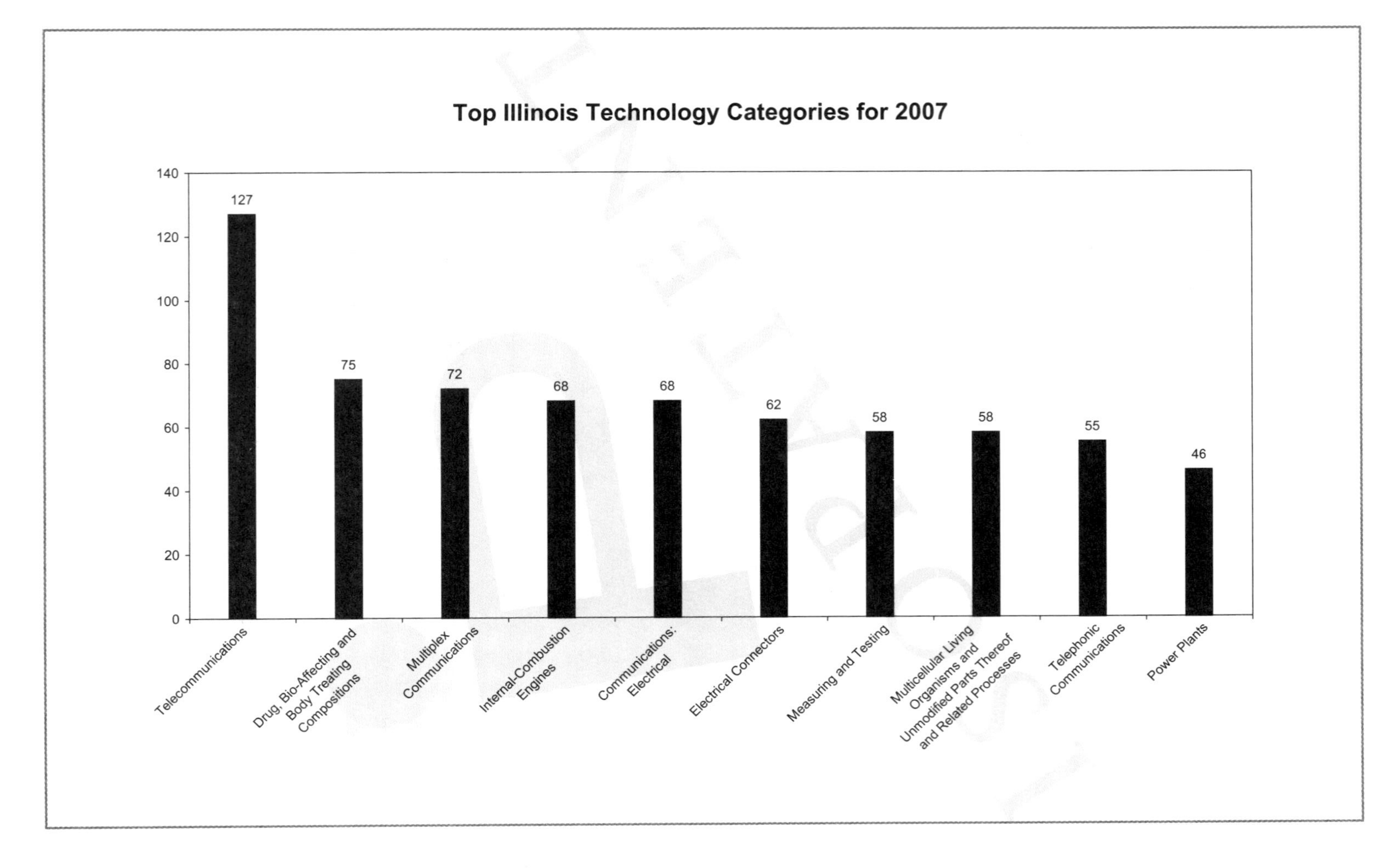
Top Illinois Technology Categories for 2007
140
120
100
80
60
40
20
0
127
75
72
68
68
62
58
58
55
46
Telecommunications
Drug, Bio-Affecting and Body Treating Compositions
Multiplex Communications
Internal-Combustion Engines
Communications: Electrical
Electrical Connectors
Measuring and Testing
Multicellular Living Organisms and Unmodified Parts Thereof and Related Processes
Telephonic Communications
Power Plants

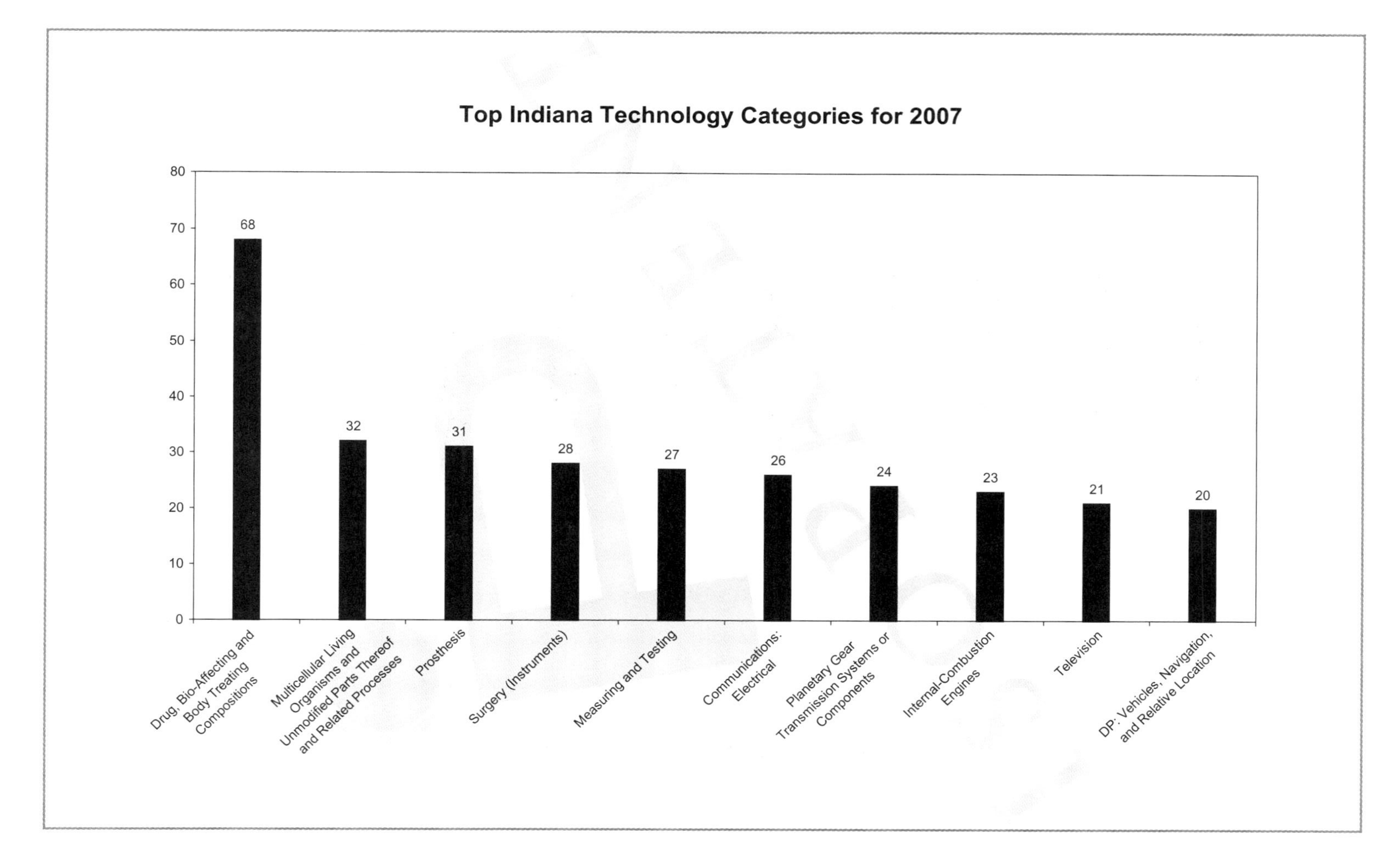

Top Indiana Technology Categories for 2007
80
70
60
50
40
30
20
10
0
68
32
31
28
27
26
24
23
21
20
Drug, Bio-Affecting and Body Treating Compositions
Multicellular Living Organisms and Unmodified Parts Thereof and Related Processes
Prosthesis
Surgery (Instruments)
Measuring and Testing
Communications: Electrical
Planetary Gear Transmission Systems or Components
Internal-Combustion Engines
Television
DP: Vehicles, Navigation, and Relative Location

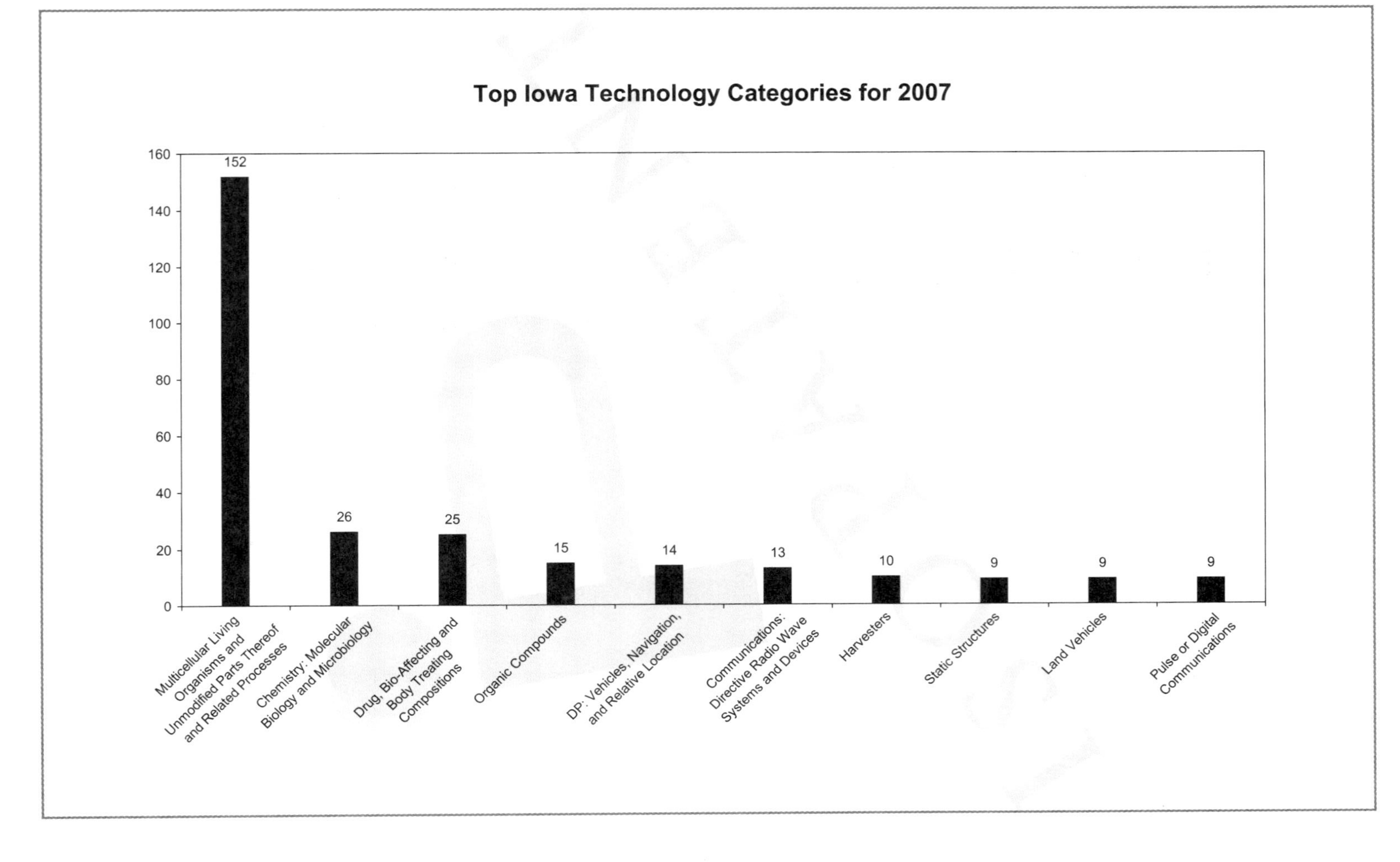
Top Iowa Technology Categories for 2007
160
140
120
100
80
60
40
20
0
152
26
25
15
14
13
10
9
9
9
Multicellular Living Organisms and Unmodified Parts Thereof and Related Processes
Chemistry: Molecular Biology and Microbiology
Drug, Bio-Affecting and Body Treating Compositions
Organic Compounds
DP: Vehicles, Navigation, and Relative Location
Communications: Directive Radio Wave Systems and Devices
Harvesters
Static Structures
Land Vehicles
Pulse or Digital Communications

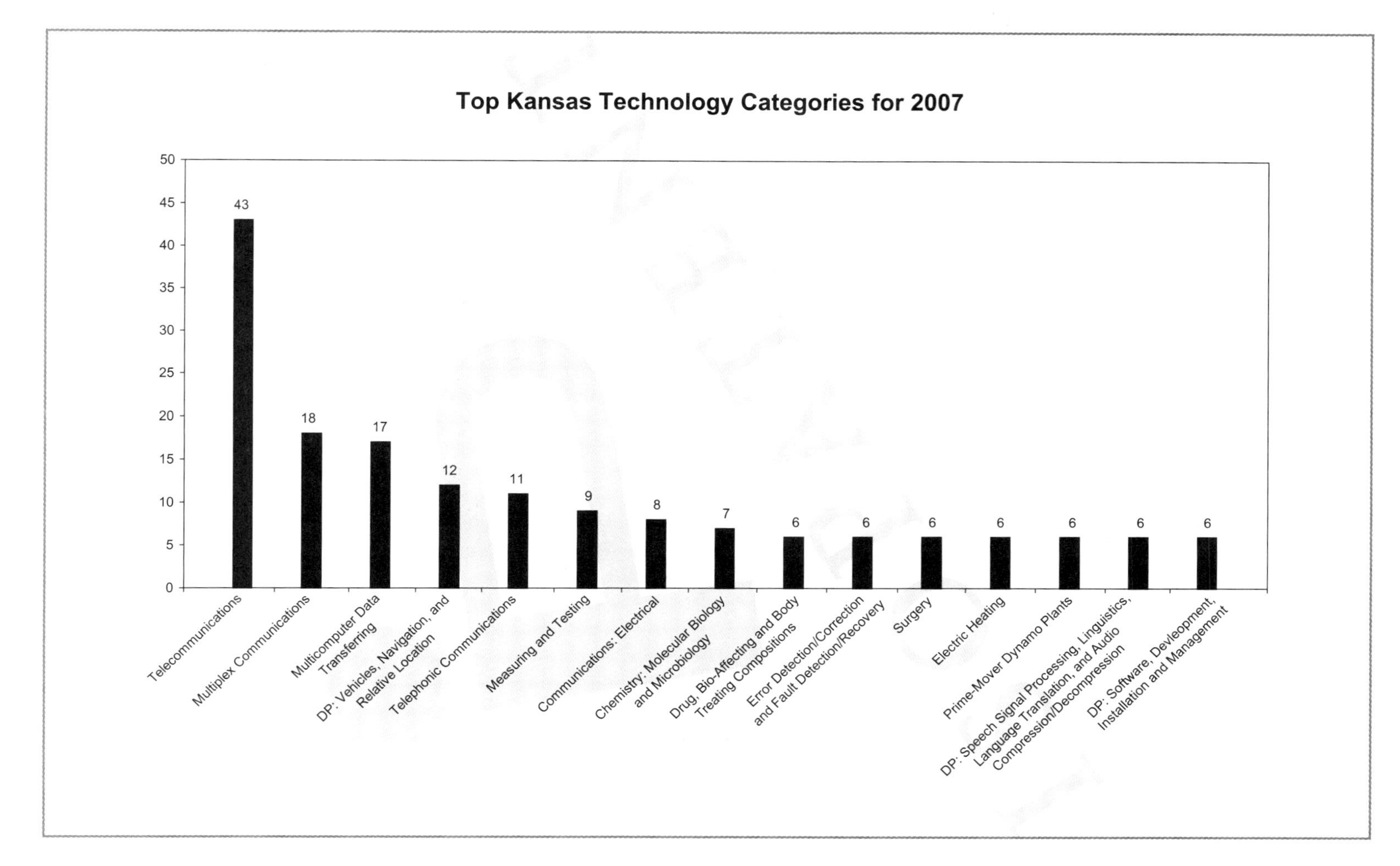
Top Kansas Technology Categories for 2007
0
5
10
15
20
25
30
35
40
45
50
Telecommunications
43
Multiplex Communications
18
Multicomputer Data Transferring
17
DP: Vehicles, Navigation, and Relative Location
12
Telephonic Communications
11
Measuring and Testing
9
Communications: Electrical
8
Chemistry: Molecular Biology and Microbiology
7
Drug, Bio-Affecting and Body Treating Compositions
6
Error Detection/Correction and Fault Detection/Recovery
6
Surgery
6
Electric Heating
6
Prime-Mover Dynamo Plants
6
DP: Speech Signal Processing, Linguistics, Language Translation, and Audio Compression/Decompression
6
DP: Software, Devleopment, Installation and Management
6

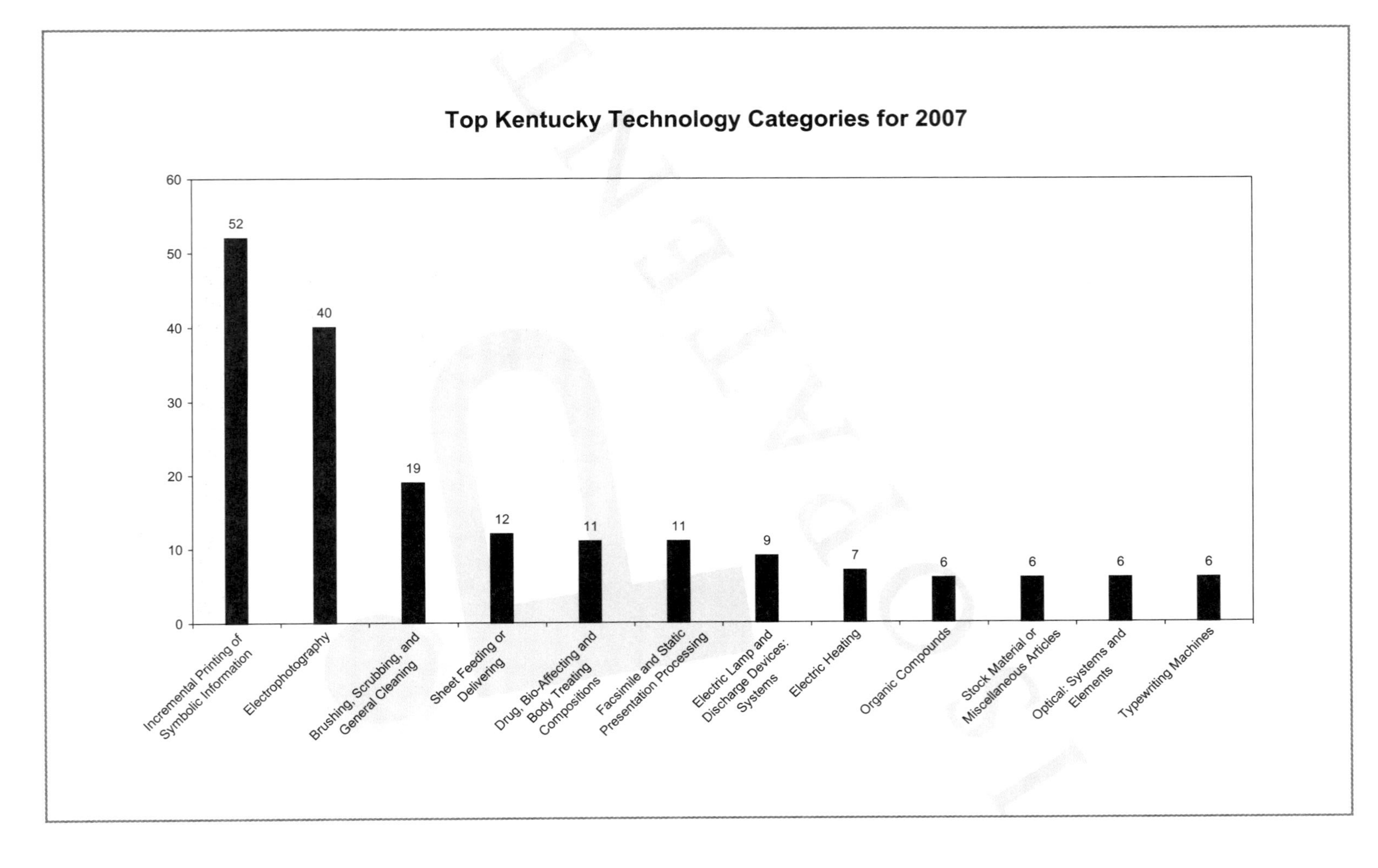
Top Kentucky Technology Categories for 2007
60
50
40
30
20
10
0
52
40
19
12
11
11
9
7
6
6
6
6
Incremental Printing of Symbolic Information
Electrophotography
Brushing, Scrubbing, and General Cleaning
Sheet Feeding or Delivering
Drug, Bio-Affecting and Body Treating Compositions
Facsimile and Static Presentation Processing
Electric Lamp and Discharge Devices: Systems
Electric Heating
Organic Compounds
Stock Material or Miscellaneous Articles
Optical: Systems and Elements
Typewriting Machines

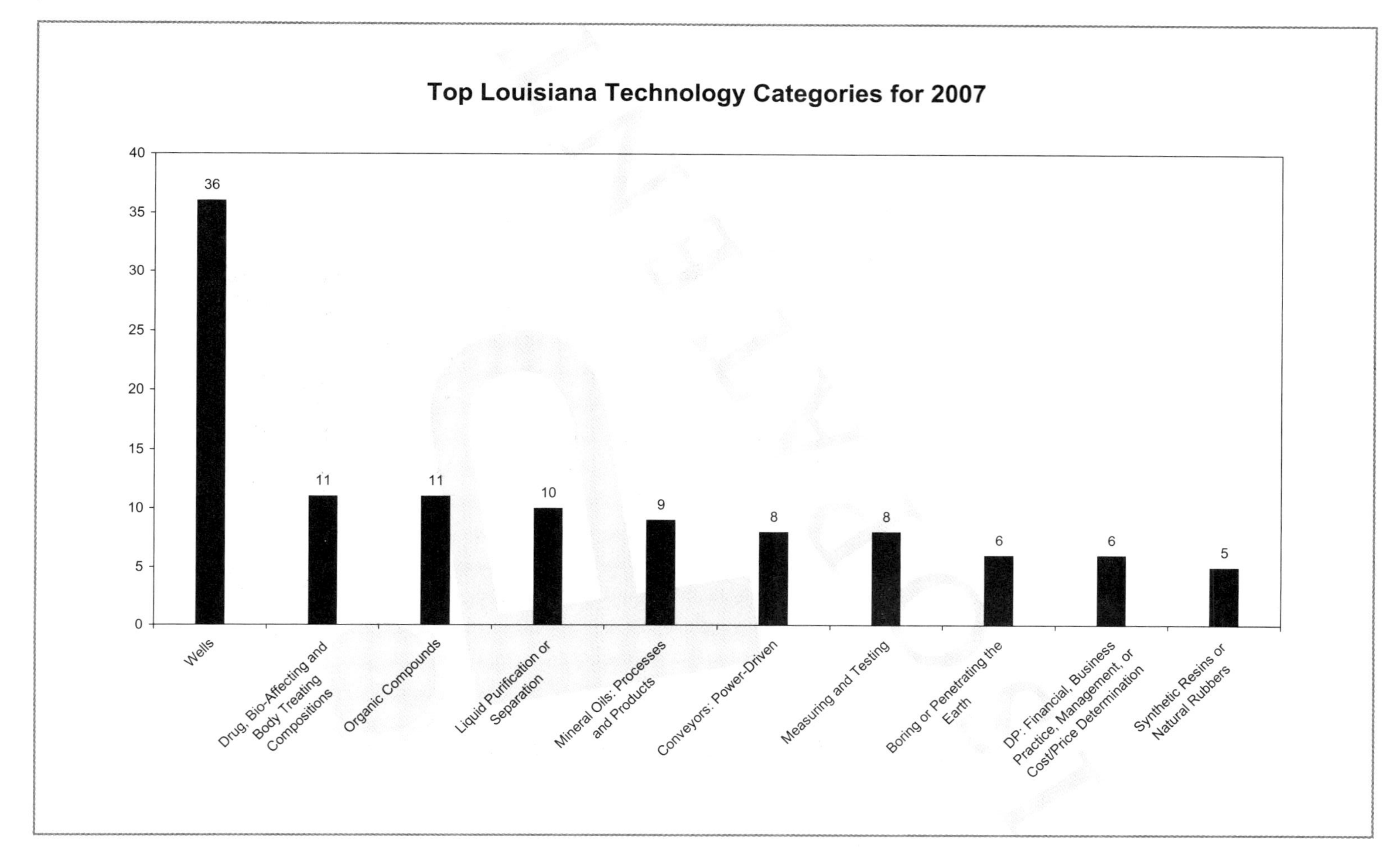
Top Louisiana Technology Categories for 2007
40
35
30
25
20
15
10
5
0
36
11
11
10
9
8
8
6
6
5
Wells
Drug, Bio-Affecting and Body Treating Compositions
Organic Compounds
Liquid Purification or Separation
Mineral Oils: Processes and Products
Conveyors: Power-Driven
Measuring and Testing
Boring or Penetrating the Earth
DP: Financial, Business Practice, Management, or Cost/Price Determination
Synthetic Resins or Natural Rubbers

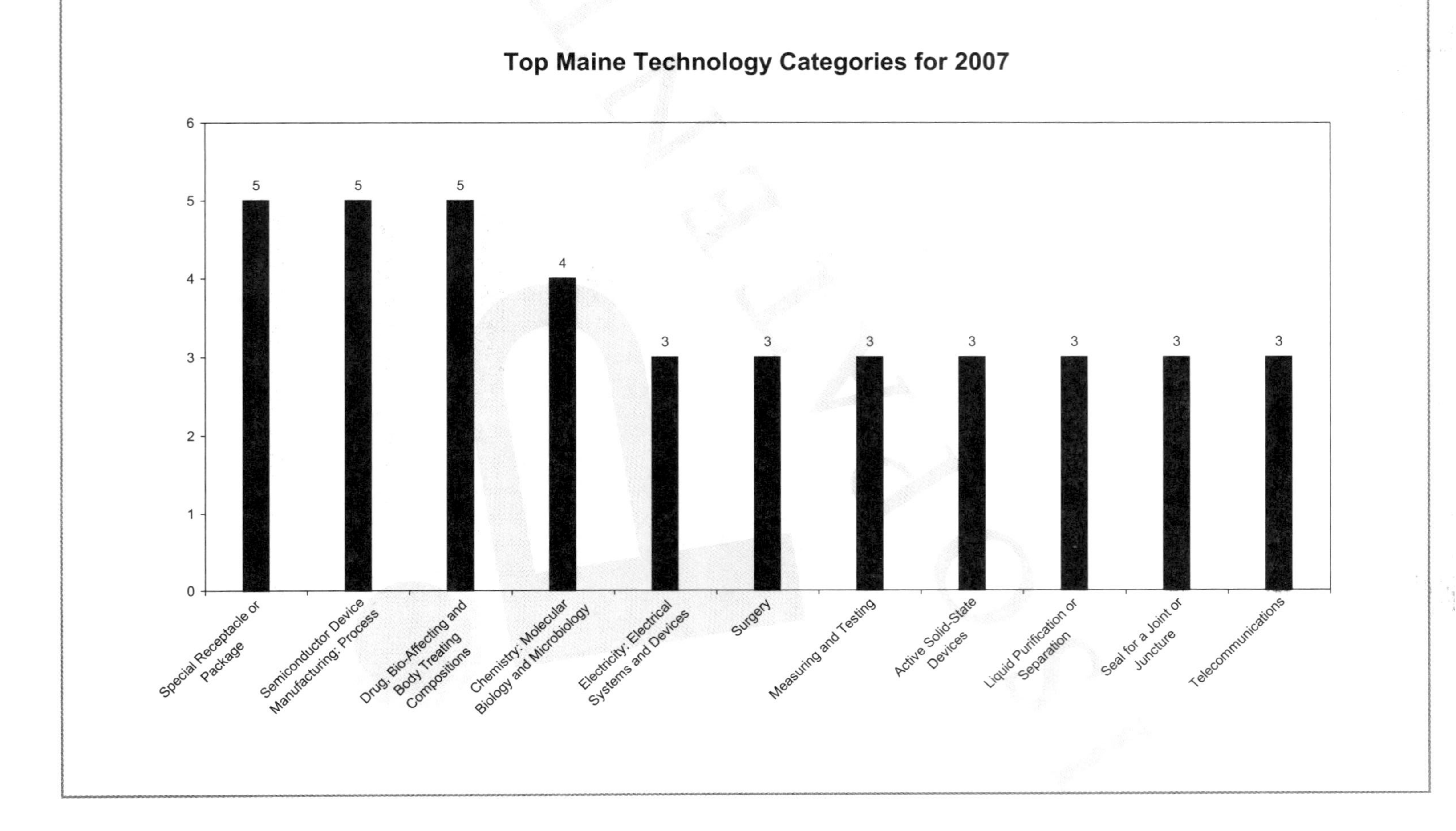
Top Maine Technology Categories for 2007
6
5
4
3
2
1
0
5
5
5
4
3
3
3
3
3
3
3
Special Receptacle or Package
Semiconductor Device Manufacturing: Process
Drug, Bio-Affecting and Body Treating Compositions
Chemistry: Molecular Biology and Microbiology
Electricity: Electrical Systems and Devices
Surgery
Measuring and Testing
Active Solid-State Devices
Liquid Purification or Separation
Seal for a Joint or Juncture
Telecommunications

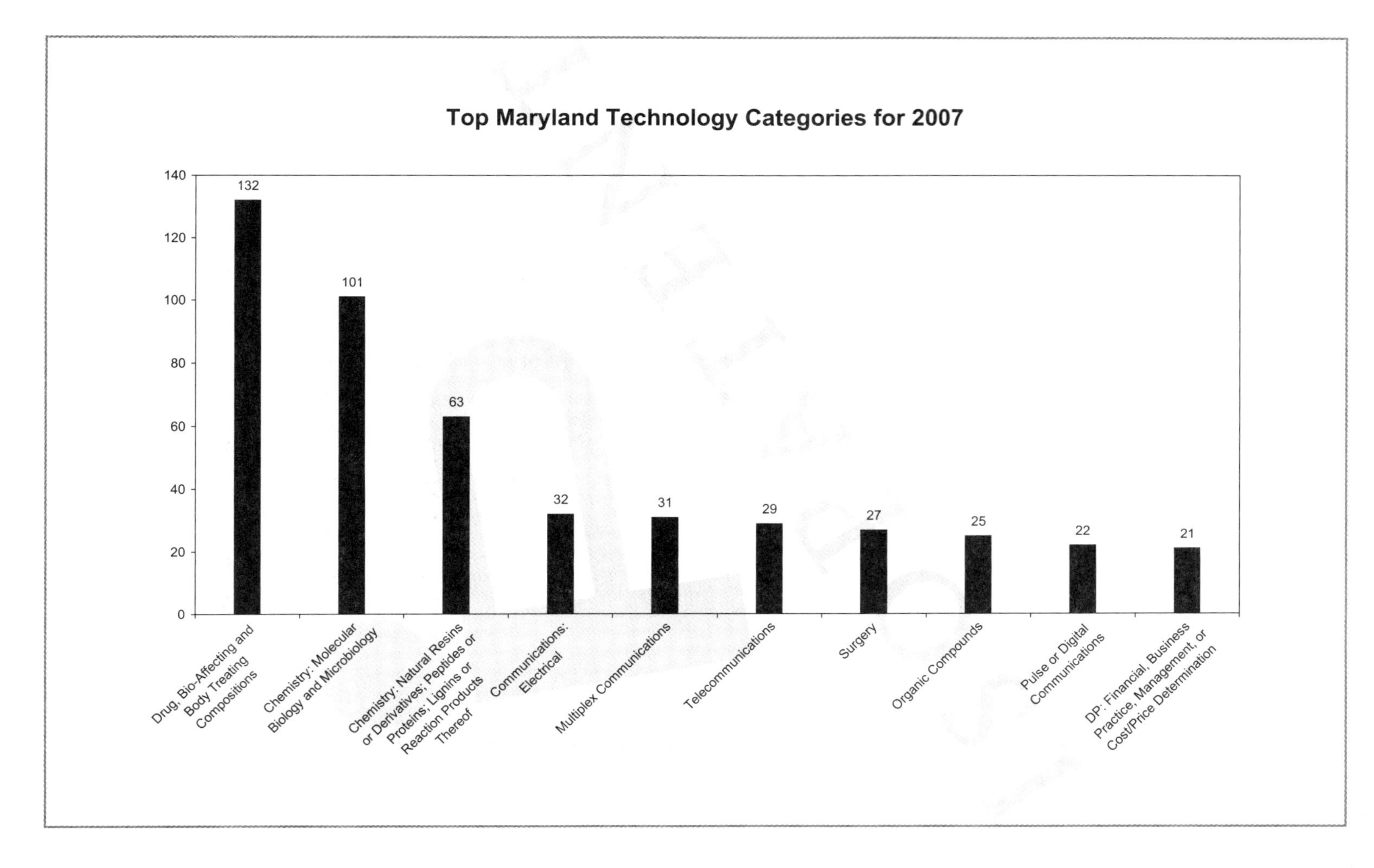

Top Maryland Technology Categories for 2007
140
120
100
80
60
40
20
0
132
101
63
32
31
29
27
25
22
21
Drug, Bio-Affecting and Body Treating Compositions
Chemistry: Molecular Biology and Microbiology
Chemistry: Natural Resins or Derivatives; Peptides or Proteins; Lignins or Reaction Products Thereof
Communications: Electrical
Multiplex Communications
Telecommunications
Surgery
Organic Compounds
Pulse or Digital Communications
DP: Financial, Business Practice, Management, or Cost/Price Determination

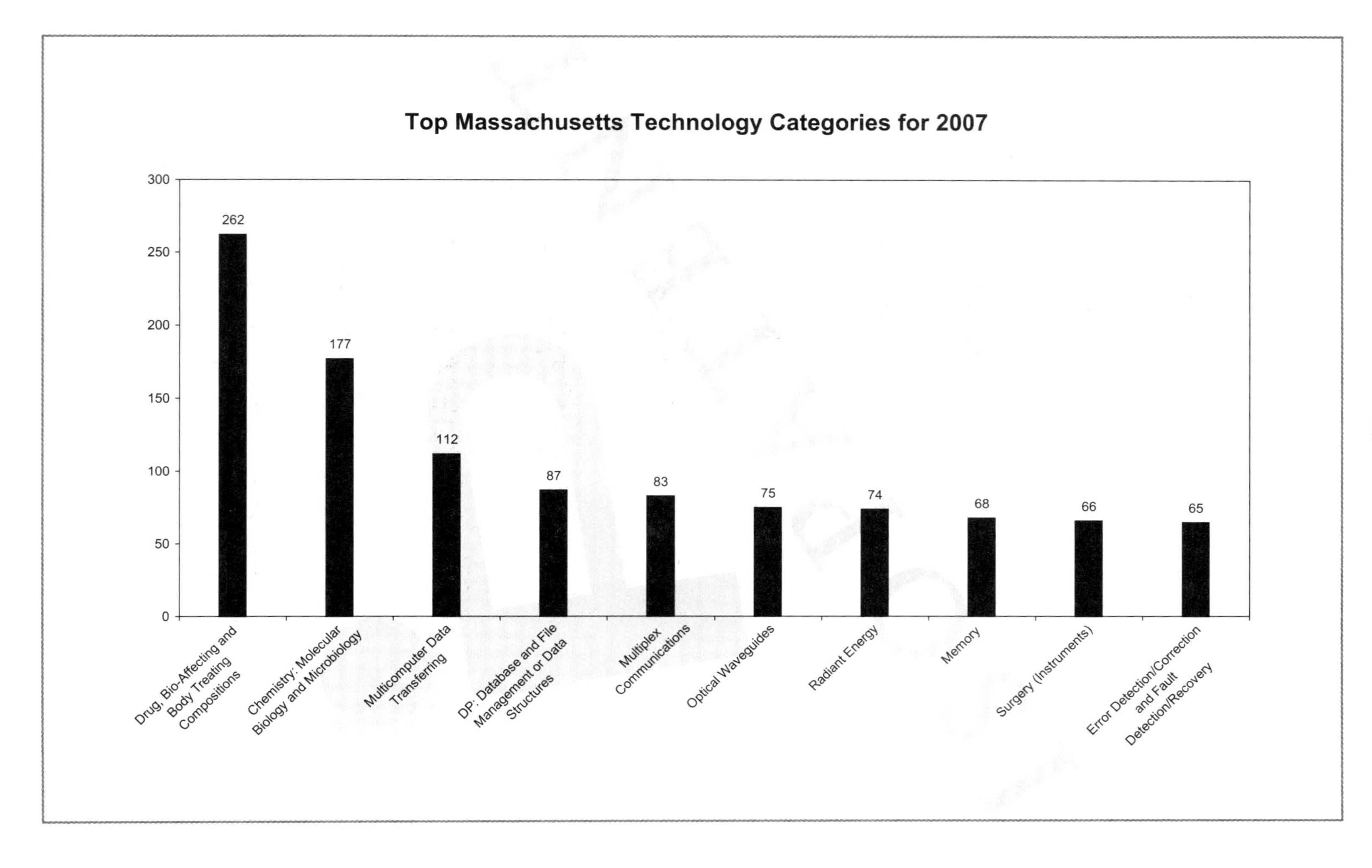

Top Massachusetts Technology Categories for 2007
300
250
200
150
100
50
0
262
177
112
87
83
75
74
68
66
65
Drug, Bio-Affecting and Body Treating Compositions
Chemistry: Molecular Biology and Microbiology
Multicomputer Data Transferring
DP: Database and File Management or Data Structures
Multiplex Communications
Optical Waveguides
Radiant Energy
Memory
Surgery (Instruments)
Error Detection/Correction and Fault Detection/Recovery

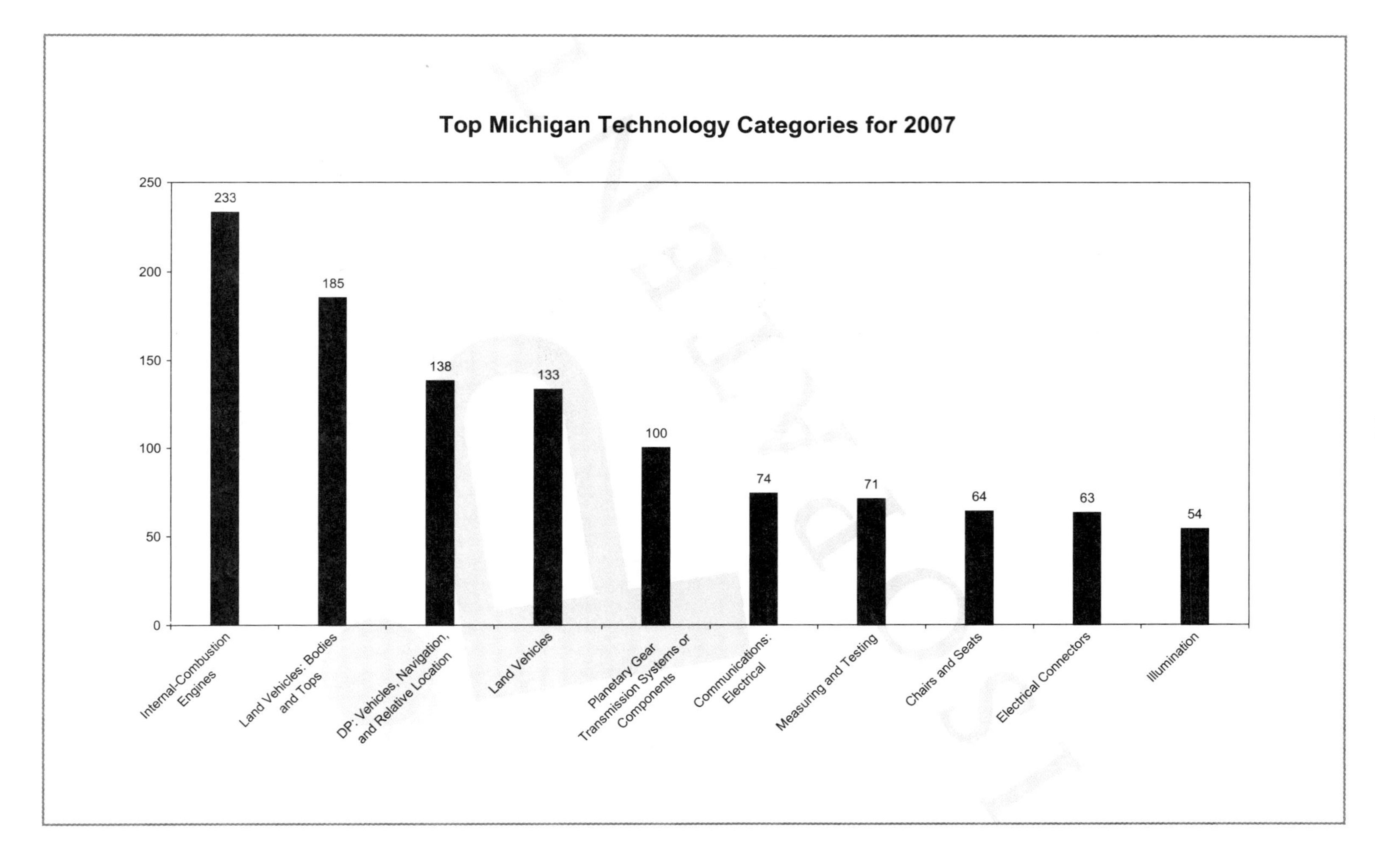
Top Michigan Technology Categories for 2007
250
200
150
100
50
0
233
185
138
133
100
74
71
64
63
54
Internal-Combustion Engines
Land Vehicles: Bodies and Tops
DP: Vehicles, Navigation, and Relative Location
Land Vehicles
Planetary Gear Transmission Systems or Components
Communications: Electrical
Measuring and Testing
Chairs and Seats
Electrical Connectors
Illumination

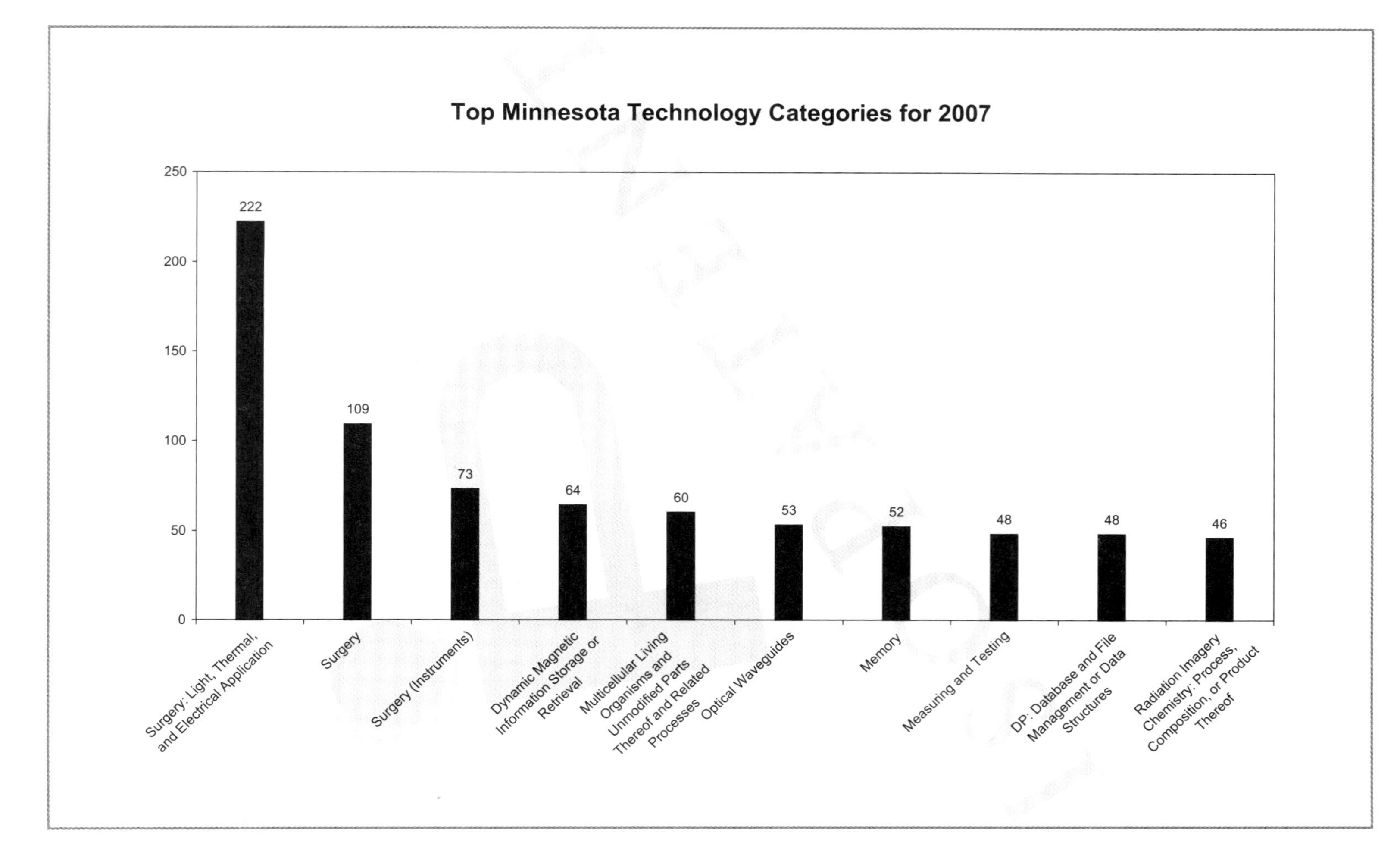
Top Minnesota Technology Categories for 2007
250
200
150
100
50
0
222
109
73
64
60
53
52
48
48
46
Surgery: Light, Thermal, and Electrical Application
Surgery
Surgery (Instruments)
Dynamic Magnetic Information Storage or Retrieval
Multicellular Living Organisms and Unmodified Parts Thereof and Related Processes
Optical Waveguides
Memory
Measuring and Testing
DP: Database and File Management or Data Structures
Radiation Imagery Chemistry: Process, Composition, or Product Thereof

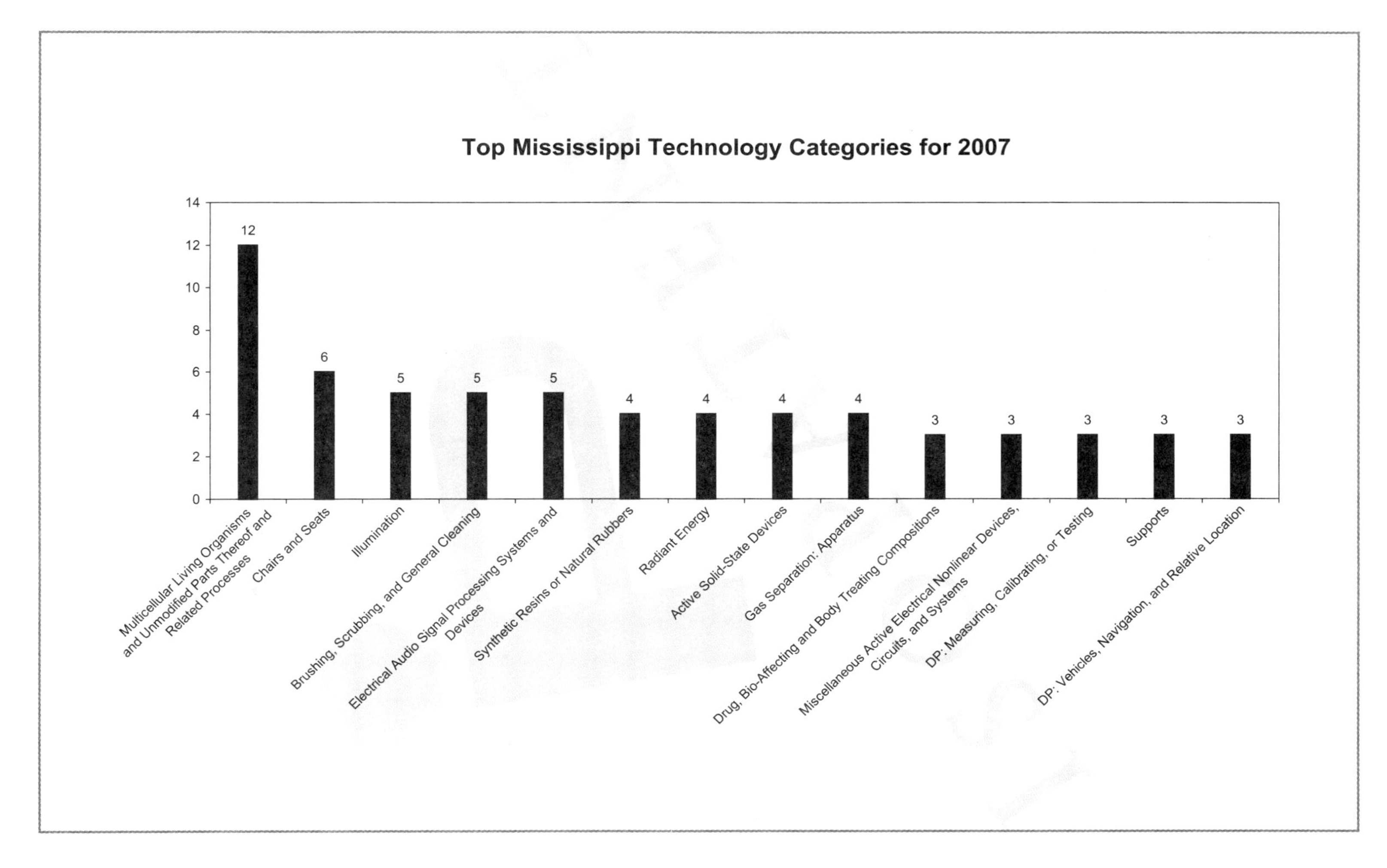

Top Mississippi Technology Categories for 2007
14
12
10
8
6
4
2
0
12
6
5
5
5
4
4
4
4
3
3
3
3
3
Multicellular Living Organisms and Unmodified Parts Thereof and Related Processes
Chairs and Seats
Illumination
Brushing, Scrubbing, and General Cleaning
Electrical Audio Signal Processing Systems and Devices
Synthetic Resins or Natural Rubbers
Radiant Energy
Active Solid-State Devices
Gas Separation: Apparatus
Drug, Bio-Affecting and Body Treating Compositions
Miscellaneous Active Electrical Nonlinear Devices, Circuits, and Systems
DP: Measuring, Calibrating, or Testing
Supports
DP: Vehicles, Navigation, and Relative Location

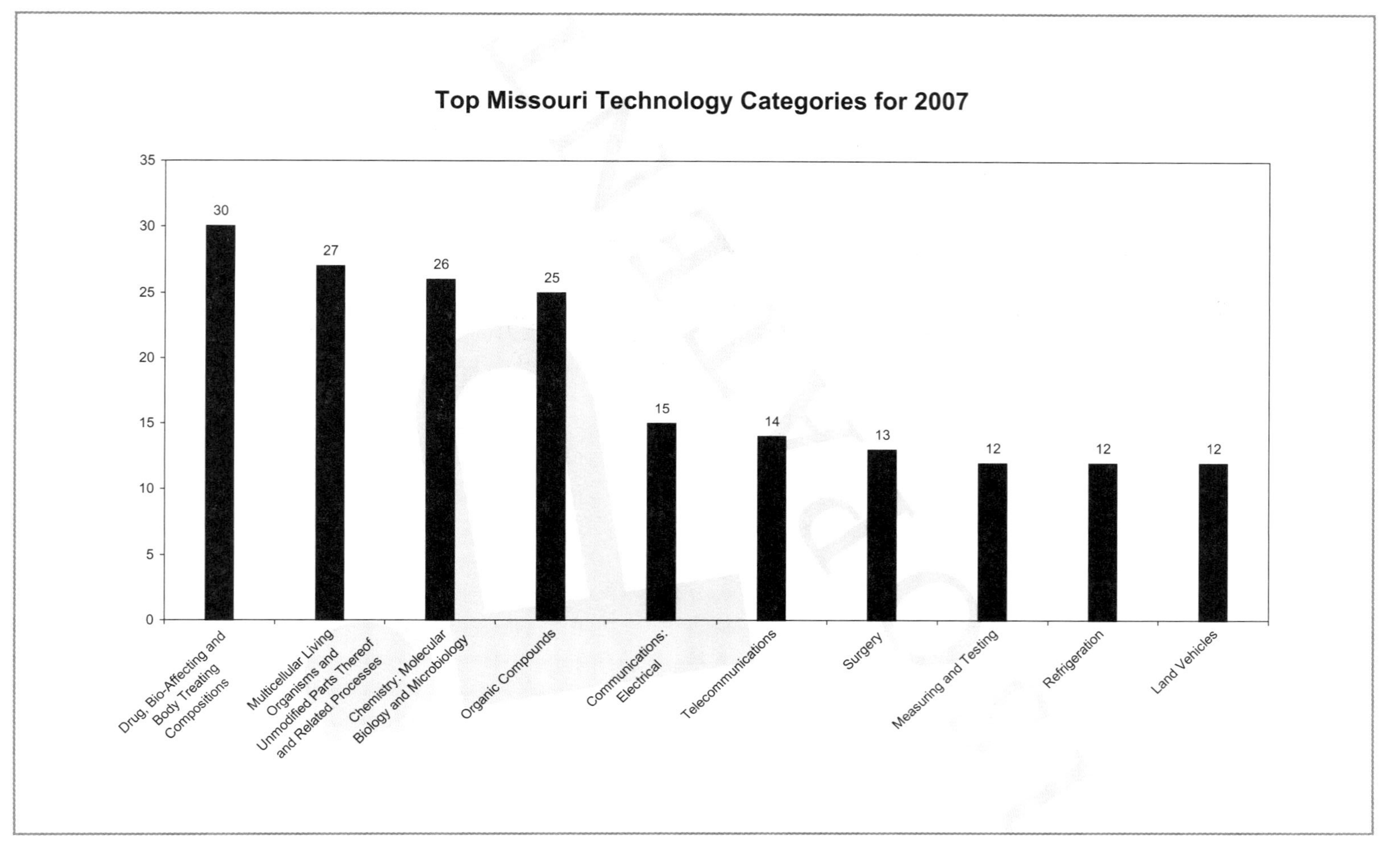
Top Missouri Technology Categories for 2007
35
30
25
20
15
10
5
0
30
27
26
25
15
14
13
12
12
12
Drug, Bio-Affecting and Body Treating Compositions
Multicellular Living Organisms and Unmodified Parts Thereof and Related Processes
Chemistry: Molecular Biology and Microbiology
Organic Compounds
Communications: Electrical
Telecommunications
Surgery
Measuring and Testing
Refrigeration
Land Vehicles

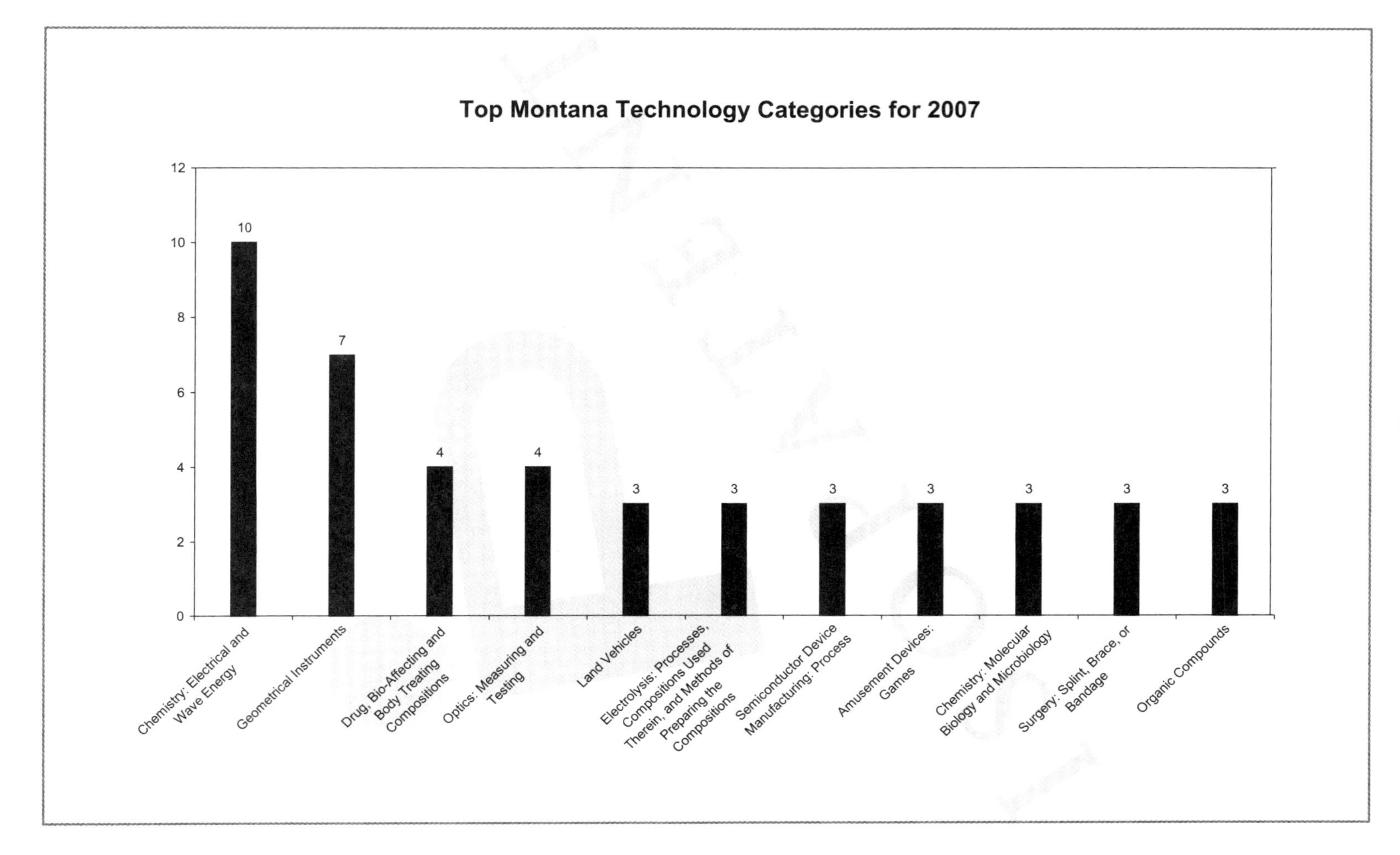

Top Montana Technology Categories for 2007
12
10
8
6
4
2
0
10
7
4
4
3
3
3
3
3
3
3
Chemistry: Electrical and Wave Energy
Geometrical Instruments
Drug, Bio-Affecting and Body Treating Compositions
Optics: Measuring and Testing
Land Vehicles
Electrolysis: Processes, Compositions Used Therein, and Methods of Preparing the Compositions
Semiconductor Device Manufacturing: Process
Amusement Devices: Games
Chemistry: Molecular Biology and Microbiology
Surgery: Splint, Brace, or Bandage
Organic Compounds

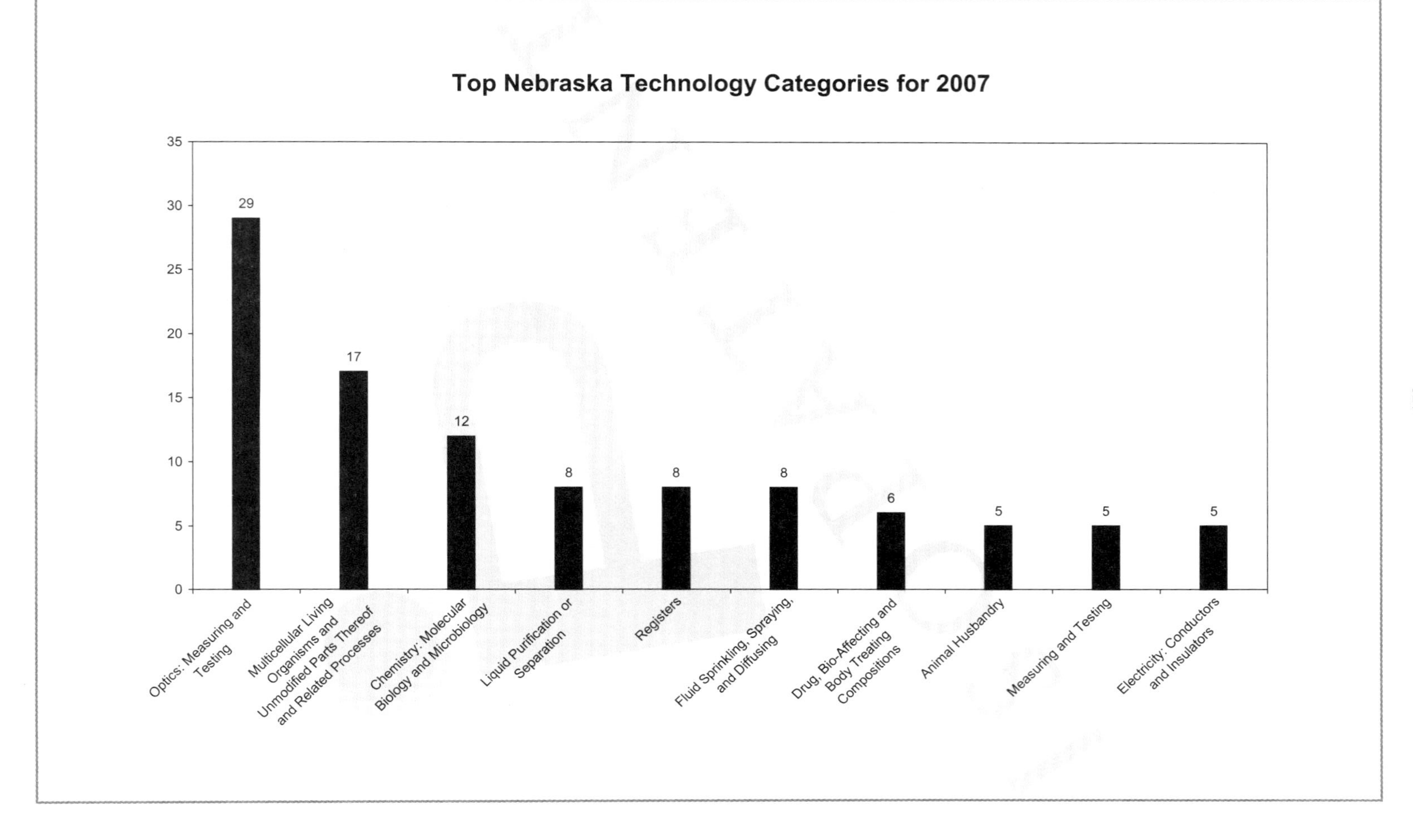
Top Nebraska Technology Categories for 2007
35
30
25
20
15
10
5
0
29
17
12
8
8
8
6
5
5
5
Optics: Measuring and Testing
Multicellular Living Organisms and Unmodified Parts Thereof and Related Processes
Chemistry: Molecular Biology and Microbiology
Liquid Purification or Separation
Registers
Fluid Sprinkling, Spraying, and Diffusing
Drug, Bio-Affecting and Body Treating Compositions
Animal Husbandry
Measuring and Testing
Electricity: Conductors and Insulators

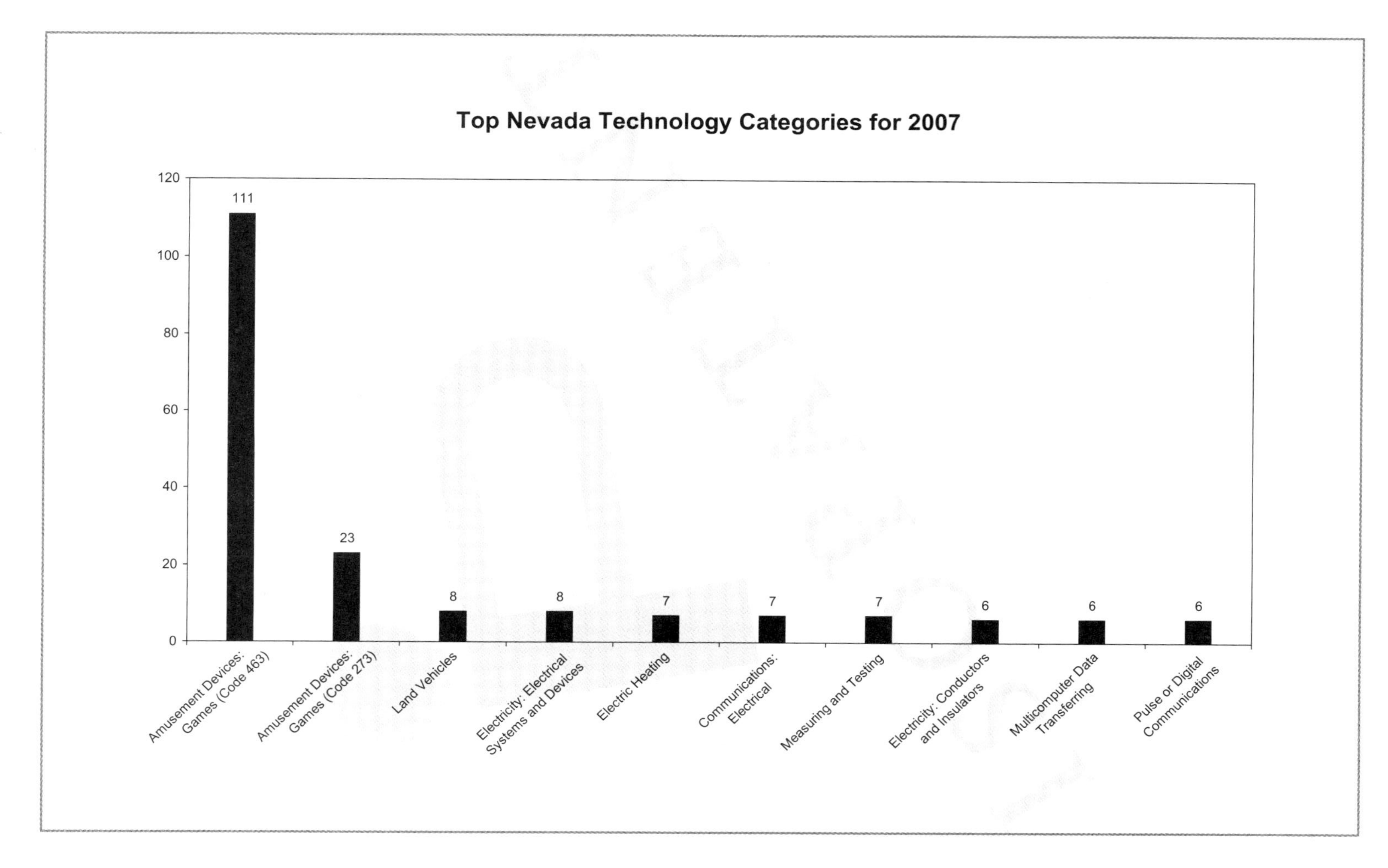

Top Nevada Technology Categories for 2007
120
100
80
60
40
20
0
111
23
8
8
7
7
7
6
6
6
Amusement Devices: Games (Code 463)
Amusement Devices: Games (Code 273)
Land Vehicles
Electricity: Electrical Systems and Devices
Electric Heating
Communications: Electrical
Measuring and Testing
Electricity: Conductors and Insulators
Multicomputer Data Transferring
Pulse or Digital Communications

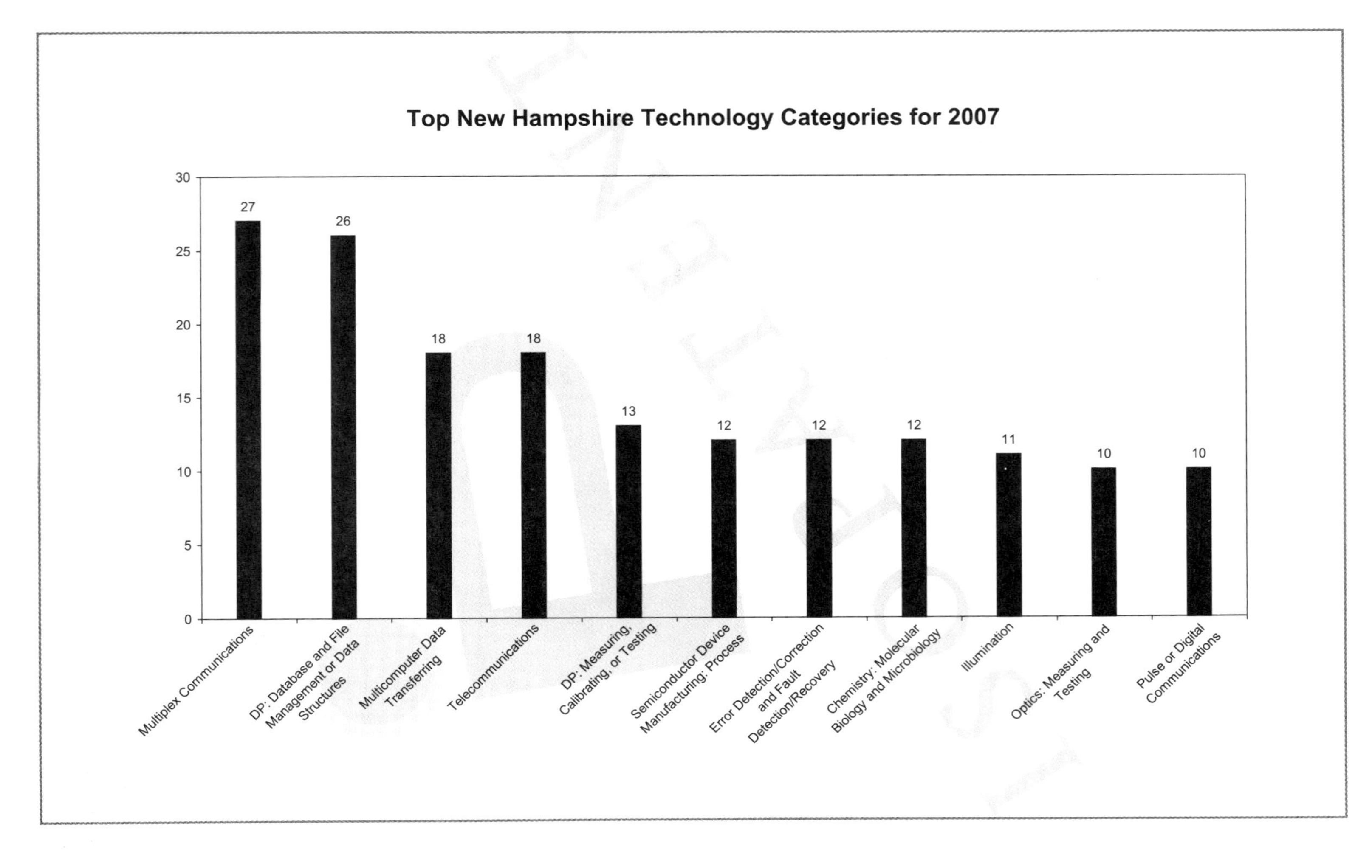

Top New Hampshire Technology Categories for 2007
30
25
20
15
10
5
0
27
26
18
18
13
12
12
12
11
10
10
Multiplex Communications
DP: Database and File Management or Data Structures
Multicomputer Data Transferring
Telecommunications
DP: Measuring, Calibrating, or Testing
Semiconductor Device Manufacturing: Process
Error Detection/Correction and Fault Detection/Recovery
Chemistry: Molecular Biology and Microbiology
Illumination
Optics: Measuring and Testing
Pulse or Digital Communications

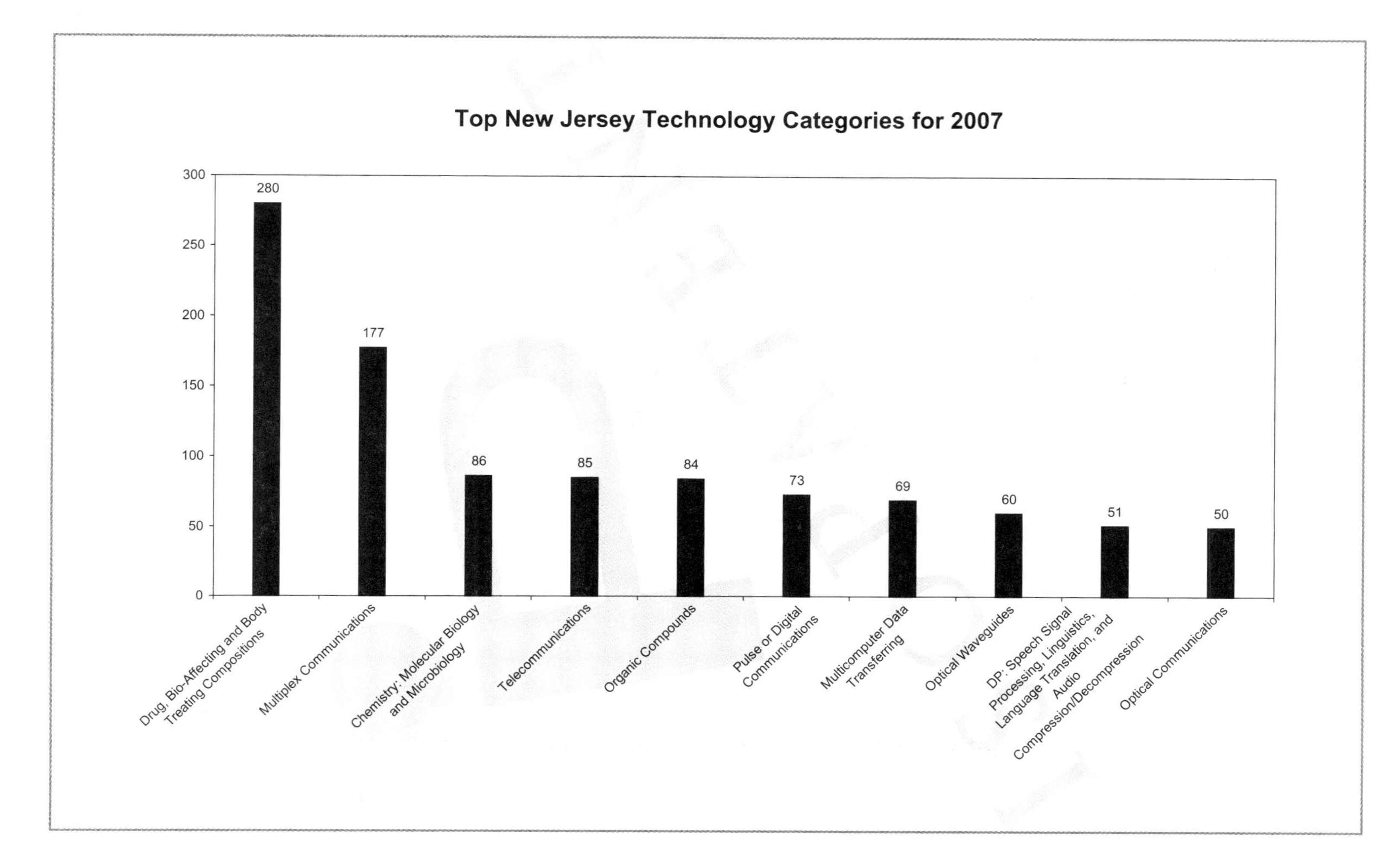

Top New Jersey Technology Categories for 2007
300
250
200
150
100
50
0
280
177
86
85
84
73
69
60
51
50
Drug, Bio-Affecting and Body Treating Compositions
Multiplex Communications
Chemistry: Molecular Biology and Microbiology
Telecommunications
Organic Compounds
Pulse or Digital Communications
Multicomputer Data Transferring
Optical Waveguides
DP: Speech Signal Processing, Linguistics, Language Translation, and Audio Compression/Decompression
Optical Communications

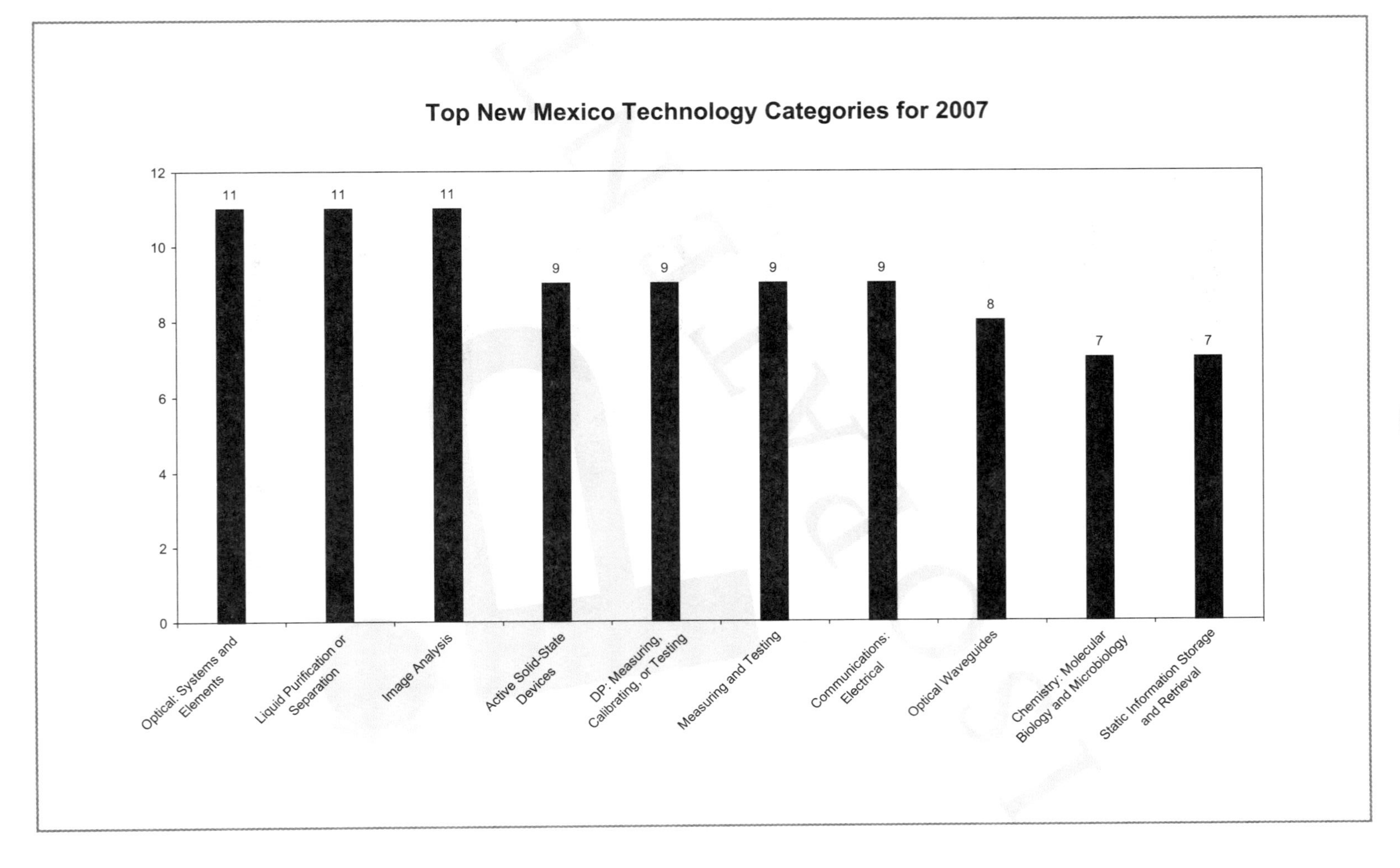

Top New Mexico Technology Categories for 2007
12
10
8
6
4
2
0
11
11
11
9
9
9
9
8
7
7
Optical: Systems and Elements
Liquid Purification or Separation
Image Analysis
Active Solid-State Devices
DP: Measuring, Calibrating, or Testing
Measuring and Testing
Communications: Electrical
Optical Waveguides
Chemistry: Molecular Biology and Microbiology
Static Information Storage and Retrieval

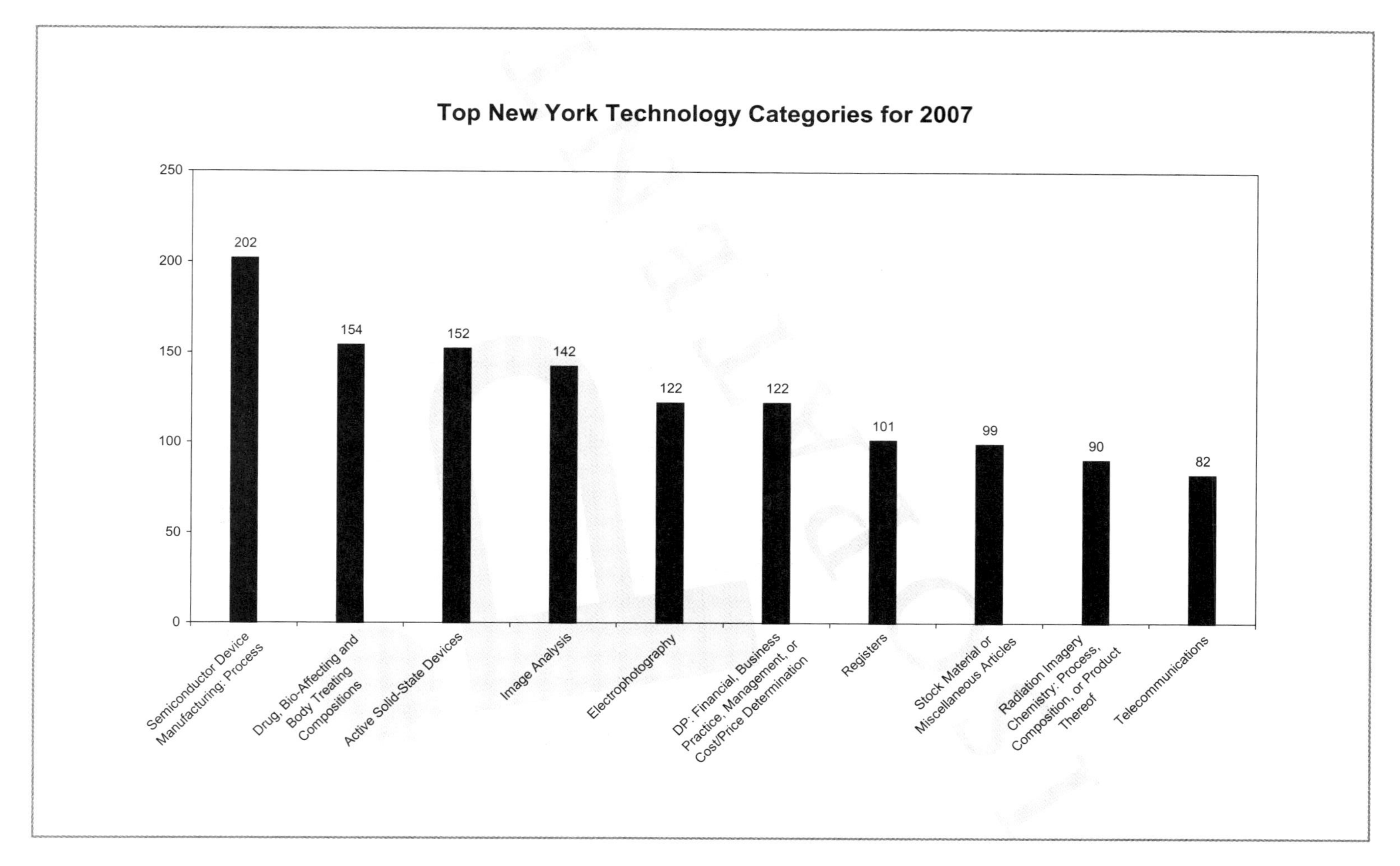
Top New York Technology Categories for 2007
250
200
150
100
50
0
202
154
152
142
122
122
101
99
90
82
Semiconductor Device Manufacturing: Process
Drug, Bio-Affecting and Body Treating Compositions
Active Solid-State Devices
Image Analysis
Electrophotography
DP: Financial, Business Practice, Management, or Cost/Price Determination
Registers
Stock Material or Miscellaneous Articles
Radiation Imagery Chemistry: Process, Composition, or Product Thereof
Telecommunications

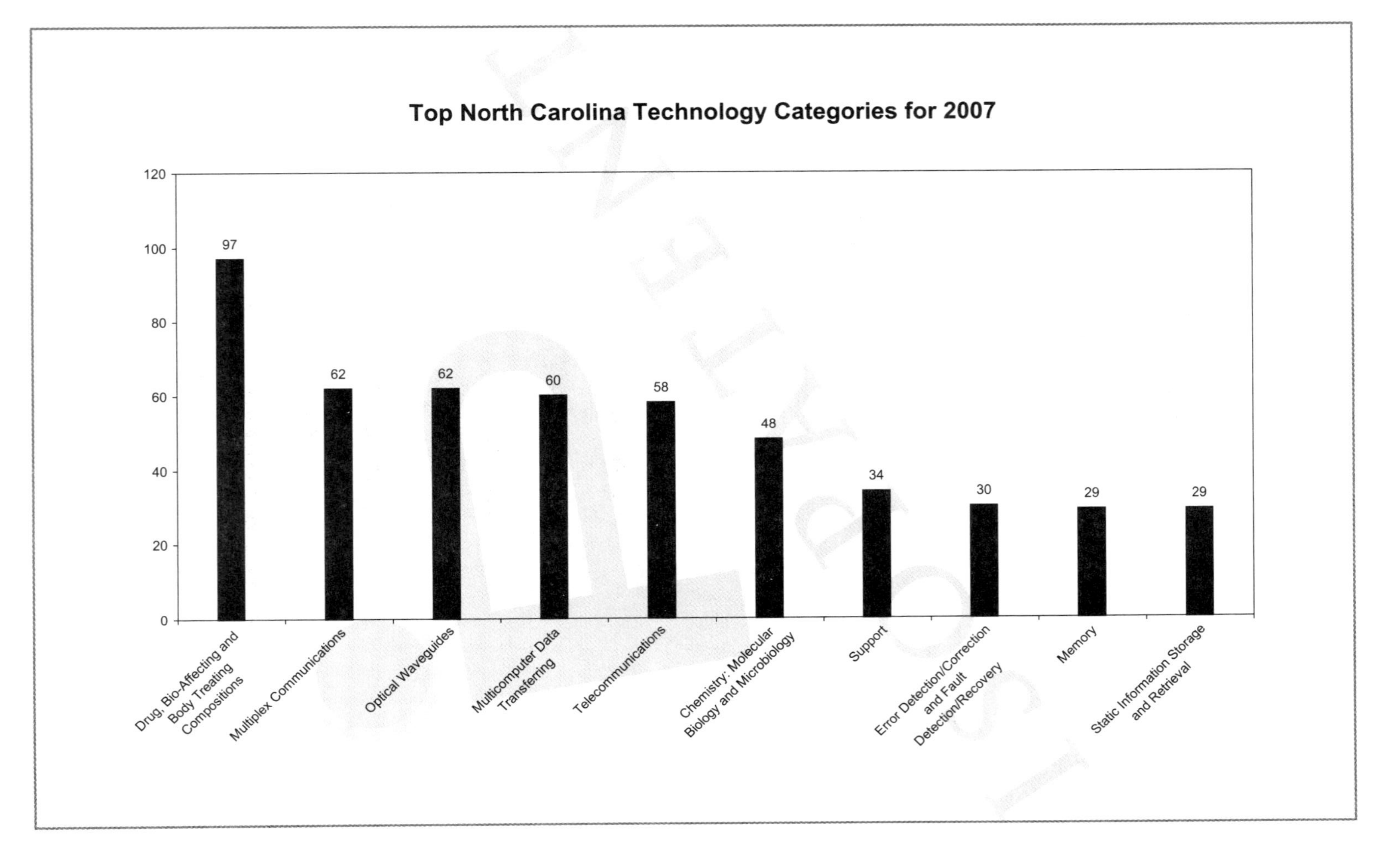
Top North Carolina Technology Categories for 2007
120
100
80
60
40
20
0
97
62
62
60
58
48
34
30
29
29
Drug, Bio-Affecting and Body Treating Compositions
Multiplex Communications
Optical Waveguides
Multicomputer Data Transferring
Telecommunications
Chemistry: Molecular Biology and Microbiology
Support
Error Detection/Correction and Fault Detection/Recovery
Memory
Static Information Storage and Retrieval

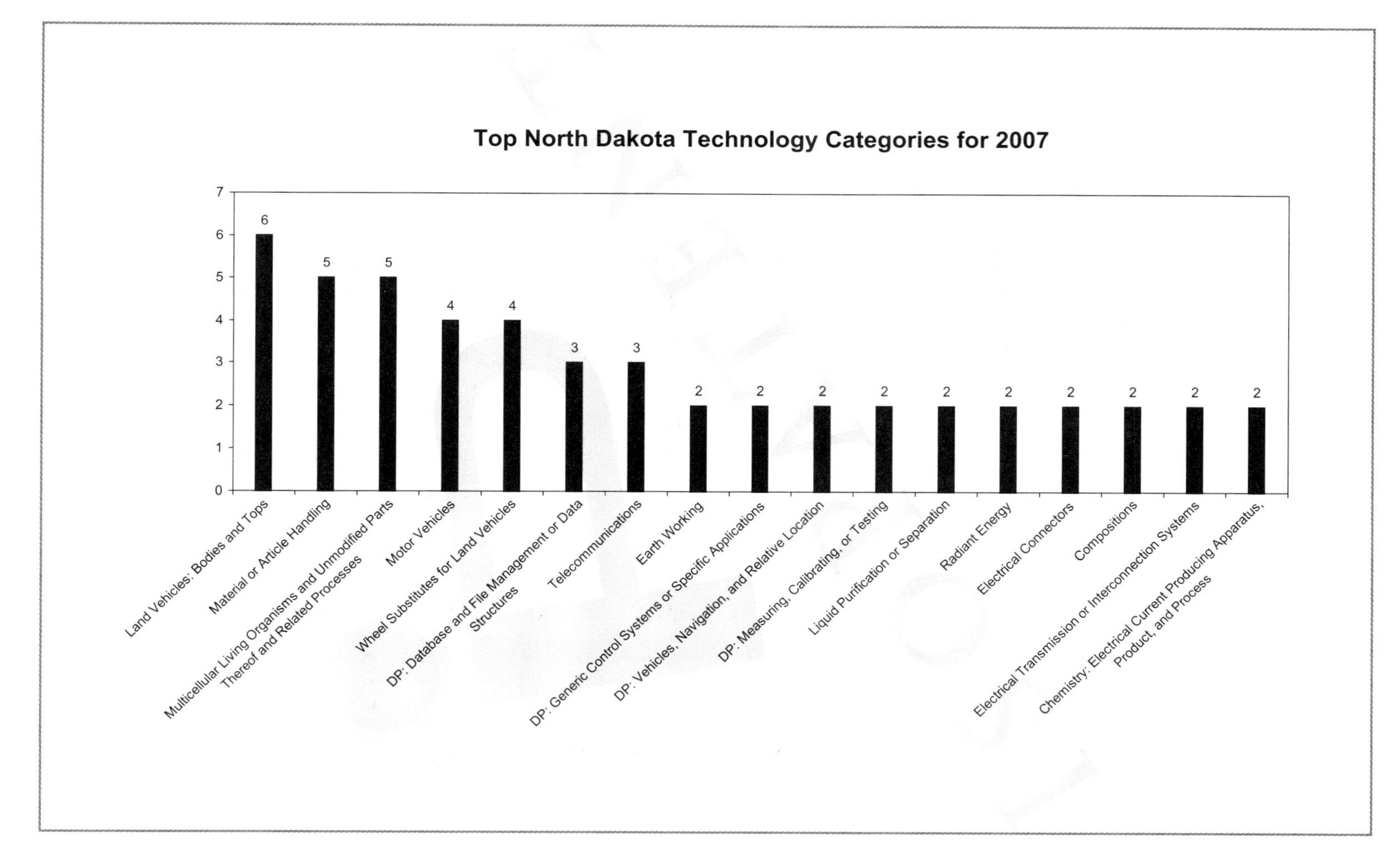
Top North Dakota Technology Categories for 2007
0
1
2
3
4
5
6
7
6
5
5
4
4
3
3
2
2
2
2
2
2
2
2
2
2
Land Vehicles: Bodies and Tops
Material or Article Handling
Multicellular Living Organisms and Unmodified Parts Thereof and Related Processes
Motor Vehicles
Wheel Substitutes for Land Vehicles
DP: Database and File Management or Data Structures
Telecommunications
Earth Working
DP: Generic Control Systems or Specific Applications
DP: Vehicles, Navigation, and Relative Location
DP: Measuring, Calibrating, or Testing
Liquid Purification or Separation
Radiant Energy
Electrical Connectors
Compositions
Electrical Transmission or Interconnection Systems
Chemistry: Electrical Current Producing Apparatus, Product, and Process

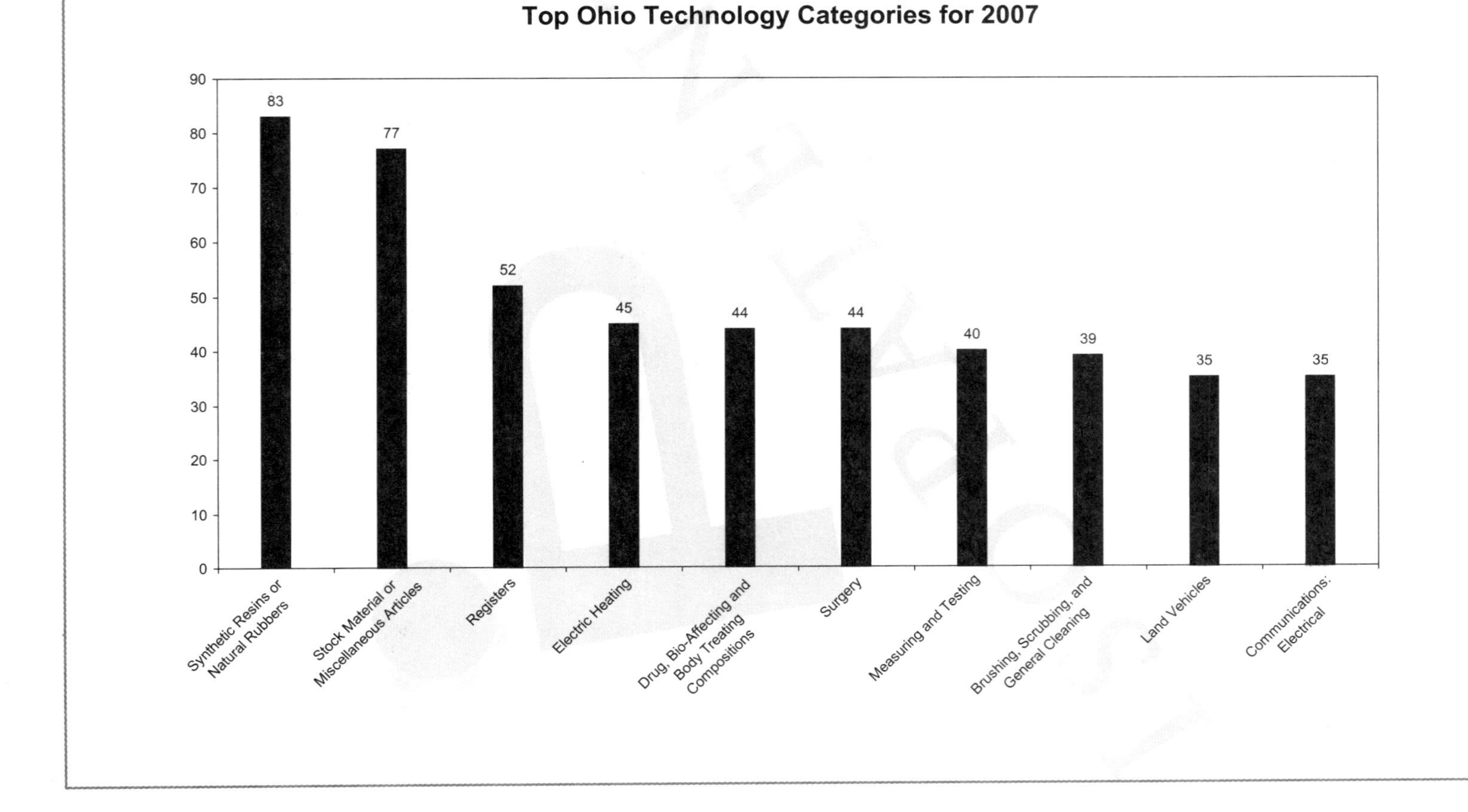

Top Ohio Technology Categories for 2007
90
80
70
60
50
40
30
20
10
0
83
77
52
45
44
44
40
39
35
35
Synthetic Resins or Natural Rubbers
Stock Material or Miscellaneous Articles
Registers
Electric Heating
Drug, Bio-Affecting and Body Treating Compositions
Surgery
Measuring and Testing
Brushing, Scrubbing, and General Cleaning
Land Vehicles
Communications: Electrical

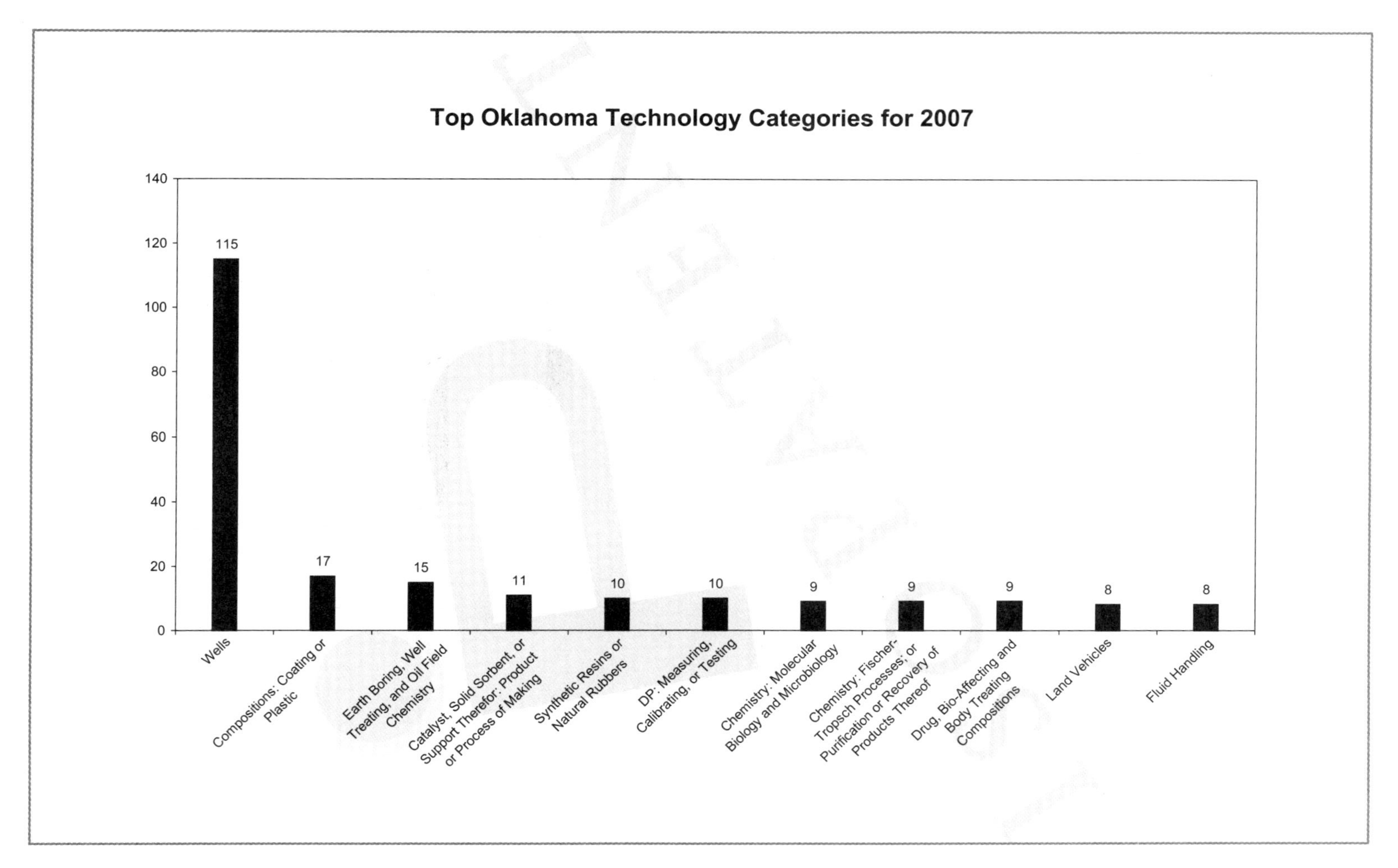
Top Oklahoma Technology Categories for 2007
140
120
100
80
60
40
20
0
115
17
15
11
10
10
9
9
9
8
8
Wells
Compositions: Coating or Plastic
Earth Boring, Well Treating, and Oil Field Chemistry
Catalyst, Solid Sorbent, or Support Therefor: Product or Process of Making
Synthetic Resins or Natural Rubbers
DP: Measuring, Calibrating, or Testing
Chemistry: Molecular Biology and Microbiology
Chemistry: Fischer-Tropsch Processes; or Purification or Recovery of Products Thereof
Drug, Bio-Affecting and Body Treating Compositions
Land Vehicles
Fluid Handling

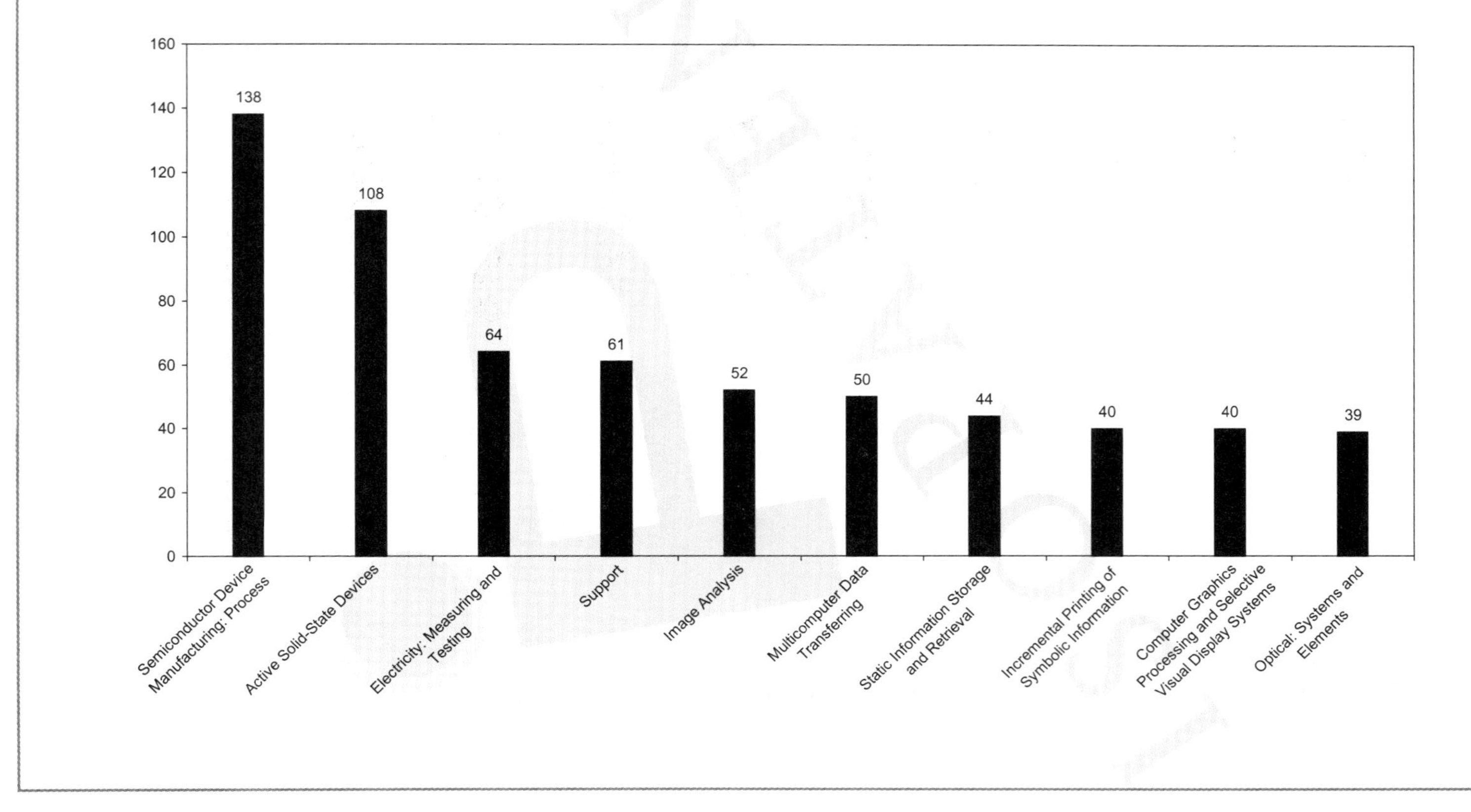
Top Oregon Technology Categories for 2007
160
140
120
100
80
60
40
20
0
138
108
64
61
52
50
44
40
40
39
Semiconductor Device Manufacturing: Process
Active Solid-State Devices
Electricity: Measuring and Testing
Support
Image Analysis
Multicomputer Data Transferring
Static Information Storage and Retrieval
Incremental Printing of Symbolic Information
Computer Graphics Processing and Selective Visual Display Systems
Optical: Systems and Elements

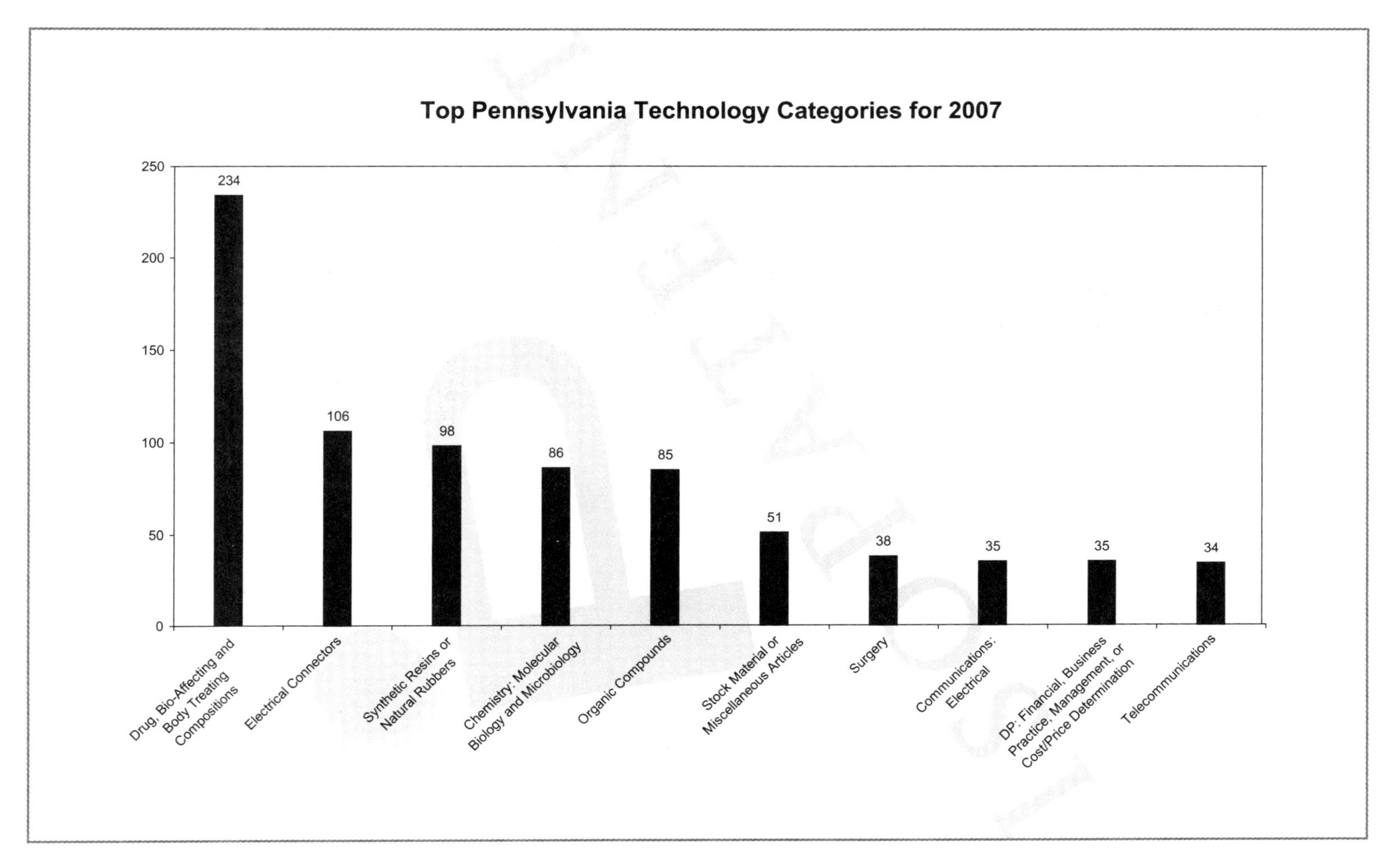

Top Pennsylvania Technology Categories for 2007
250
200
150
100
50
0
234
106
98
86
85
51
38
35
35
34
Drug, Bio-Affecting and Body Treating Compositions
Electrical Connectors
Synthetic Resins or Natural Rubbers
Chemistry: Molecular Biology and Microbiology
Organic Compounds
Stock Material or Miscellaneous Articles
Surgery
Communications: Electrical
DP: Financial, Business Practice, Management, or Cost/Price Determination
Telecommunications

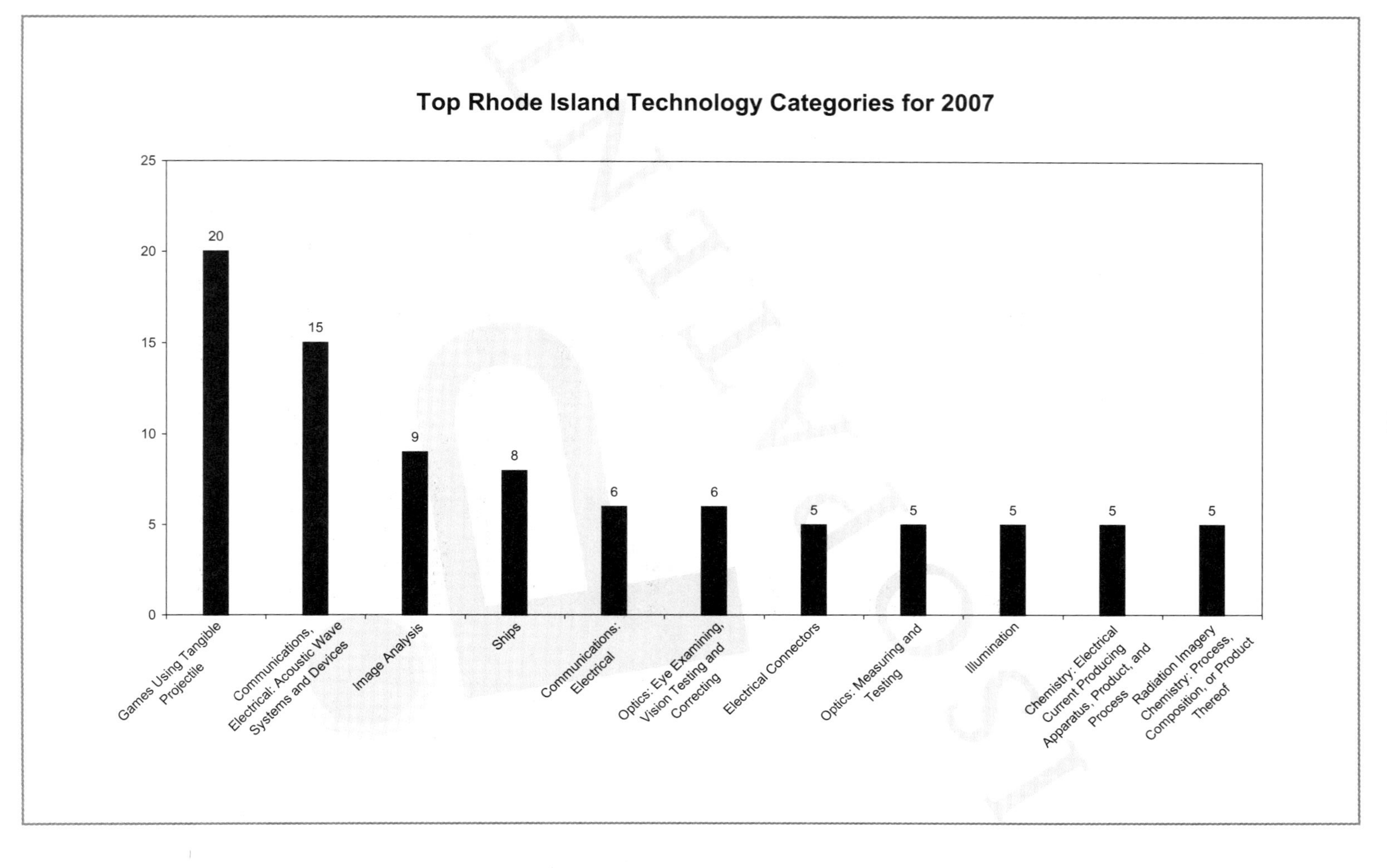
Top Rhode Island Technology Categories for 2007
25
20
15
10
5
0
20
15
9
8
6
6
5
5
5
5
5
Games Using Tangible Projectile
Communications, Electrical: Acoustic Wave Systems and Devices
Image Analysis
Ships
Communications: Electrical
Optics: Eye Examining, Vision Testing and Correcting
Electrical Connectors
Optics: Measuring and Testing
Illumination
Chemistry: Electrical Current Producing Apparatus, Product, and Process
Radiation Imagery Chemistry: Process, Composition, or Product Thereof

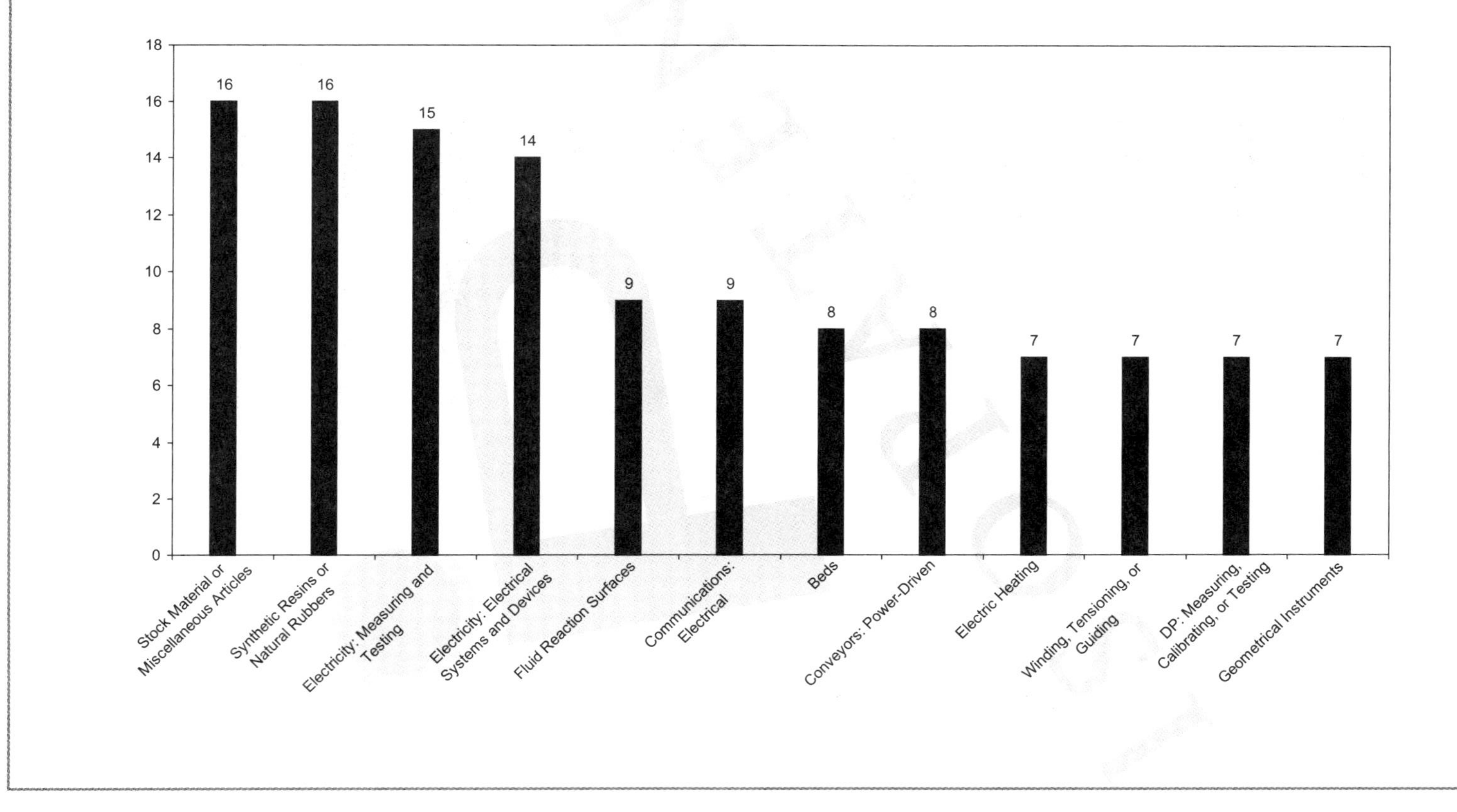
Top South Carolina Technology Categories for 2007
0
2
4
6
8
10
12
14
16
18
Stock Material or Miscellaneous Articles
16
Synthetic Resins or Natural Rubbers
16
Electricity: Measuring and Testing
15
Electricity: Electrical Systems and Devices
14
Fluid Reaction Surfaces
9
Communications: Electrical
9
Beds
8
Conveyors: Power-Driven
8
Electric Heating
7
Winding, Tensioning, or Guiding
7
DP: Measuring, Calibrating, or Testing
7
Geometrical Instruments
7

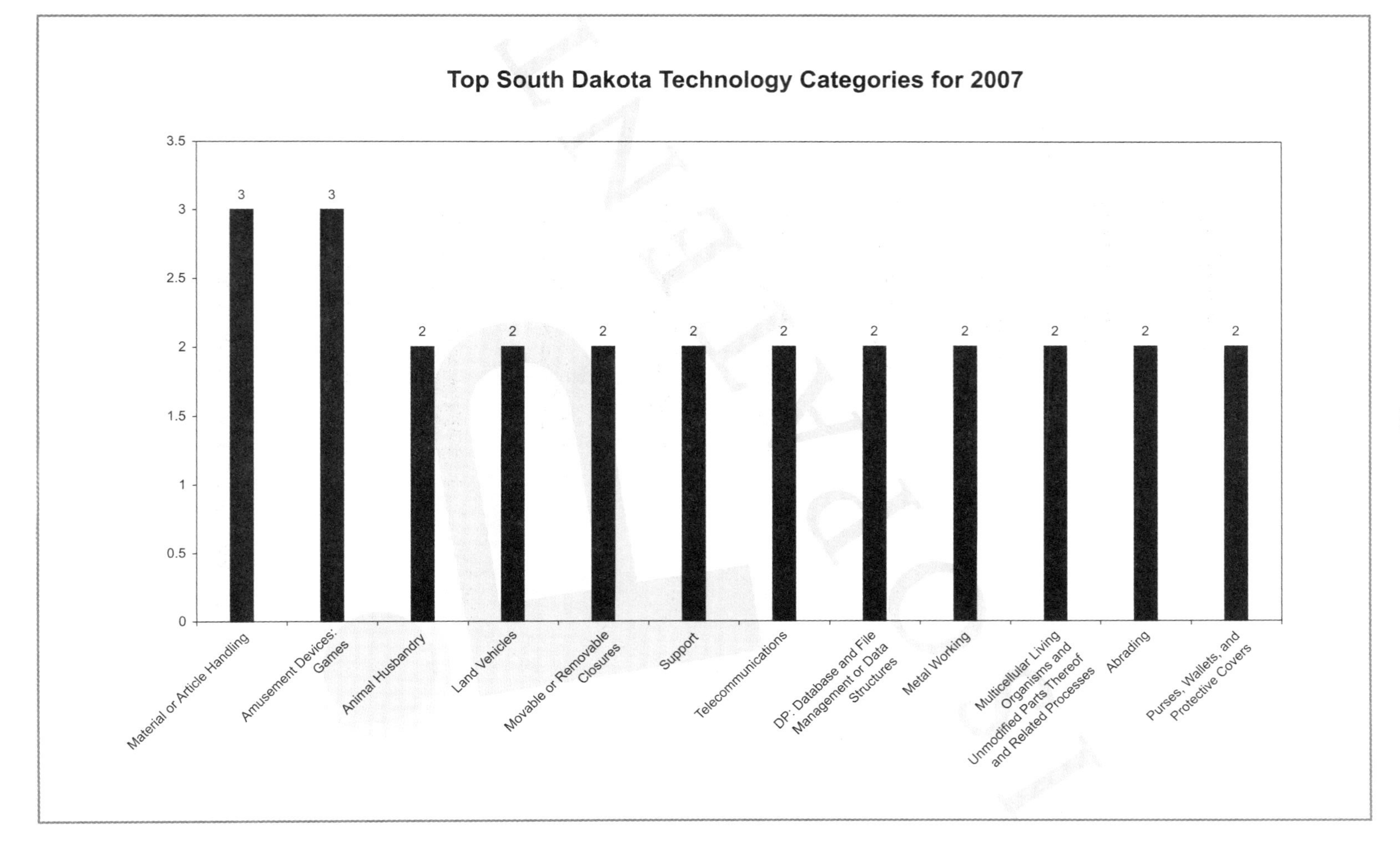
Top South Dakota Technology Categories for 2007
3.5
3
2.5
2
1.5
1
0.5
0
3
3
2
2
2
2
2
2
2
2
2
2
Material or Article Handling
Amusement Devices: Games
Animal Husbandry
Land Vehicles
Movable or Removable Closures
Support
Telecommunications
DP: Database and File Management or Data Structures
Metal Working
Multicellular Living Organisms and Unmodified Parts Thereof and Related Processes
Abrading
Purses, Wallets, and Protective Covers

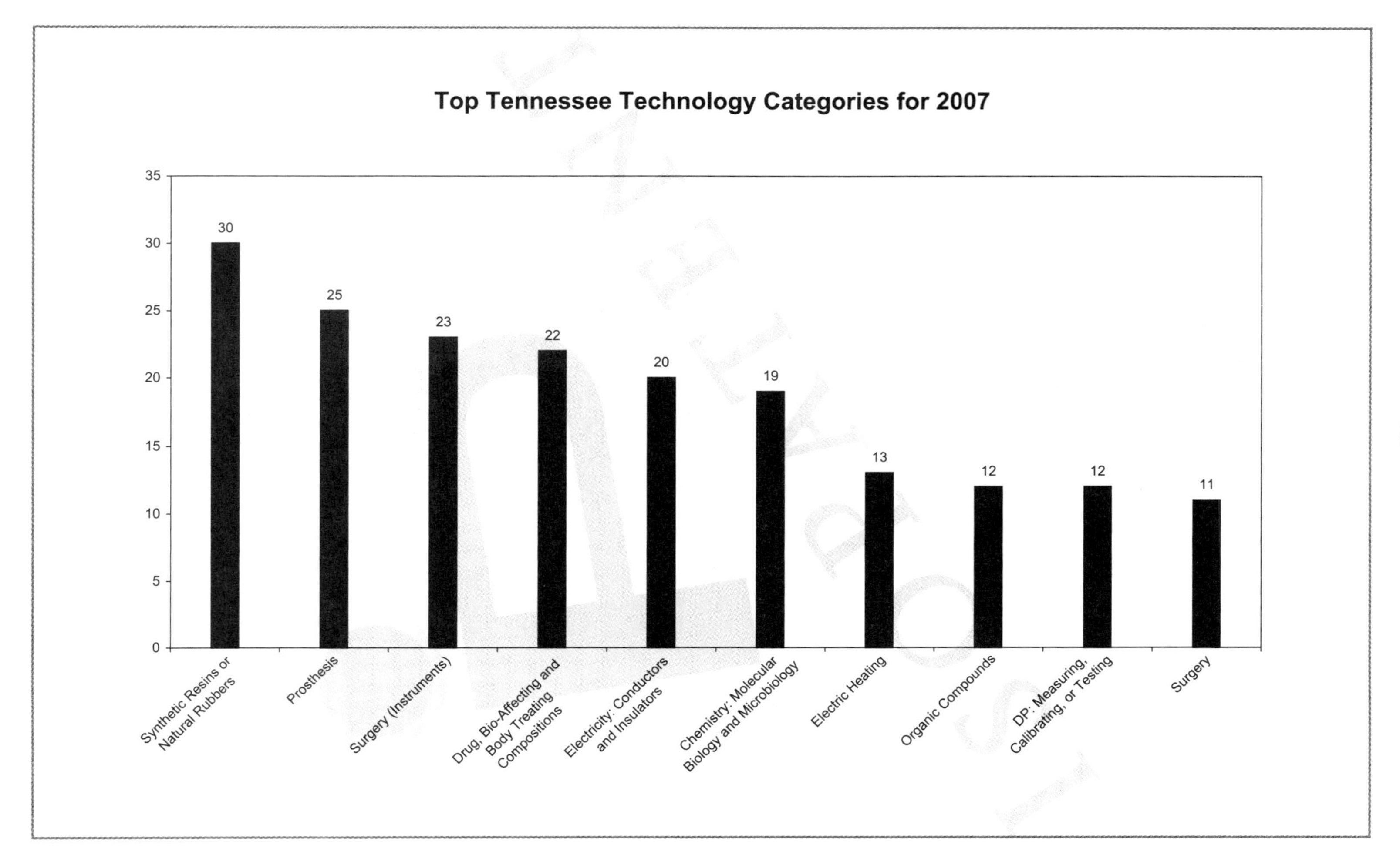
Top Tennessee Technology Categories for 2007
35
30
25
20
15
10
5
0
30
25
23
22
20
19
13
12
12
11
Synthetic Resins or Natural Rubbers
Prosthesis
Surgery (Instruments)
Drug, Bio-Affecting and Body Treating Compositions
Electricity: Conductors and Insulators
Chemistry: Molecular Biology and Microbiology
Electric Heating
Organic Compounds
DP: Measuring, Calibrating, or Testing
Surgery

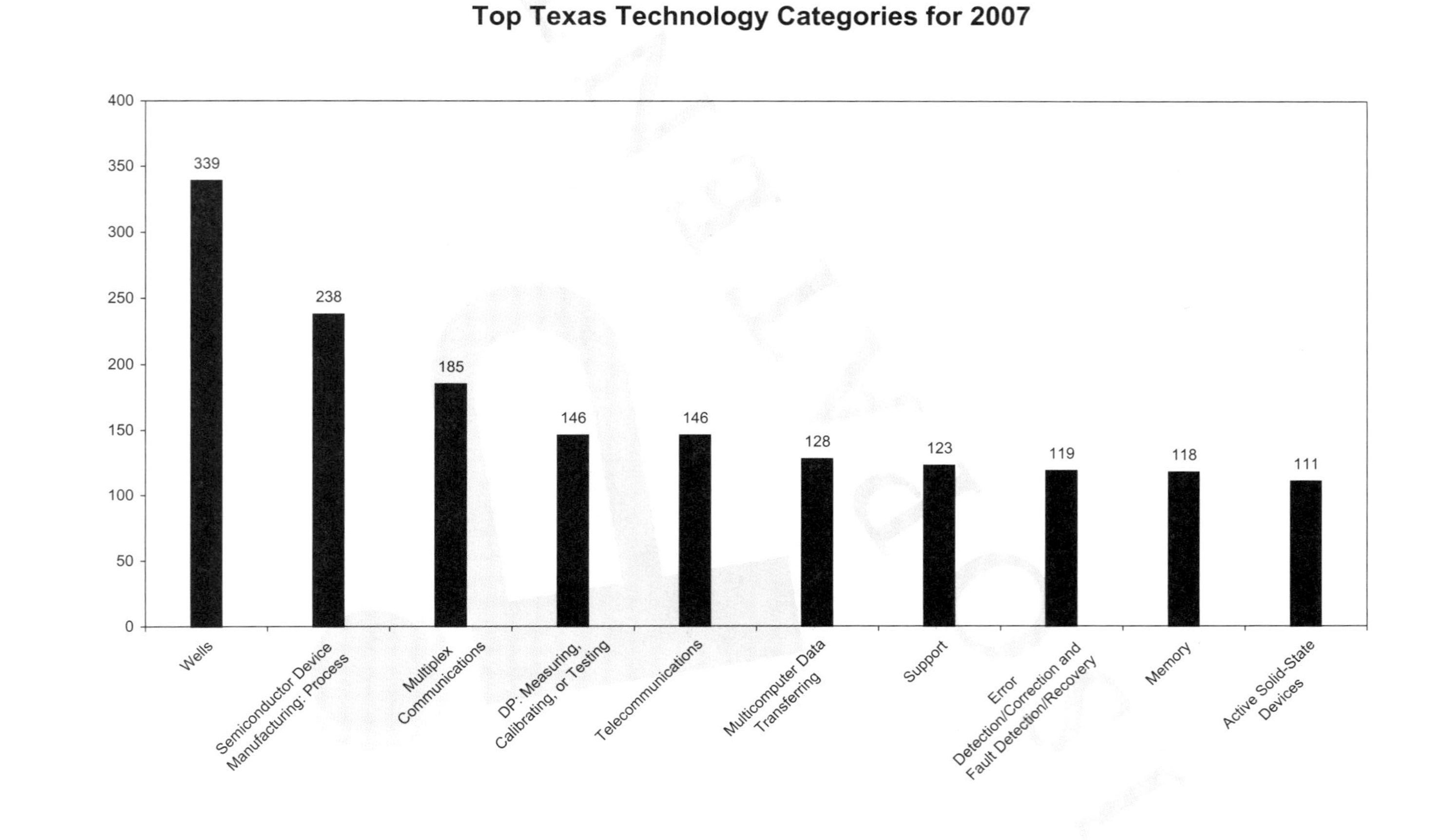

Top Texas Technology Categories for 2007
400
350
300
250
200
150
100
50
0
339
238
185
146
146
128
123
119
118
111
Wells
Semiconductor Device Manufacturing: Process
Multiplex Communications
DP: Measuring, Calibrating, or Testing
Telecommunications
Multicomputer Data Transferring
Support
Error Detection/Correction and Fault Detection/Recovery
Memory
Active Solid-State Devices

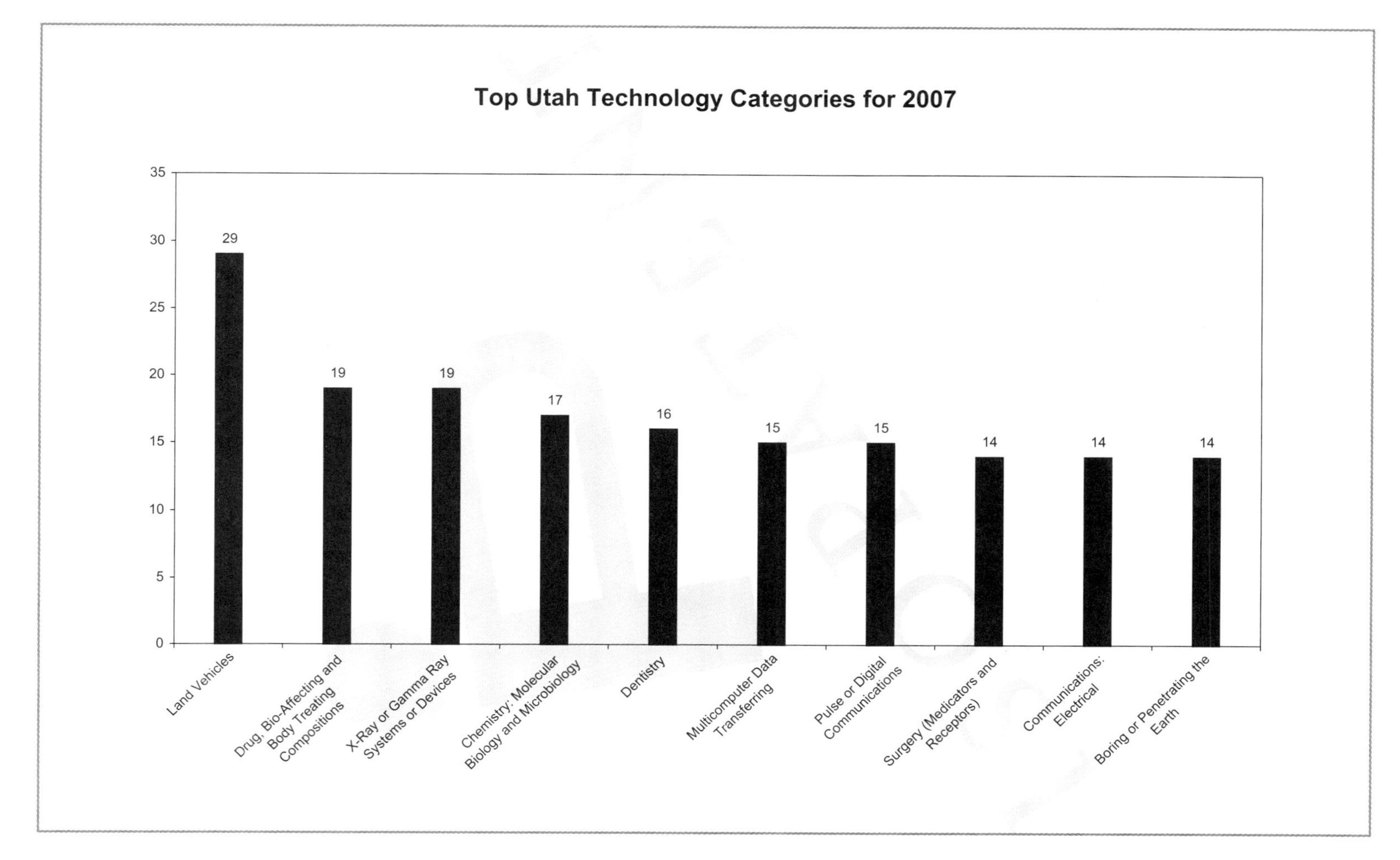
Top Utah Technology Categories for 2007
35
30
25
20
15
10
5
0
29
19
19
17
16
15
15
14
14
14
Land Vehicles
Drug, Bio-Affecting and Body Treating Compositions
X-Ray or Gamma Ray Systems or Devices
Chemistry: Molecular Biology and Microbiology
Dentistry
Multicomputer Data Transferring
Pulse or Digital Communications
Surgery (Medicators and Receptors)
Communications: Electrical
Boring or Penetrating the Earth

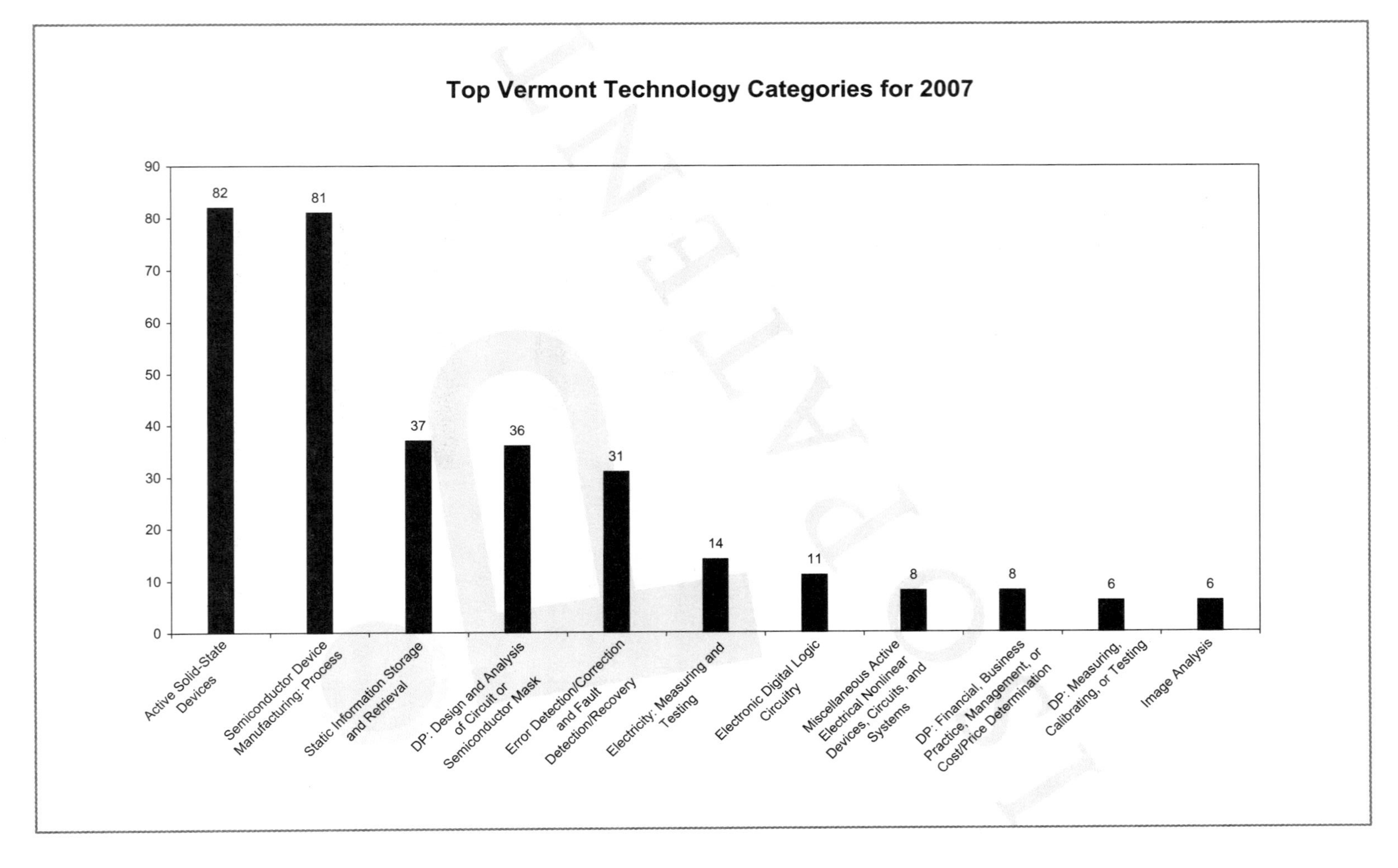
Top Vermont Technology Categories for 2007
0
10
20
30
40
50
60
70
80
90
82
81
37
36
31
14
11
8
8
6
6
Active Solid-State Devices
Semiconductor Device Manufacturing: Process
Static Information Storage and Retrieval
DP: Design and Analysis of Circuit or Semiconductor Mask
Error Detection/Correction and Fault Detection/Recovery
Electricity: Measuring and Testing
Electronic Digital Logic Circuitry
Miscellaneous Active Electrical Nonlinear Devices, Circuits, and Systems
DP: Financial, Business Practice, Management, or Cost/Price Determination
DP: Measuring, Calibrating, or Testing
Image Analysis

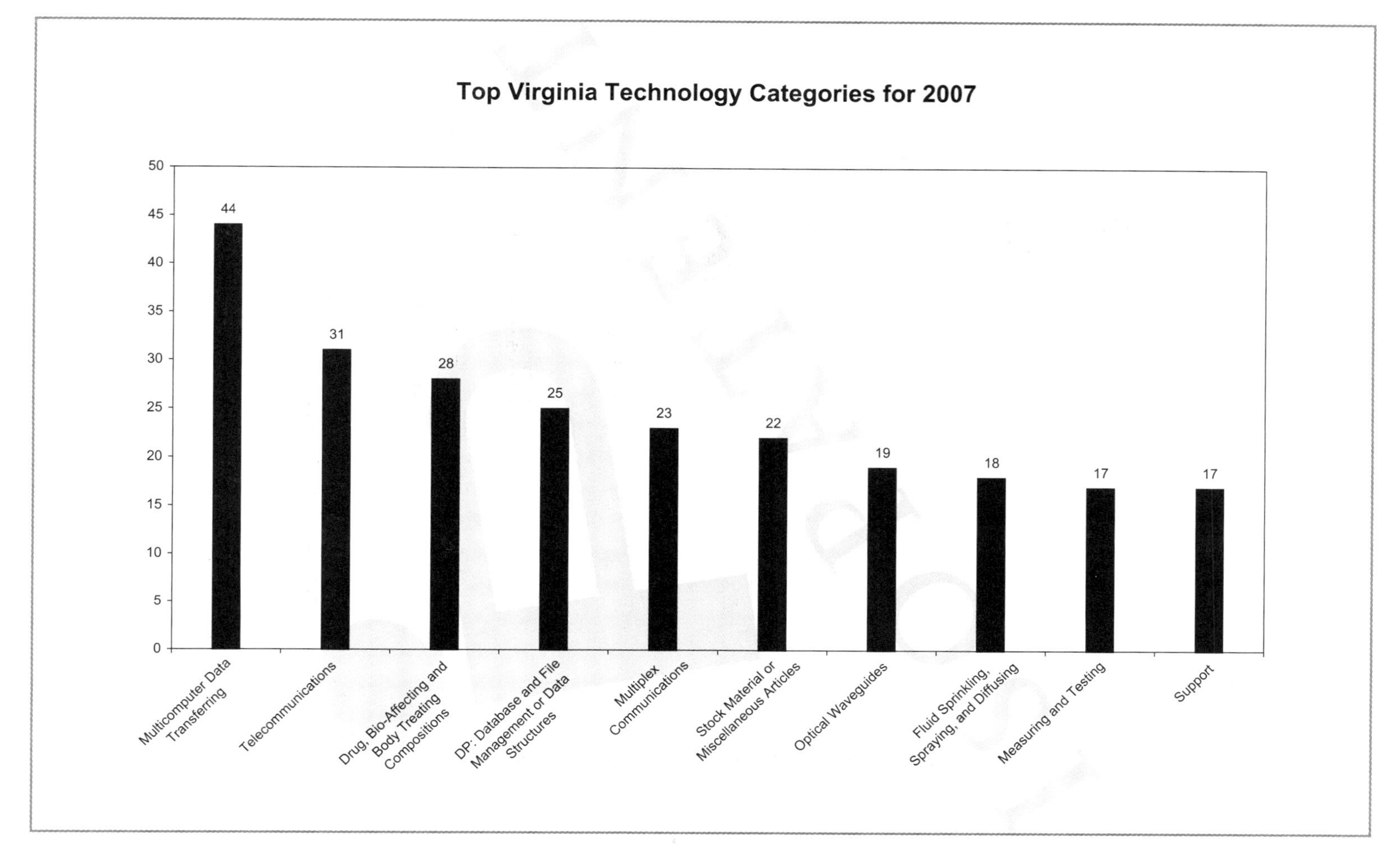
Top Virginia Technology Categories for 2007
50
45
40
35
30
25
20
15
10
5
0
44
31
28
25
23
22
19
18
17
17
Multicomputer Data Transferring
Telecommunications
Drug, Bio-Affecting and Body Treating Compositions
DP: Database and File Management or Data Structures
Multiplex Communications
Stock Material or Miscellaneous Articles
Optical Waveguides
Fluid Sprinkling, Spraying, and Diffusing
Measuring and Testing
Support

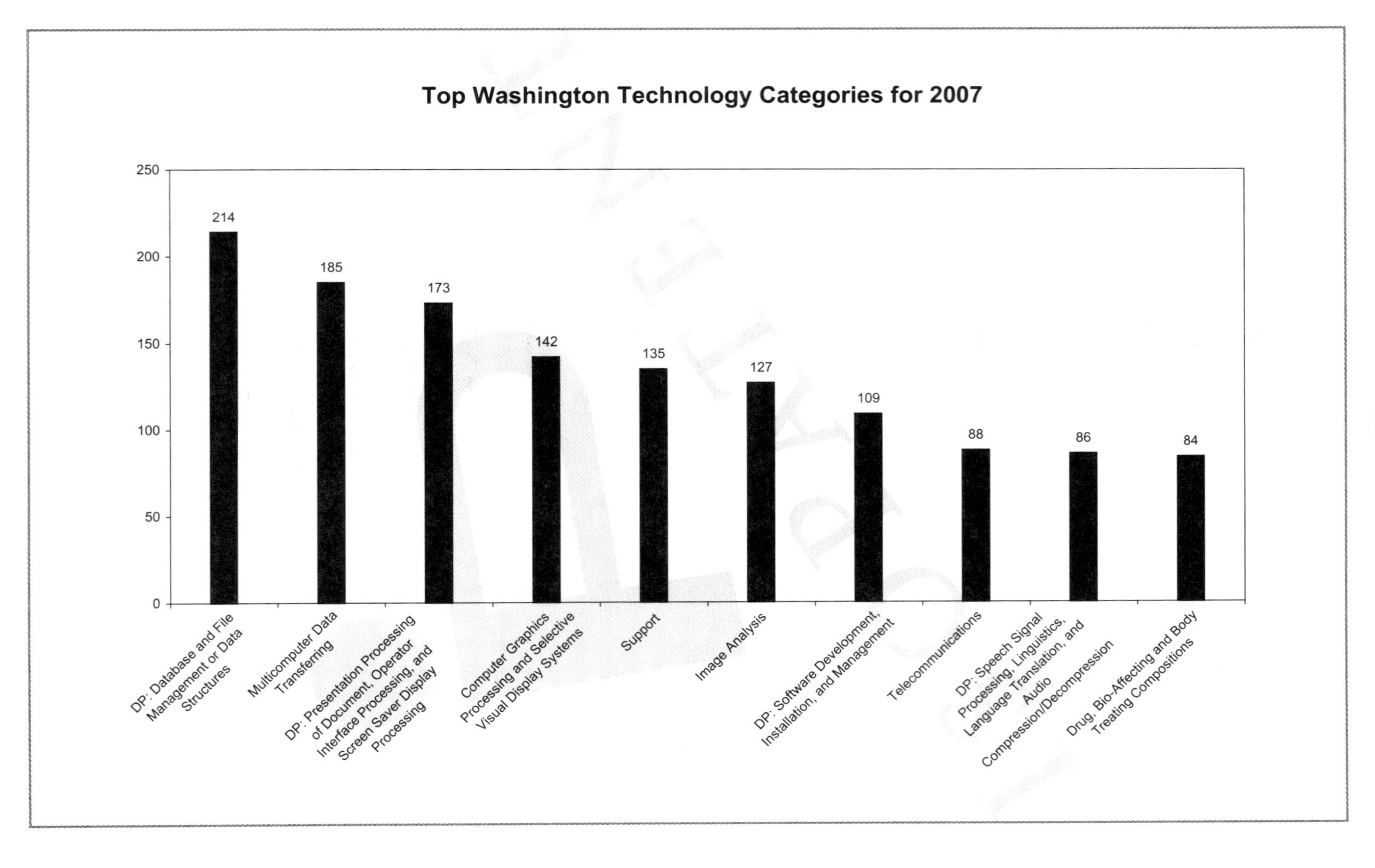
Top Washington Technology Categories for 2007
250
200
150
100
50
0
214
185
173
142
135
127
109
88
86
84
DP: Database and File Management or Data Structures
Multicomputer Data Transferring
DP: Presentation Processing of Document, Operator Interface Processing, and Screen Saver Display Processing
Computer Graphics Processing and Selective Visual Display Systems
Support
Image Analysis
DP: Software Development, Installation, and Management
Telecommunications
DP: Speech Signal Processing, Linguistics, Language Translation, and Audio Compression/Decompression
Drug, Bio-Affecting and Body Treating Compositions

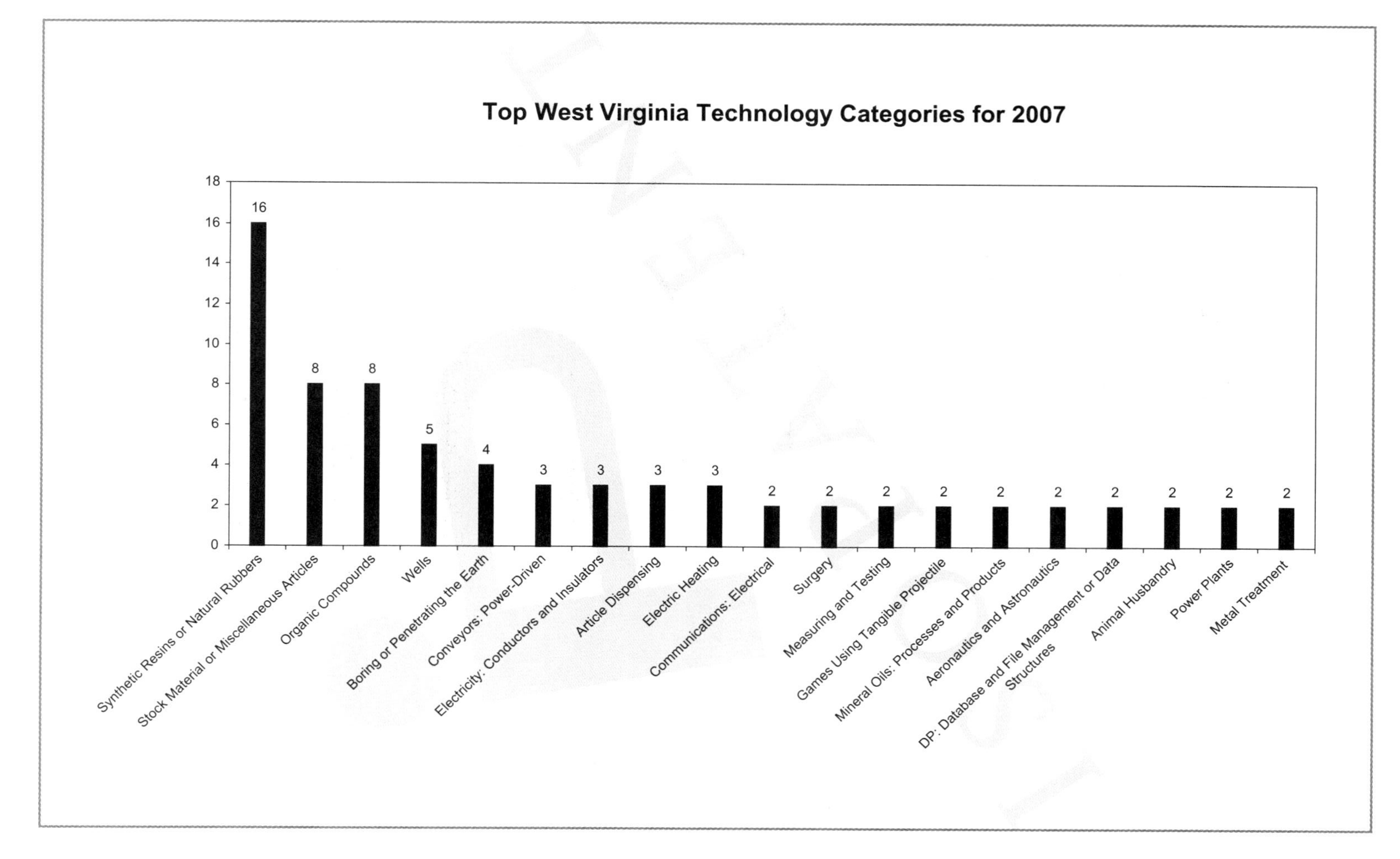

Top West Virginia Technology Categories for 2007
18
16
14
12
10
8
6
4
2
0
16
8
8
5
4
3
3
3
3
2
2
2
2
2
2
2
2
2
2
Synthetic Resins or Natural Rubbers
Stock Material or Miscellaneous Articles
Organic Compounds
Wells
Boring or Penetrating the Earth
Conveyors: Power-Driven
Electricity: Conductors and Insulators
Article Dispensing
Electric Heating
Communications: Electrical
Surgery
Measuring and Testing
Games Using Tangible Projectile
Mineral Oils: Processes and Products
Aeronautics and Astronautics
DP: Database and File Management or Data Structures
Animal Husbandry
Power Plants
Metal Treatment

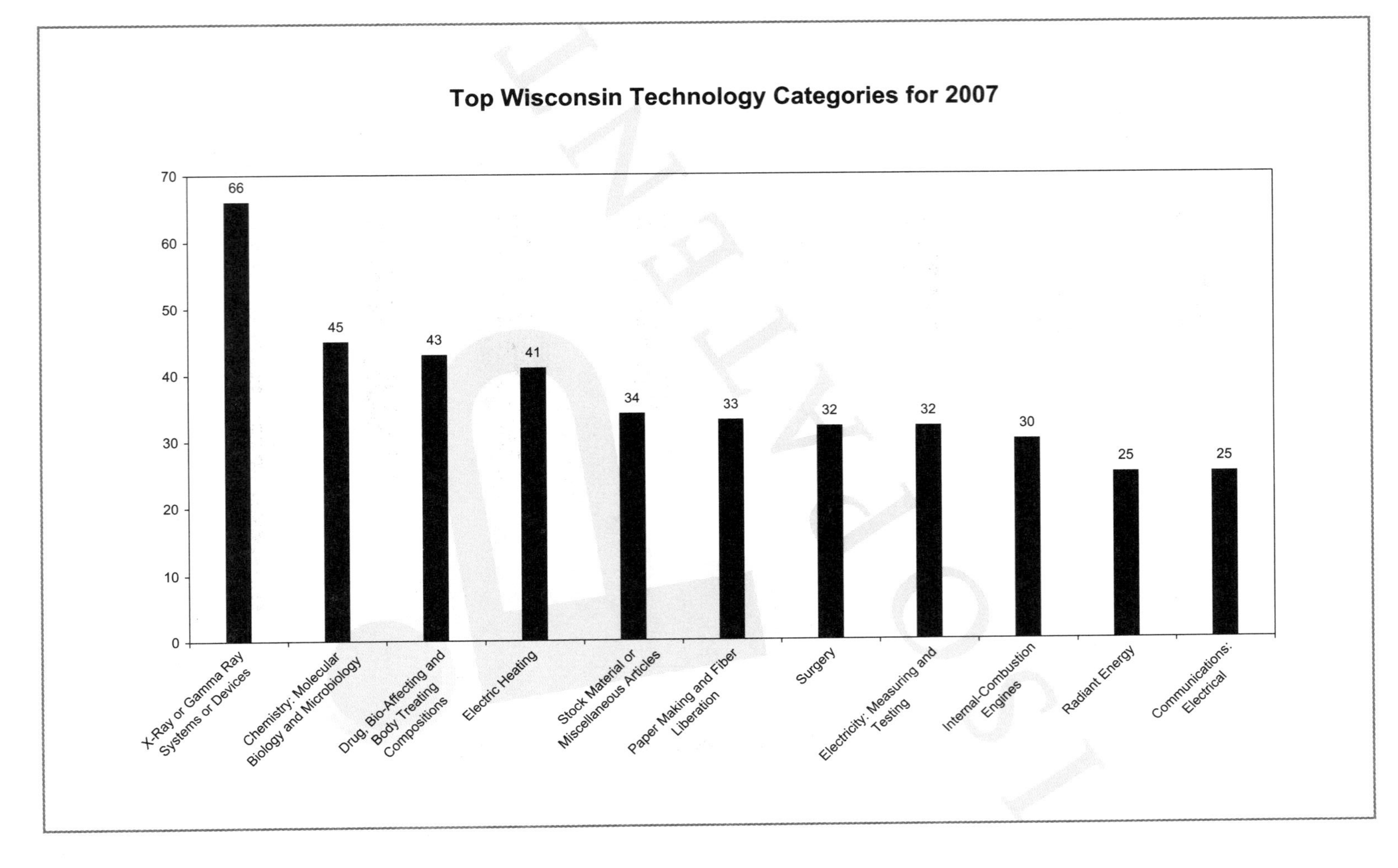

Top Wisconsin Technology Categories for 2007
70
60
50
40
30
20
10
0
66
45
43
41
34
33
32
32
30
25
25
X-Ray or Gamma Ray Systems or Devices
Chemistry: Molecular Biology and Microbiology
Drug, Bio-Affecting and Body Treating Compositions
Electric Heating
Stock Material or Miscellaneous Articles
Paper Making and Fiber Liberation
Surgery
Electricity: Measuring and Testing
Internal-Combustion Engines
Radiant Energy
Communications: Electrical

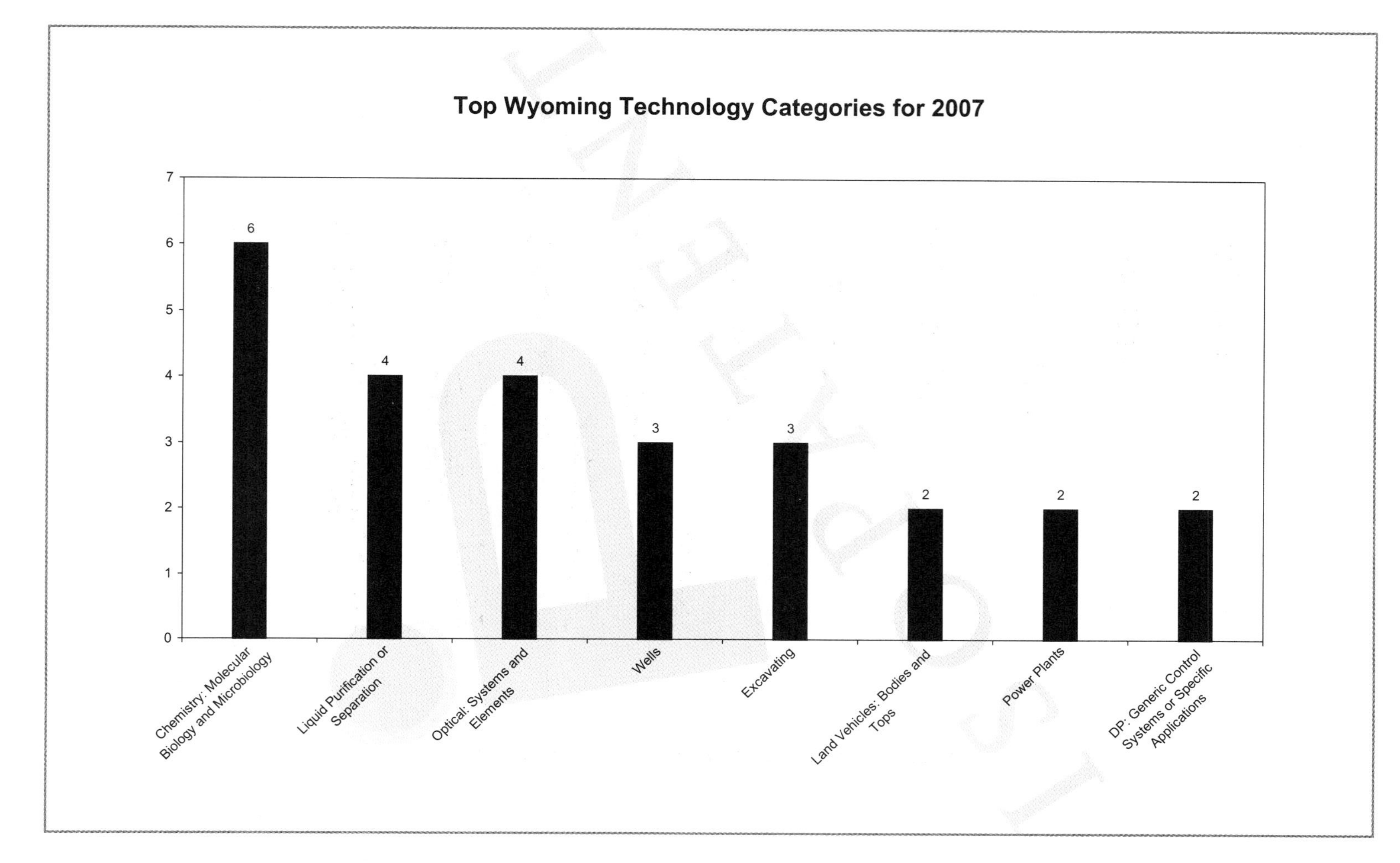

Top Wyoming Technology Categories for 2007
7
6
5
4
3
2
1
0
6
4
4
3
3
2
2
2
Chemistry: Molecular Biology and Microbiology
Liquid Purification or Separation
Optical: Systems and Elements
Wells
Excavating
Land Vehicles: Bodies and Tops
Power Plants
DP: Generic Control Systems or Specific Applications

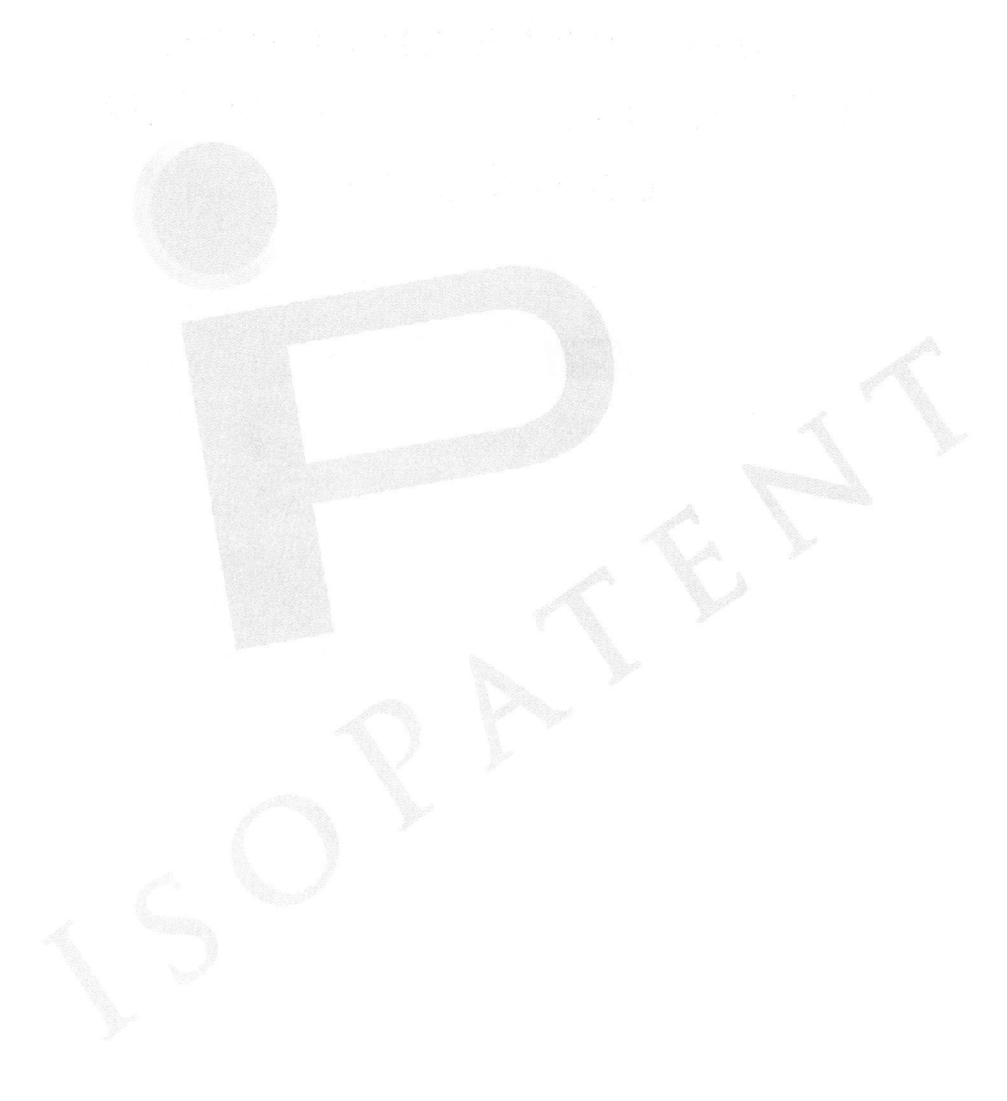

Chapter 17
Top United States Patent Recipients for Each of the Fifty United States

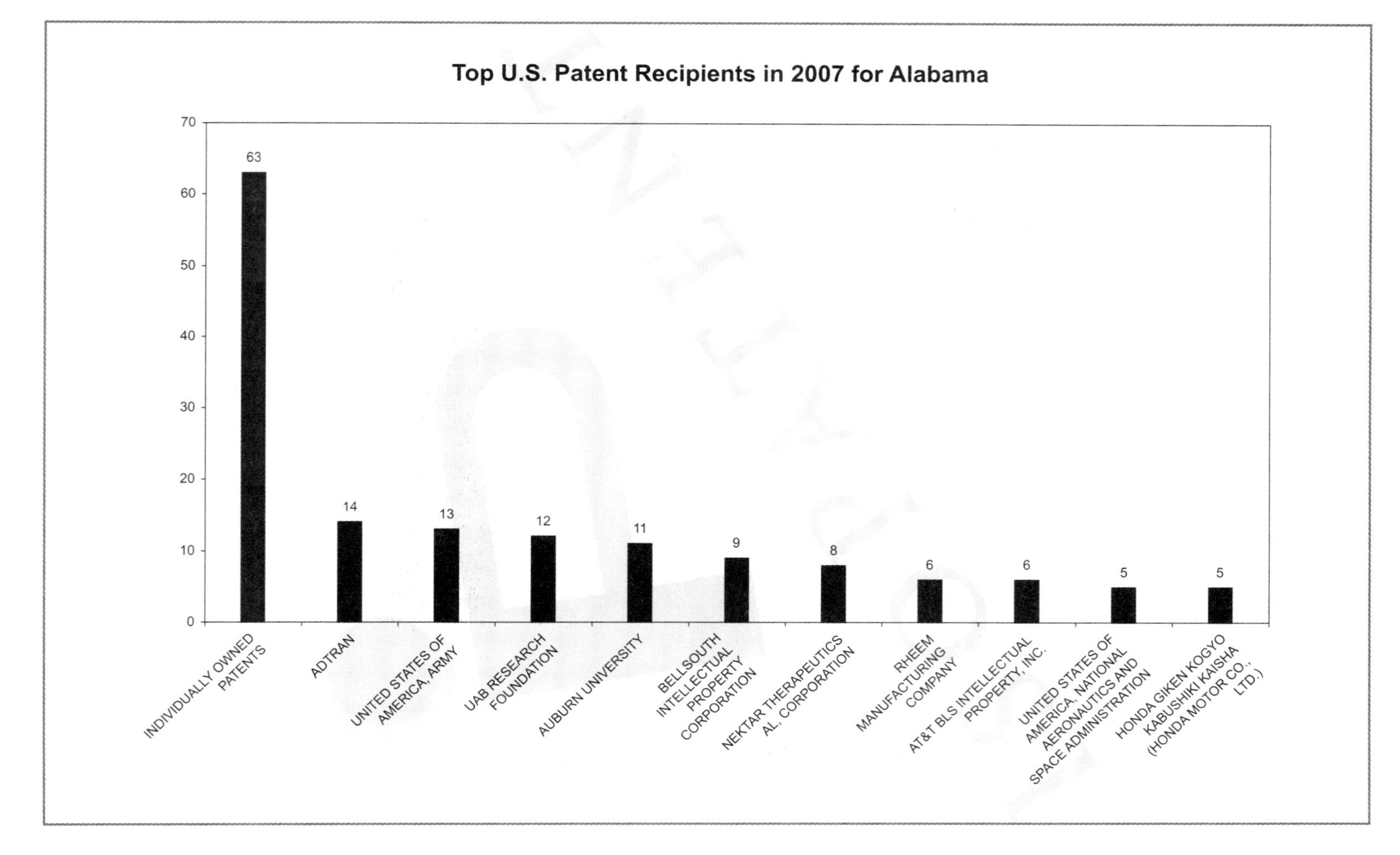

Top U.S. Patent Recipients in 2007 for Alabama
70
60
50
40
30
20
10
0
63
14
13
12
11
9
8
6
6
5
5
INDIVIDUALLY OWNED PATENTS
ADTRAN
UNITED STATES OF AMERICA, ARMY
UAB RESEARCH FOUNDATION
AUBURN UNIVERSITY
BELLSOUTH INTELLECTUAL PROPERTY CORPORATION
NEKTAR THERAPEUTICS AL, CORPORATION
RHEEM MANUFACTURING COMPANY
AT&T BLS INTELLECTUAL PROPERTY, INC.
UNITED STATES OF AMERICA, NATIONAL AERONAUTICS AND SPACE ADMINISTRATION
HONDA GIKEN KOGYO KABUSHIKI KAISHA (HONDA MOTOR CO., LTD.)

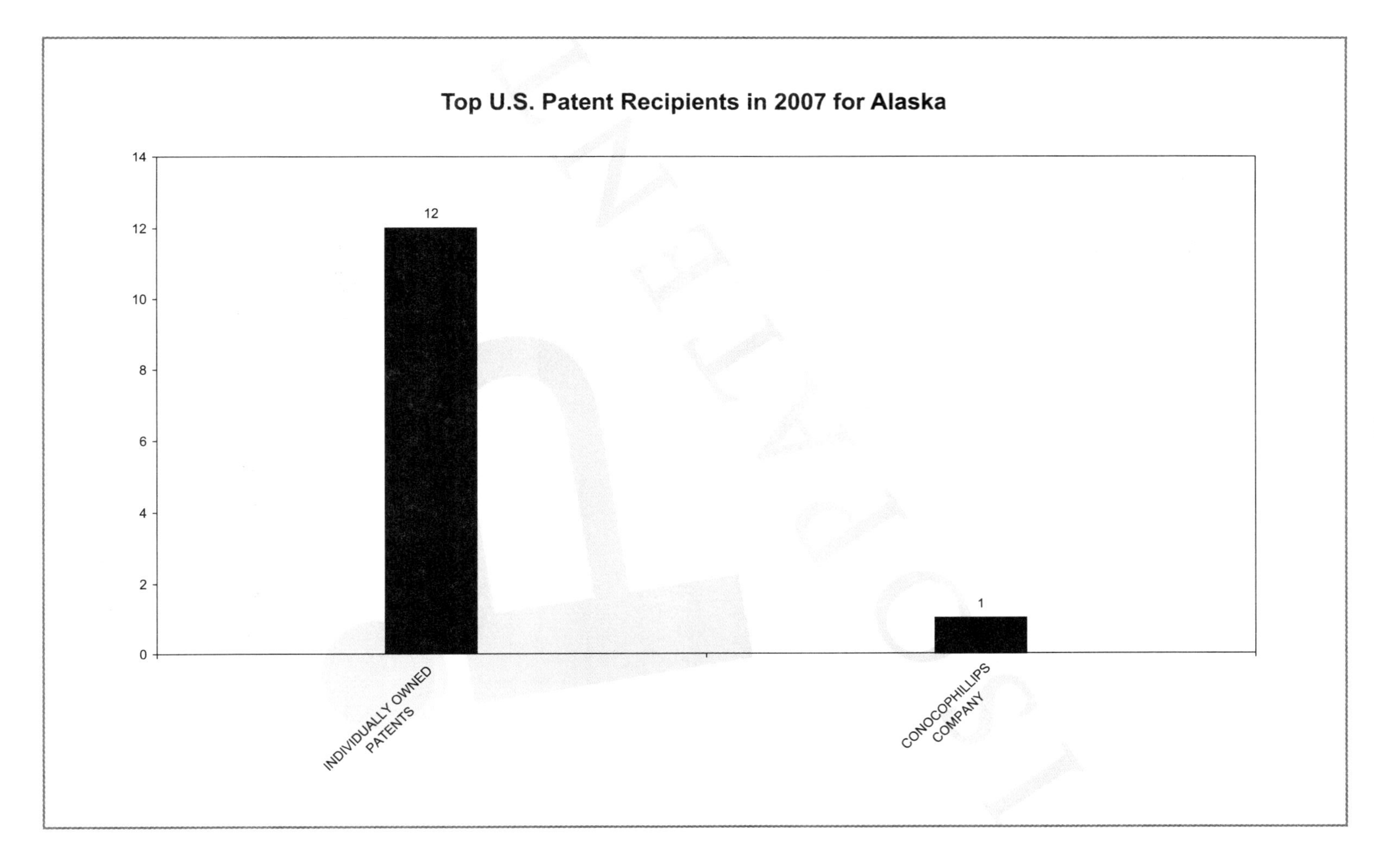
Top U.S. Patent Recipients in 2007 for Alaska
14
12
10
8
6
4
2
0
12
1
INDIVIDUALLY OWNED PATENTS
CONOCOPHILLIPS COMPANY

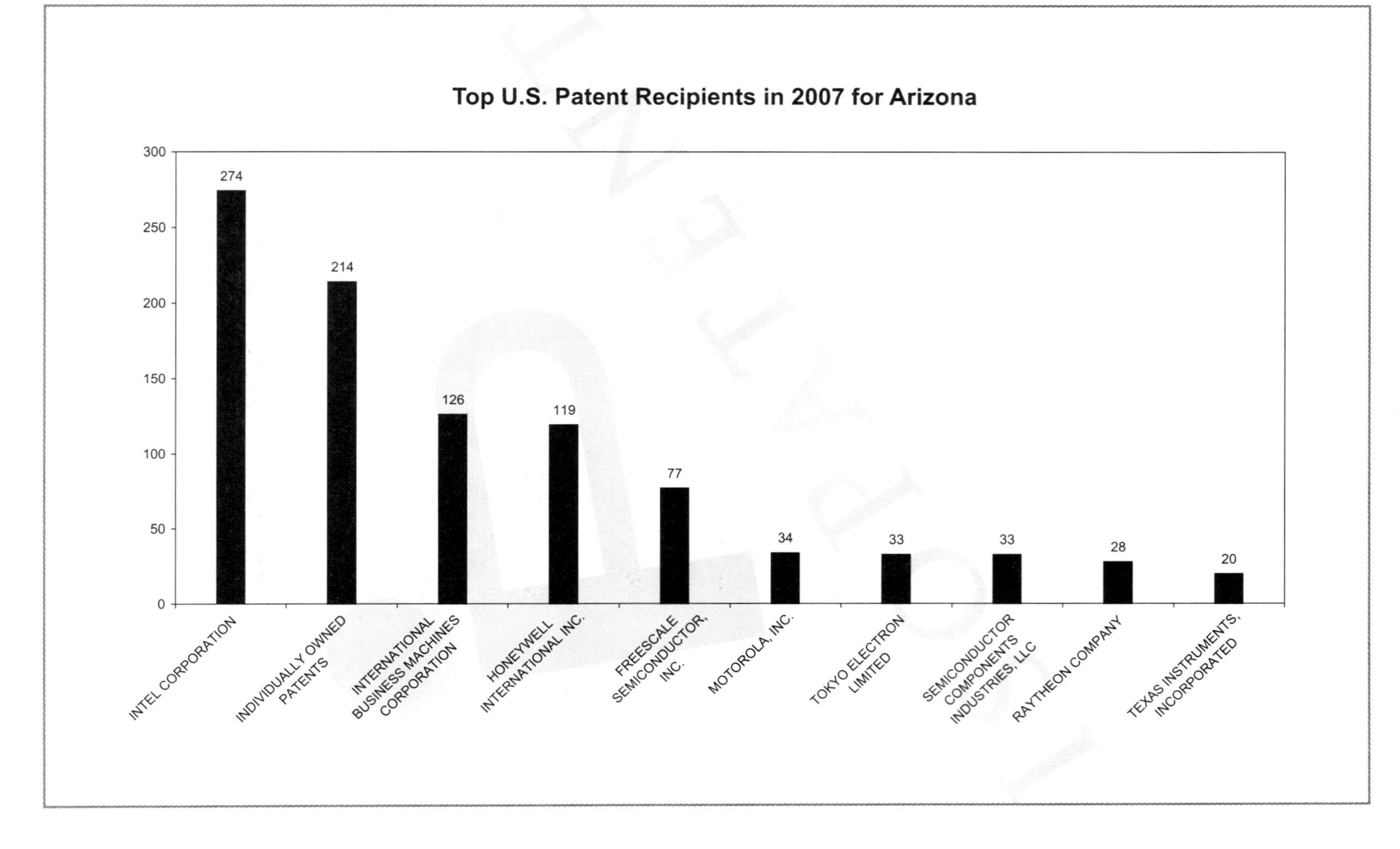
Top U.S. Patent Recipients in 2007 for Arizona
300
250
200
150
100
50
0
274
214
126
119
77
34
33
33
28
20
INTEL CORPORATION
INDIVIDUALLY OWNED PATENTS
INTERNATIONAL BUSINESS MACHINES CORPORATION
HONEYWELL INTERNATIONAL INC.
FREESCALE SEMICONDUCTOR, INC.
MOTOROLA, INC.
TOKYO ELECTRON LIMITED
SEMICONDUCTOR COMPONENTS INDUSTRIES, LLC
RAYTHEON COMPANY
TEXAS INSTRUMENTS, INCORPORATED

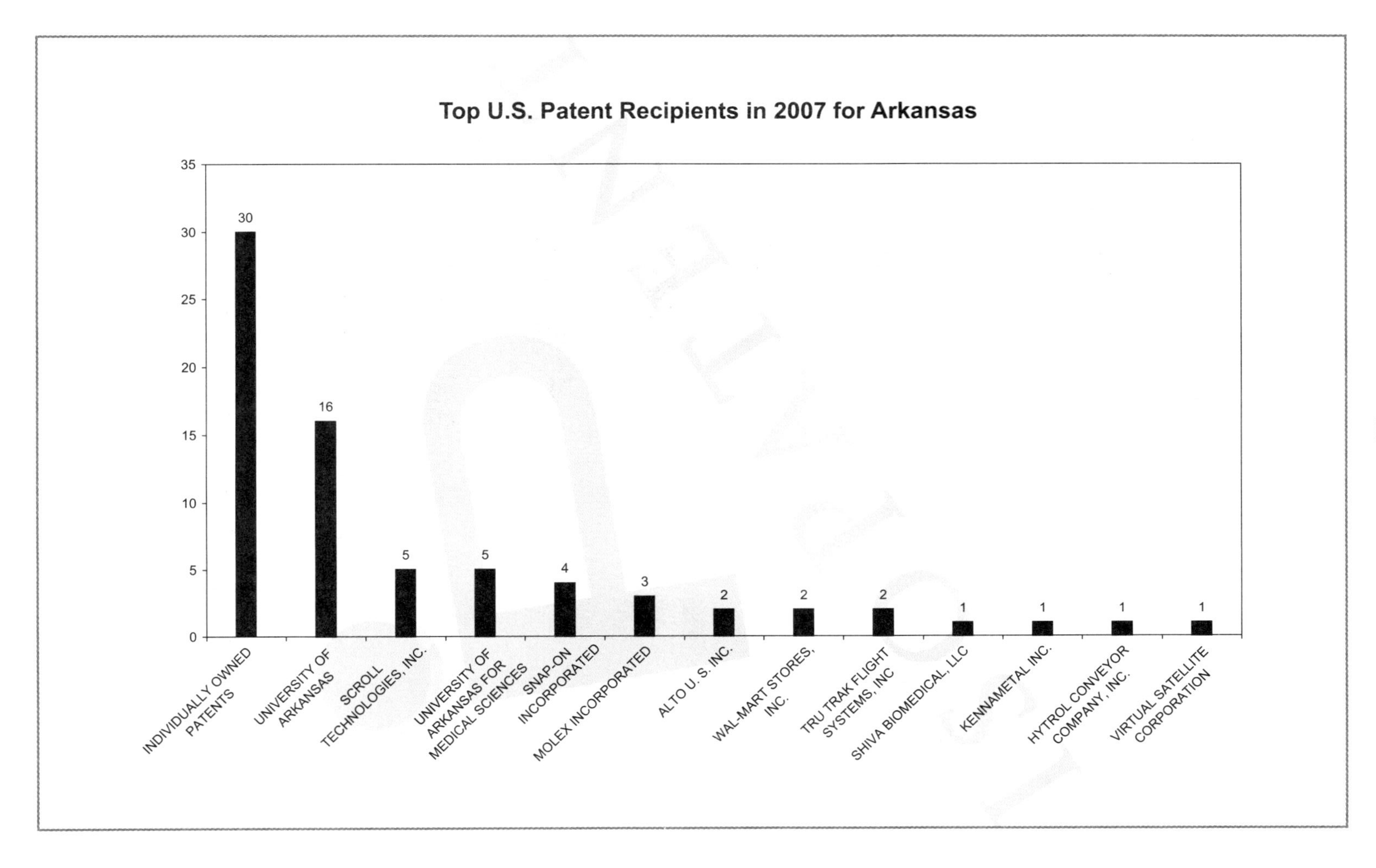

Top U.S. Patent Recipients in 2007 for Arkansas
35
30
25
20
15
10
5
0
30
16
5
5
4
3
2
2
2
1
1
1
1
INDIVIDUALLY OWNED PATENTS
UNIVERSITY OF ARKANSAS
SCROLL TECHNOLOGIES, INC.
UNIVERSITY OF ARKANSAS FOR MEDICAL SCIENCES
SNAP-ON INCORPORATED
MOLEX INCORPORATED
ALTO U. S. INC.
WAL-MART STORES, INC.
TRU TRAK FLIGHT SYSTEMS, INC
SHIVA BIOMEDICAL, LLC
KENNAMETAL INC.
HYTROL CONVEYOR COMPANY, INC.
VIRTUAL SATELLITE CORPORATION

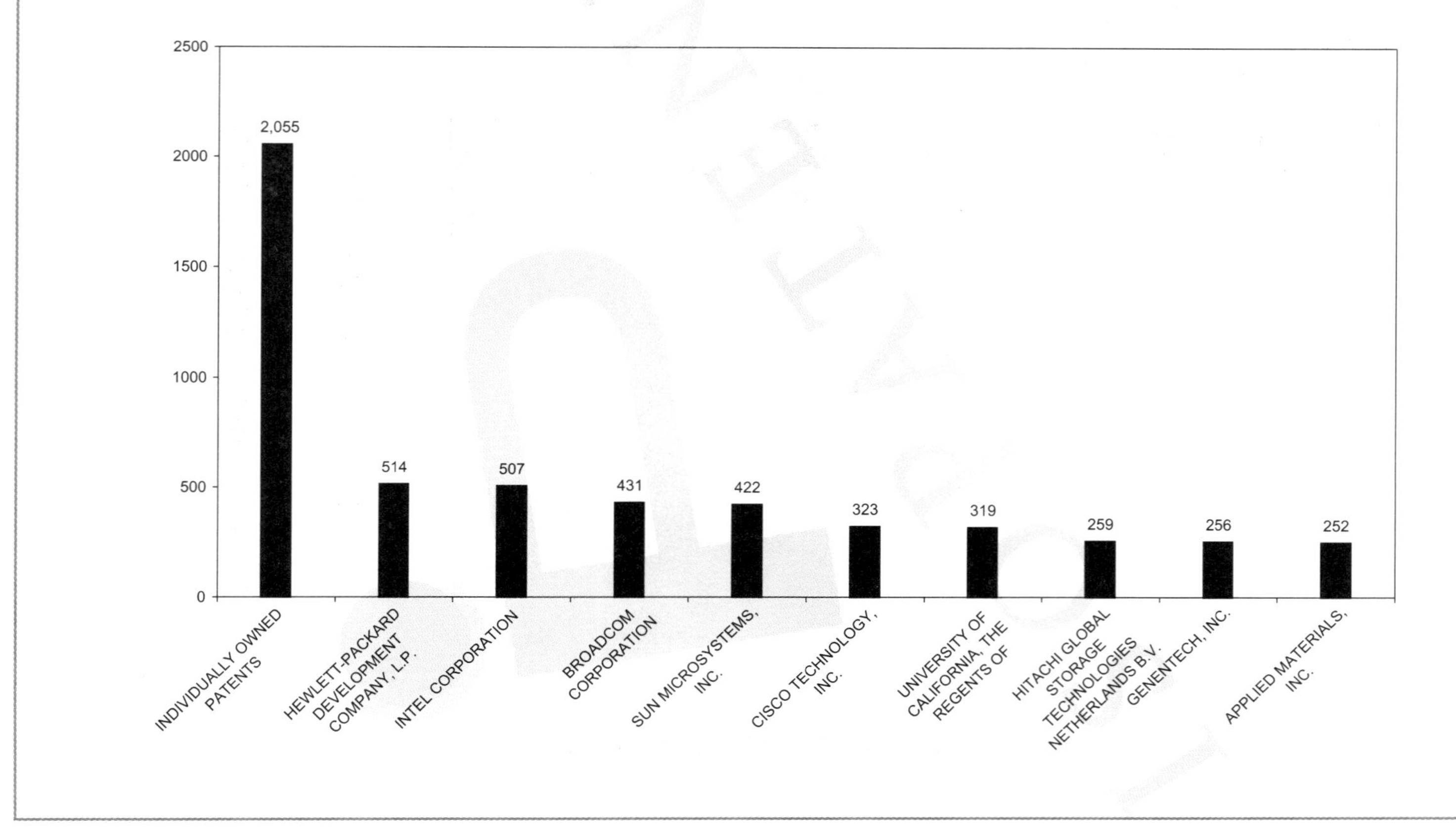
Top U.S. Patent Recipients in 2007 for California
2500
2000
1500
1000
500
0
2,055
514
507
431
422
323
319
259
256
252
INDIVIDUALLY OWNED PATENTS
HEWLETT-PACKARD DEVELOPMENT COMPANY, L.P.
INTEL CORPORATION
BROADCOM CORPORATION
SUN MICROSYSTEMS, INC.
CISCO TECHNOLOGY, INC.
UNIVERSITY OF CALIFORNIA, THE REGENTS OF
HITACHI GLOBAL STORAGE TECHNOLOGIES NETHERLANDS B.V.
GENENTECH, INC.
APPLIED MATERIALS, INC.

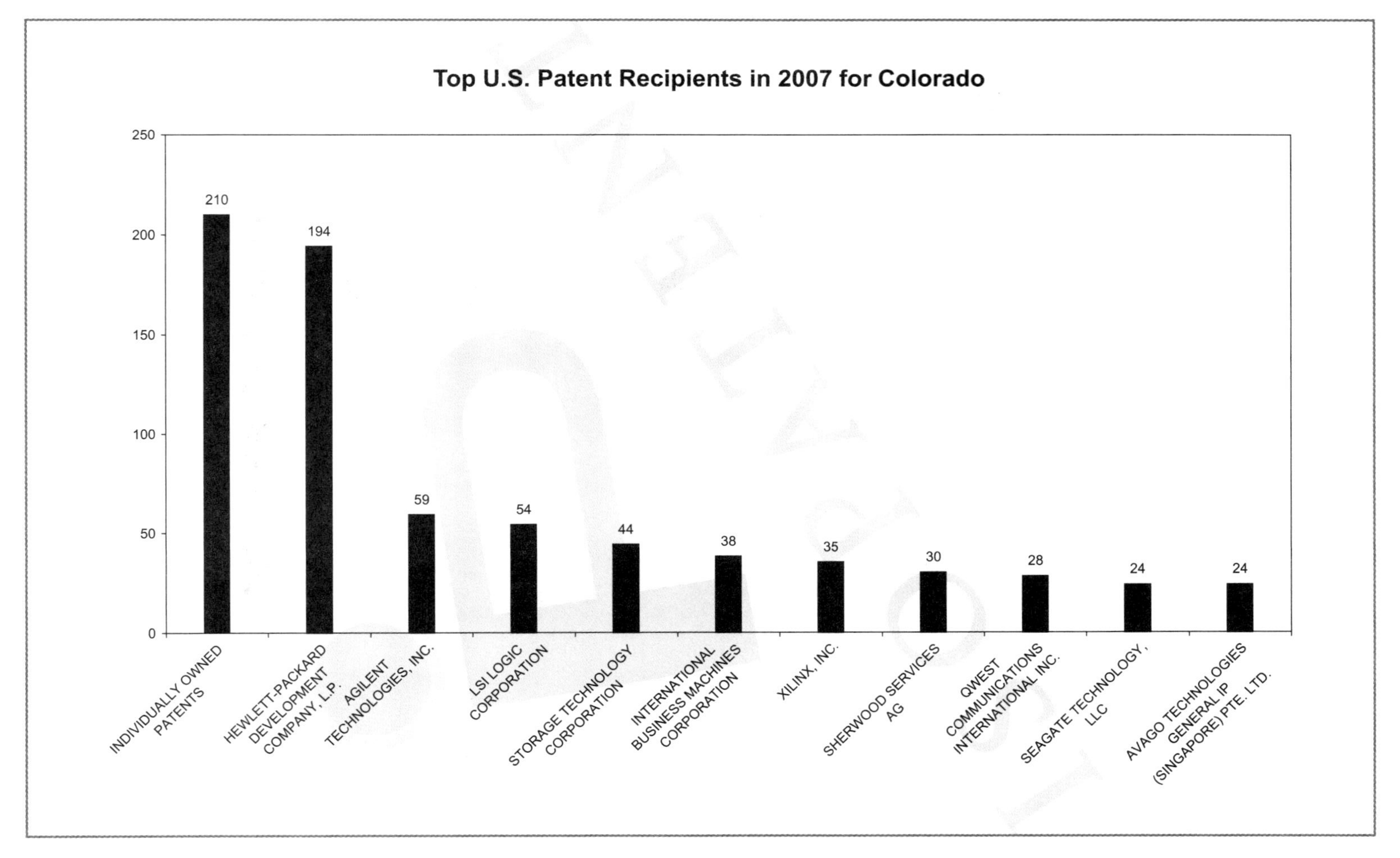
Top U.S. Patent Recipients in 2007 for Colorado
250
200
150
100
50
0
210
194
59
54
44
38
35
30
28
24
24
INDIVIDUALLY OWNED PATENTS
HEWLETT-PACKARD DEVELOPMENT COMPANY, L.P.
AGILENT TECHNOLOGIES, INC.
LSI LOGIC CORPORATION
STORAGE TECHNOLOGY CORPORATION
INTERNATIONAL BUSINESS MACHINES CORPORATION
XILINX, INC.
SHERWOOD SERVICES AG
QWEST COMMUNICATIONS INTERNATIONAL INC.
SEAGATE TECHNOLOGY, LLC
AVAGO TECHNOLOGIES GENERAL IP (SINGAPORE) PTE. LTD.

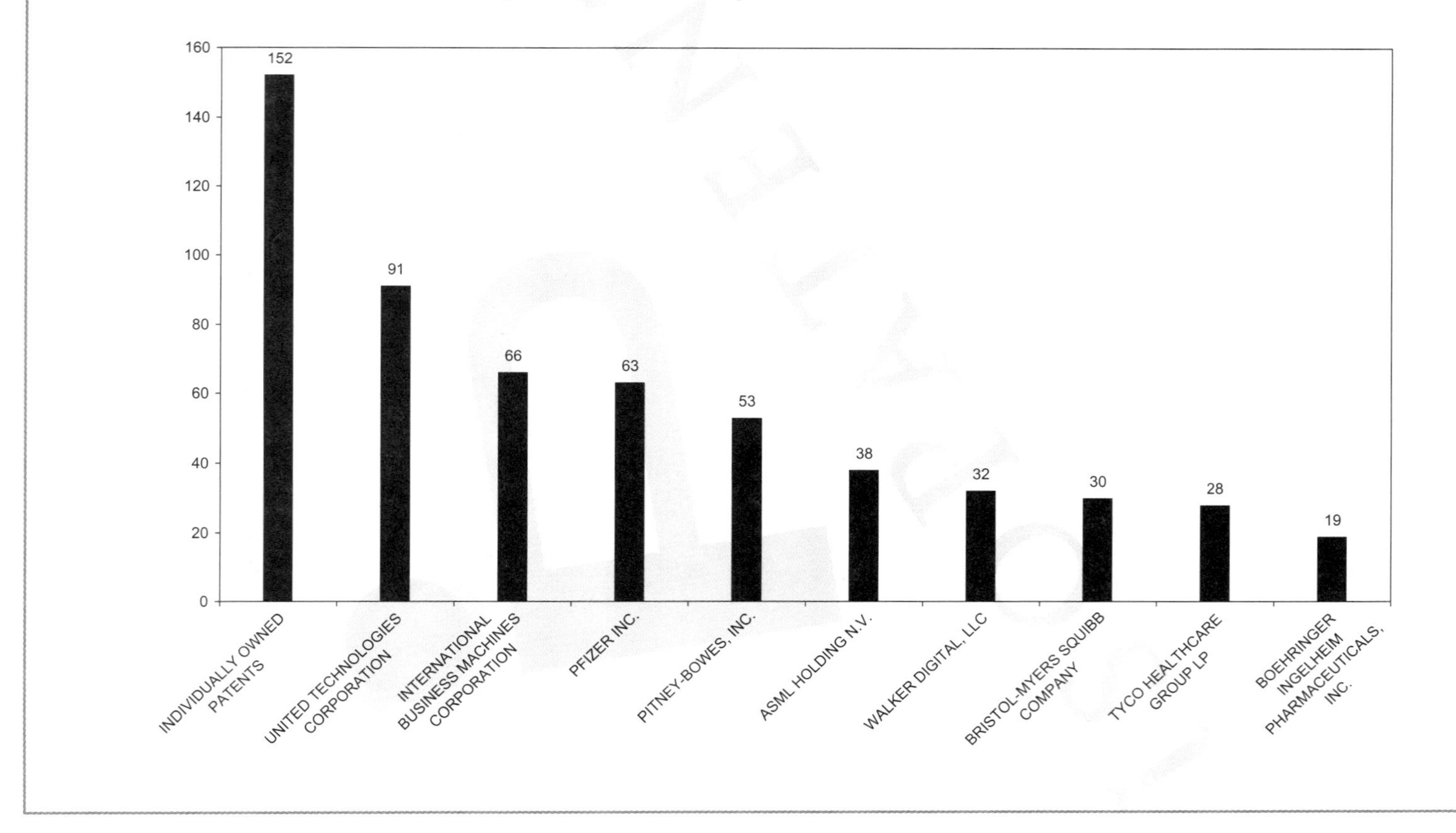

Top U.S. Patent Recipients in 2007 for Connecticut
160
140
120
100
80
60
40
20
0
152
91
66
63
53
38
32
30
28
19
INDIVIDUALLY OWNED PATENTS
UNITED TECHNOLOGIES CORPORATION
INTERNATIONAL BUSINESS MACHINES CORPORATION
PFIZER INC.
PITNEY-BOWES, INC.
ASML HOLDING N.V.
WALKER DIGITAL, LLC
BRISTOL-MYERS SQUIBB COMPANY
TYCO HEALTHCARE GROUP LP
BOEHRINGER INGELHEIM PHARMACEUTICALS, INC.

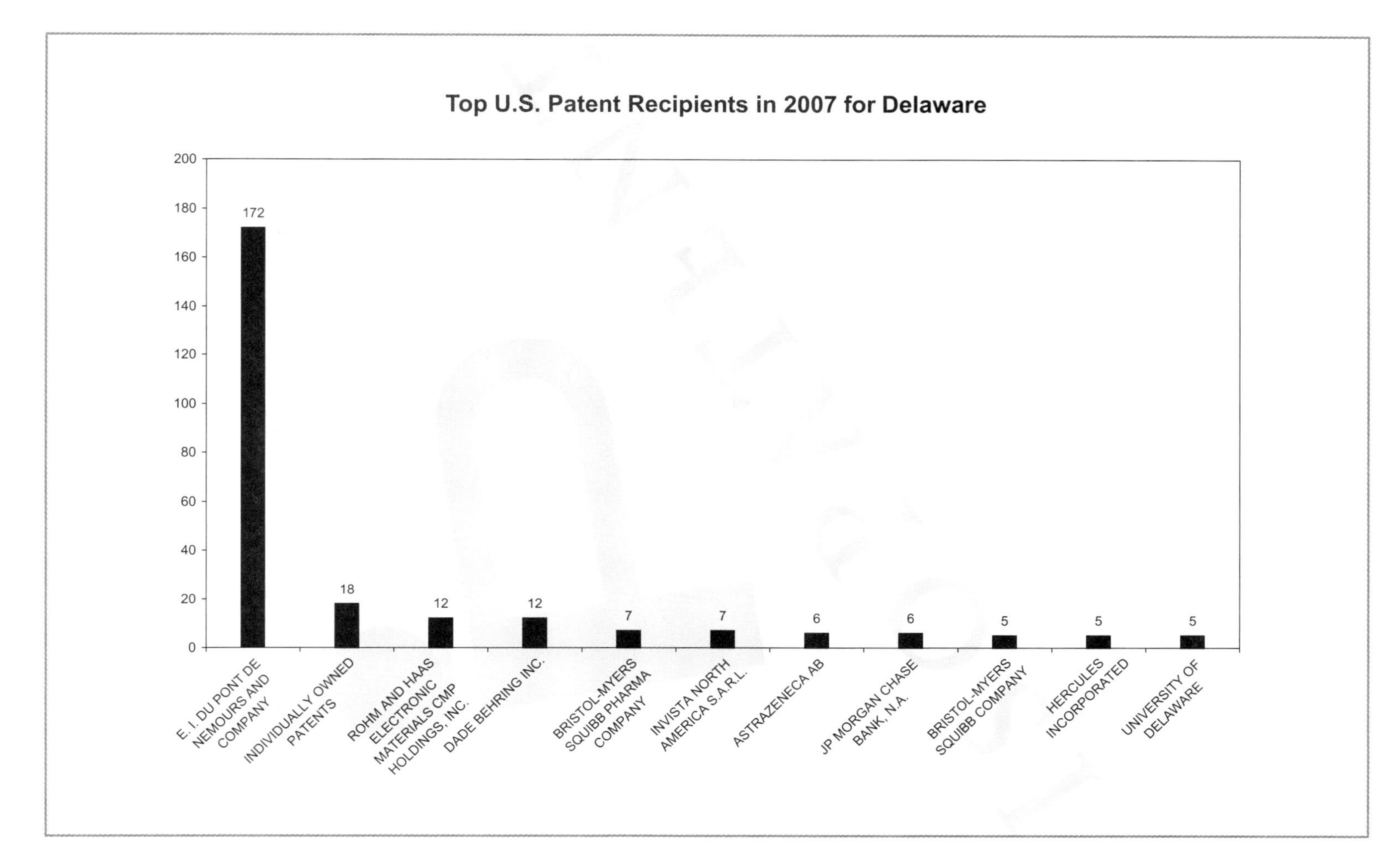
Top U.S. Patent Recipients in 2007 for Delaware
200
180
160
140
120
100
80
60
40
20
0
172
18
12
12
7
7
6
6
5
5
5
E. I. DU PONT DE NEMOURS AND COMPANY
INDIVIDUALLY OWNED PATENTS
ROHM AND HAAS ELECTRONIC MATERIALS CMP HOLDINGS, INC.
DADE BEHRING INC.
BRISTOL-MYERS SQUIBB PHARMA COMPANY
INVISTA NORTH AMERICA S.A.R.L.
ASTRAZENECA AB
JP MORGAN CHASE BANK, N.A.
BRISTOL-MYERS SQUIBB COMPANY
HERCULES INCORPORATED
UNIVERSITY OF DELAWARE

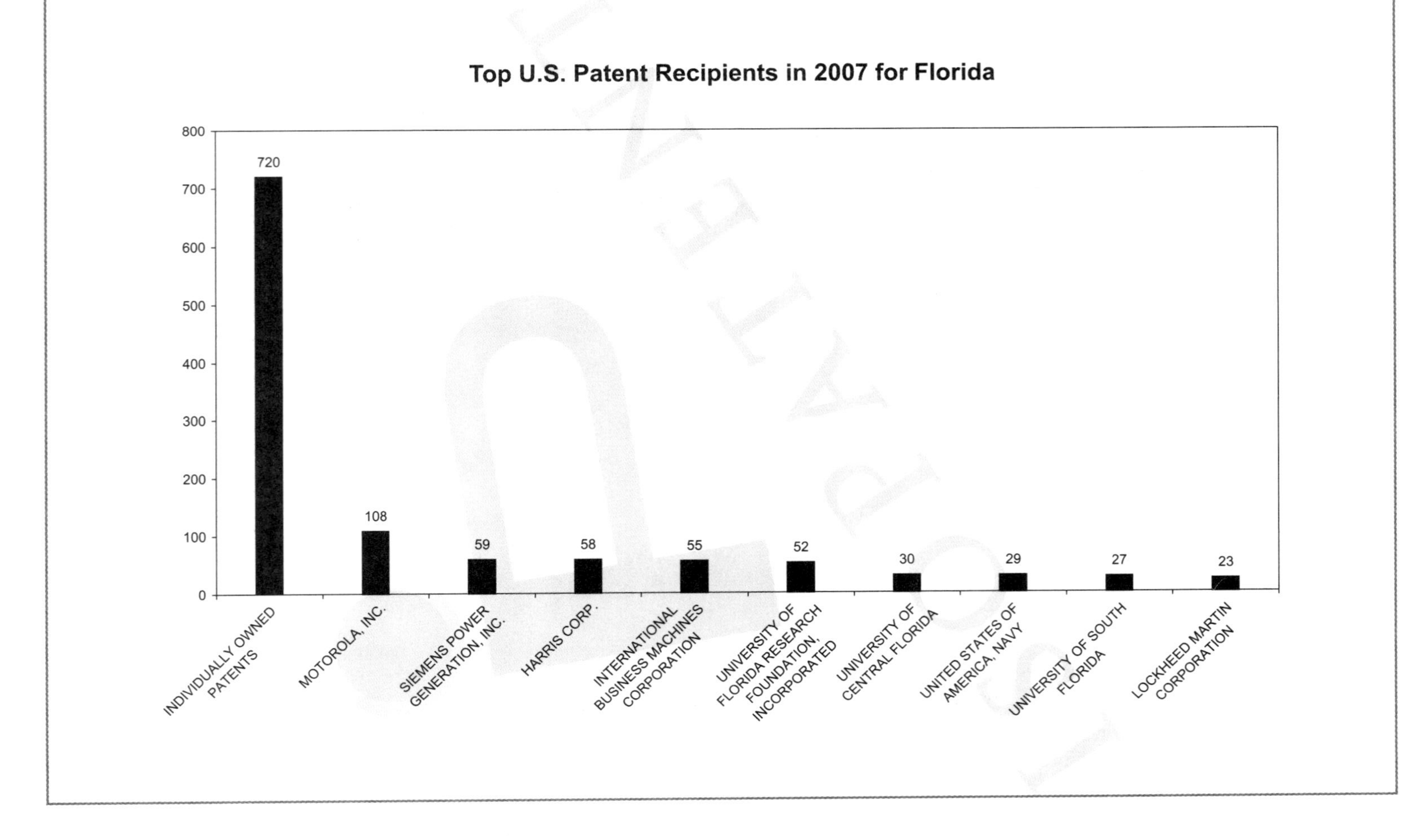
Top U.S. Patent Recipients in 2007 for Florida
800
700
600
500
400
300
200
100
0
720
108
59
58
55
52
30
29
27
23
INDIVIDUALLY OWNED PATENTS
MOTOROLA, INC.
SIEMENS POWER GENERATION, INC.
HARRIS CORP.
INTERNATIONAL BUSINESS MACHINES CORPORATION
UNIVERSITY OF FLORIDA RESEARCH FOUNDATION, INCORPORATED
UNIVERSITY OF CENTRAL FLORIDA
UNITED STATES OF AMERICA, NAVY
UNIVERSITY OF SOUTH FLORIDA
LOCKHEED MARTIN CORPORATION

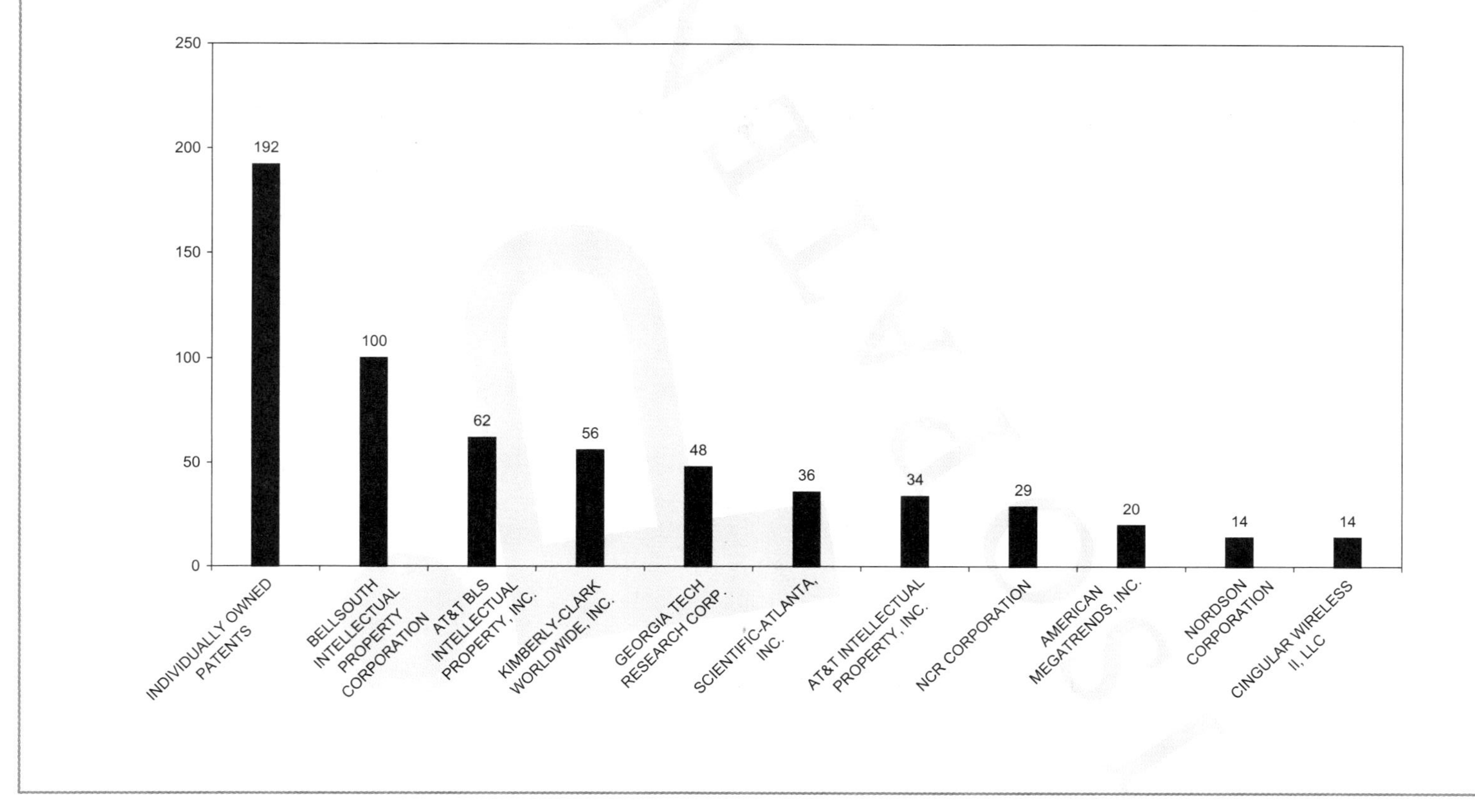
Top U.S. Patent Recipients in 2007 for Georgia
250
200
150
100
50
0
192
100
62
56
48
36
34
29
20
14
14
INDIVIDUALLY OWNED PATENTS
BELLSOUTH INTELLECTUAL PROPERTY CORPORATION
AT&T BLS INTELLECTUAL PROPERTY, INC.
KIMBERLY-CLARK WORLDWIDE, INC.
GEORGIA TECH RESEARCH CORP.
SCIENTIFIC-ATLANTA, INC.
AT&T INTELLECTUAL PROPERTY, INC.
NCR CORPORATION
AMERICAN MEGATRENDS, INC.
NORDSON CORPORATION
CINGULAR WIRELESS II, LLC

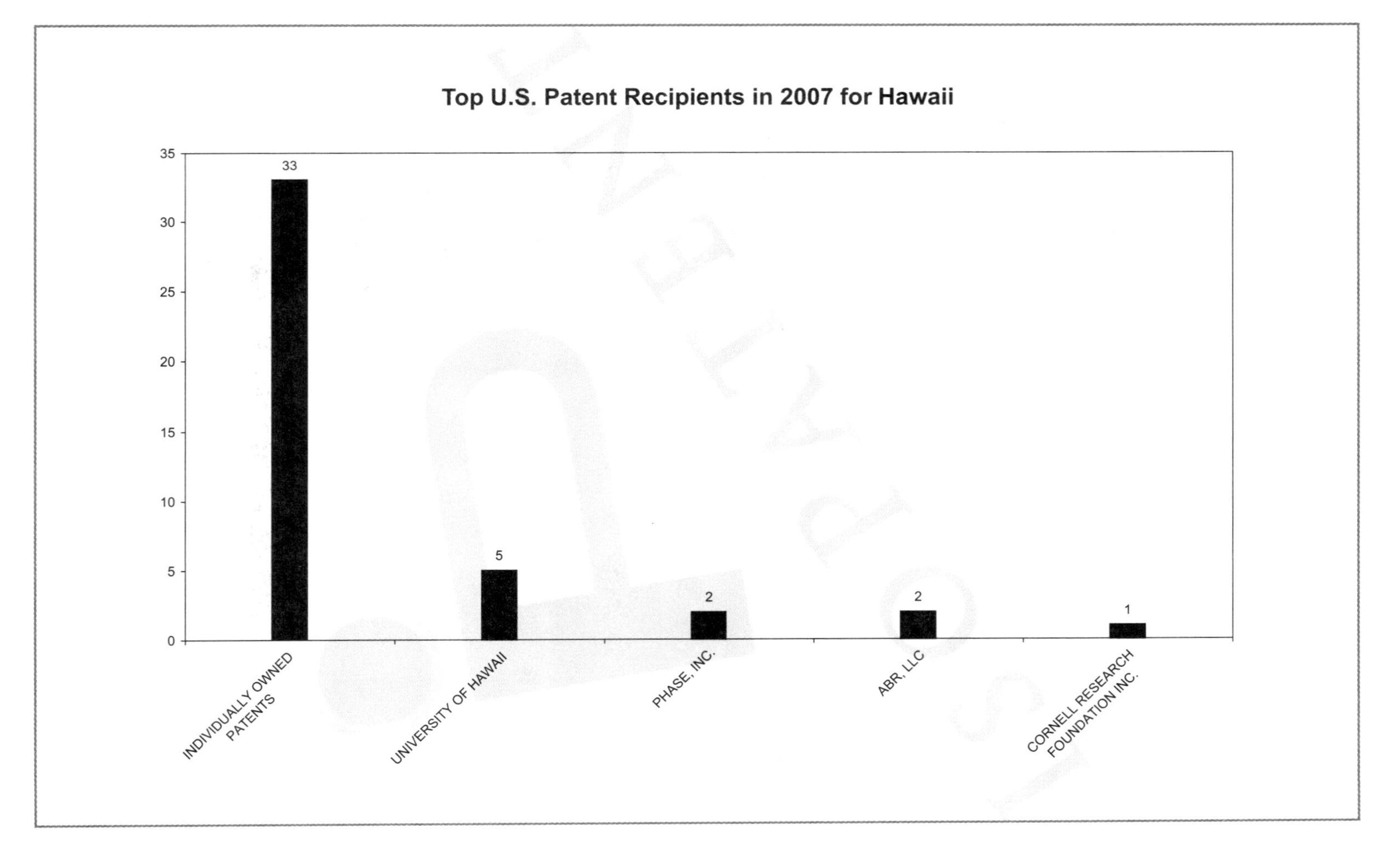
Top U.S. Patent Recipients in 2007 for Hawaii
35
30
25
20
15
10
5
0
33
5
2
2
1
INDIVIDUALLY OWNED PATENTS
UNIVERSITY OF HAWAII
PHASE, INC.
ABR, LLC
CORNELL RESEARCH FOUNDATION INC.

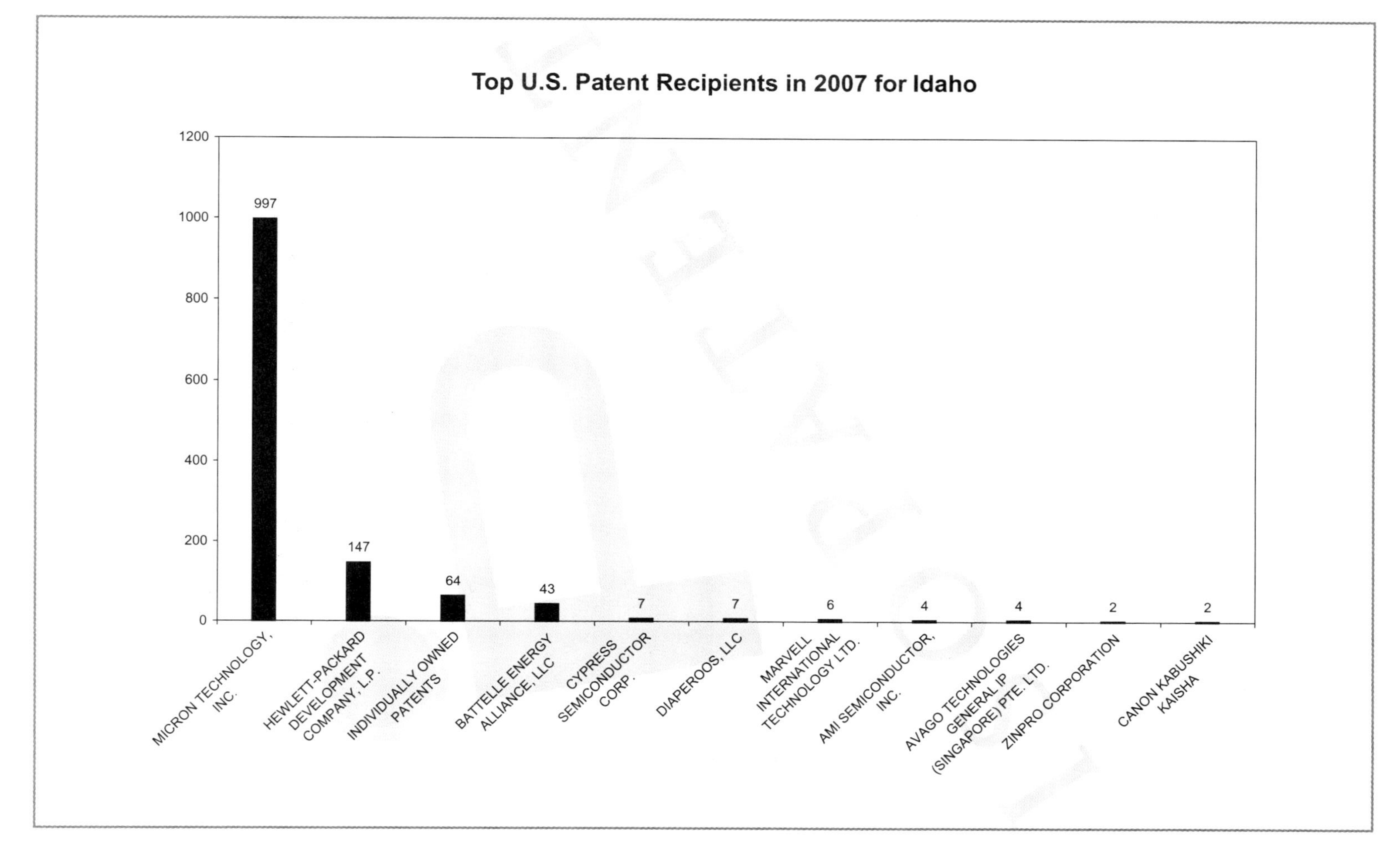

Top U.S. Patent Recipients in 2007 for Idaho
1200
1000
800
600
400
200
0
997
147
64
43
7
7
6
4
4
2
2
MICRON TECHNOLOGY, INC.
HEWLETT-PACKARD DEVELOPMENT COMPANY, L.P.
INDIVIDUALLY OWNED PATENTS
BATTELLE ENERGY ALLIANCE, LLC
CYPRESS SEMICONDUCTOR CORP.
DIAPEROOS, LLC
MARVELL INTERNATIONAL TECHNOLOGY LTD.
AMI SEMICONDUCTOR, INC.
AVAGO TECHNOLOGIES GENERAL IP (SINGAPORE) PTE. LTD.
ZINPRO CORPORATION
CANON KABUSHIKI KAISHA

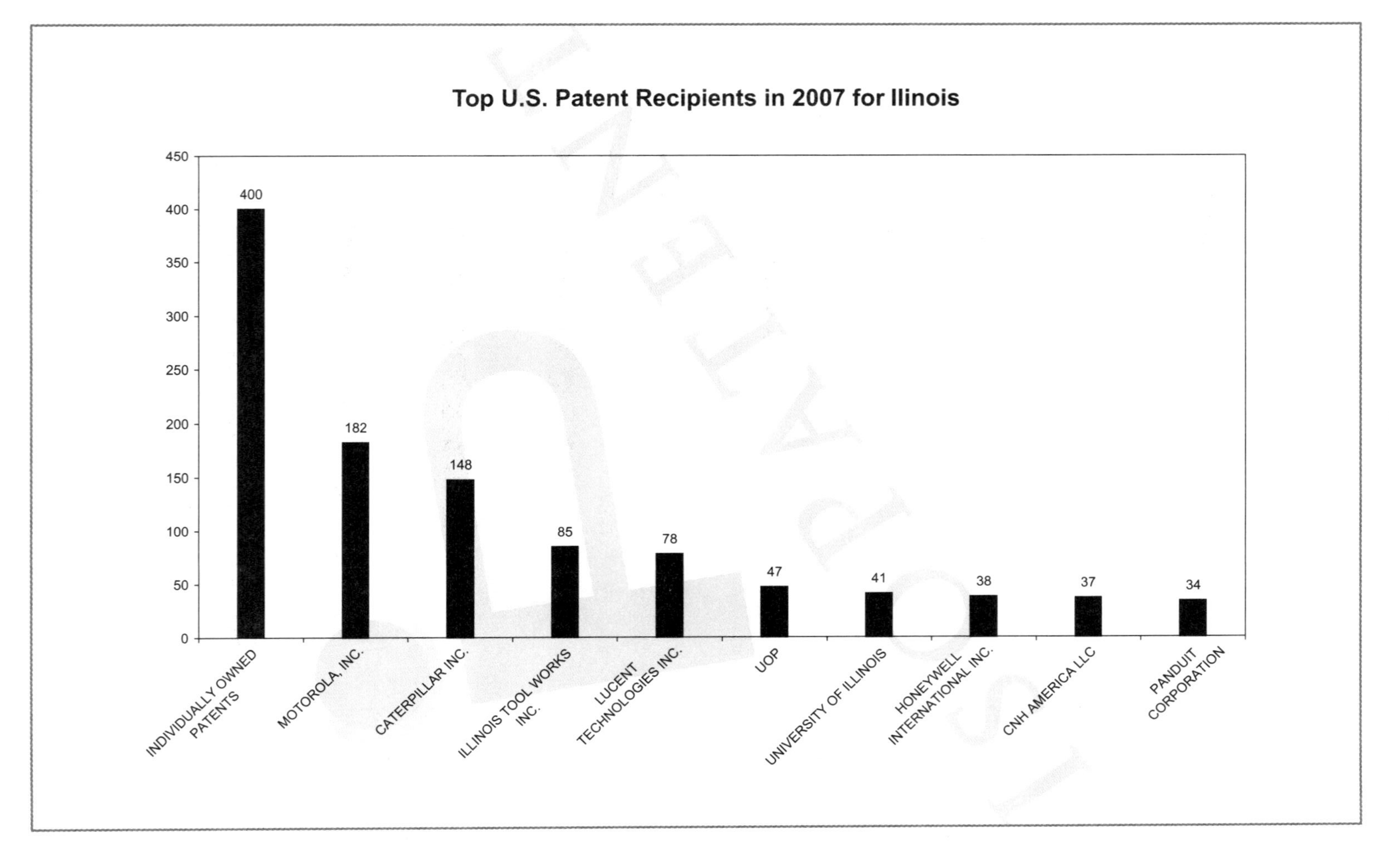
Top U.S. Patent Recipients in 2007 for Ilinois
450
400
350
300
250
200
150
100
50
0
400
182
148
85
78
47
41
38
37
34
INDIVIDUALLY OWNED PATENTS
MOTOROLA, INC.
CATERPILLAR INC.
ILLINOIS TOOL WORKS INC.
LUCENT TECHNOLOGIES INC.
UOP
UNIVERSITY OF ILLINOIS
HONEYWELL INTERNATIONAL INC.
CNH AMERICA LLC
PANDUIT CORPORATION

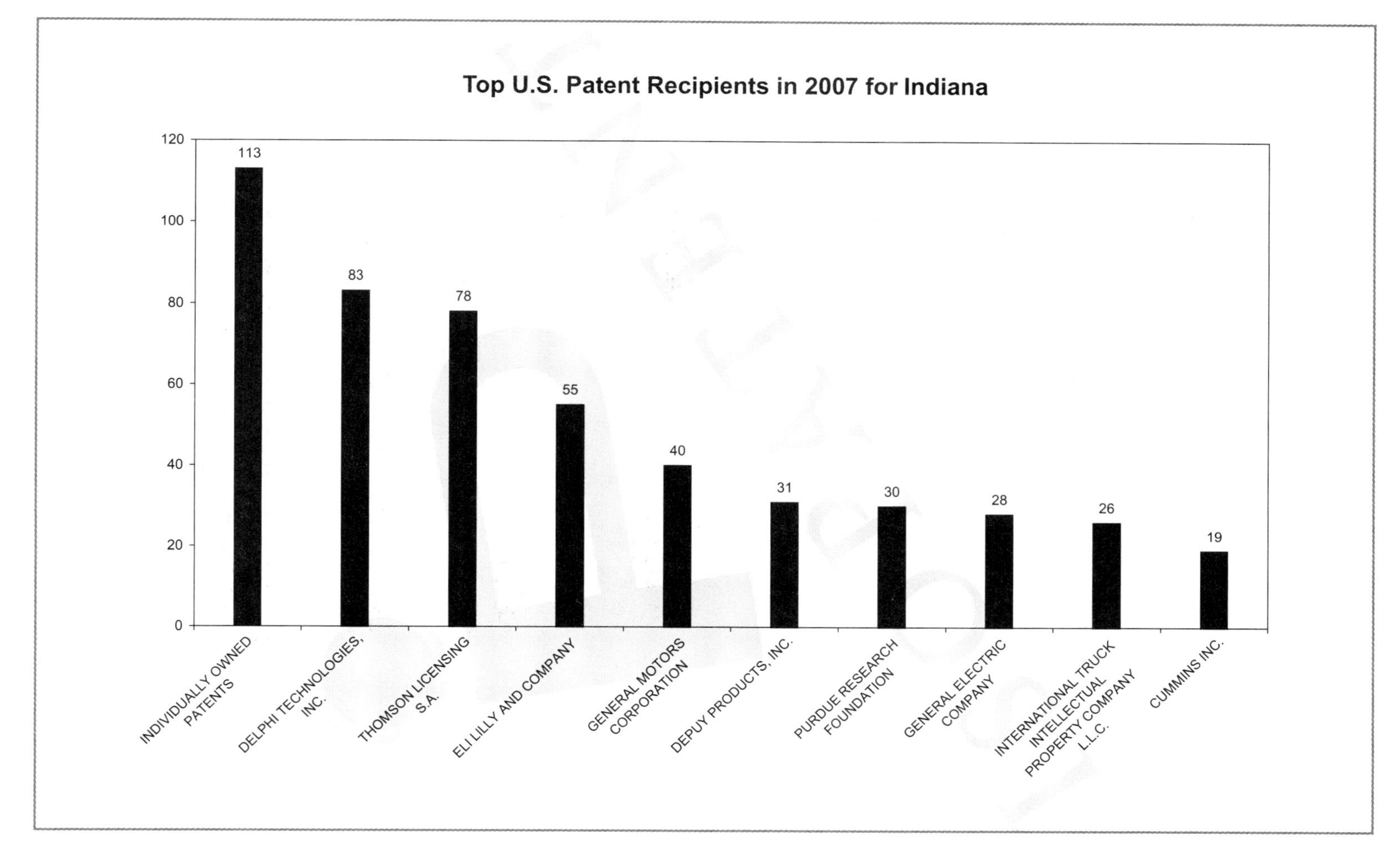

Top U.S. Patent Recipients in 2007 for Indiana
120
100
80
60
40
20
0
113
83
78
55
40
31
30
28
26
19
INDIVIDUALLY OWNED PATENTS
DELPHI TECHNOLOGIES, INC.
THOMSON LICENSING S.A.
ELI LILLY AND COMPANY
GENERAL MOTORS CORPORATION
DEPUY PRODUCTS, INC.
PURDUE RESEARCH FOUNDATION
GENERAL ELECTRIC COMPANY
INTERNATIONAL TRUCK INTELLECTUAL PROPERTY COMPANY L.L.C.
CUMMINS INC.

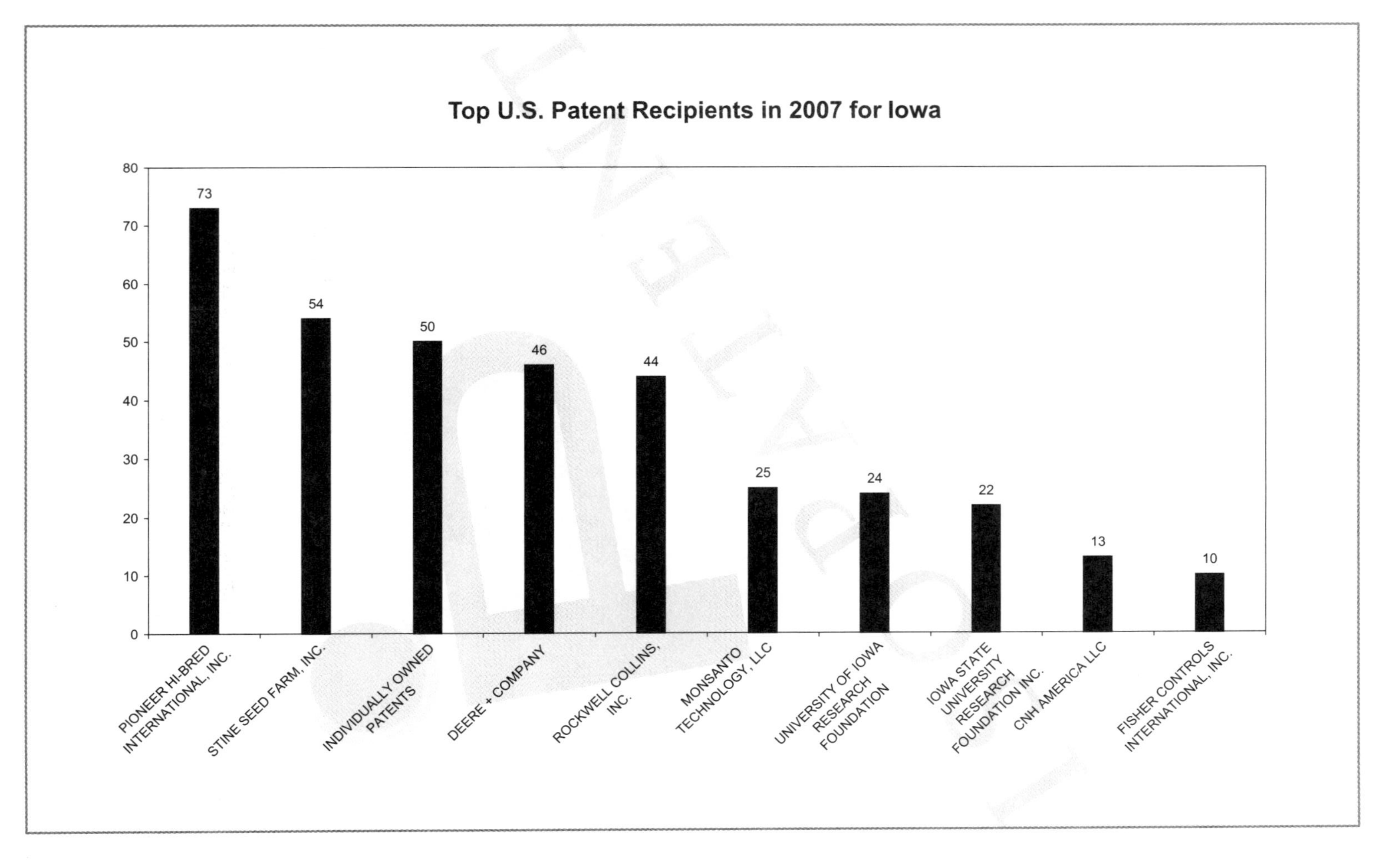
Top U.S. Patent Recipients in 2007 for Iowa
80
70
60
50
40
30
20
10
0
73
54
50
46
44
25
24
22
13
10
PIONEER HI-BRED INTERNATIONAL, INC.
STINE SEED FARM, INC.
INDIVIDUALLY OWNED PATENTS
DEERE + COMPANY
ROCKWELL COLLINS, INC.
MONSANTO TECHNOLOGY, LLC
UNIVERSITY OF IOWA RESEARCH FOUNDATION
IOWA STATE UNIVERSITY RESEARCH FOUNDATION INC.
CNH AMERICA LLC
FISHER CONTROLS INTERNATIONAL, INC.

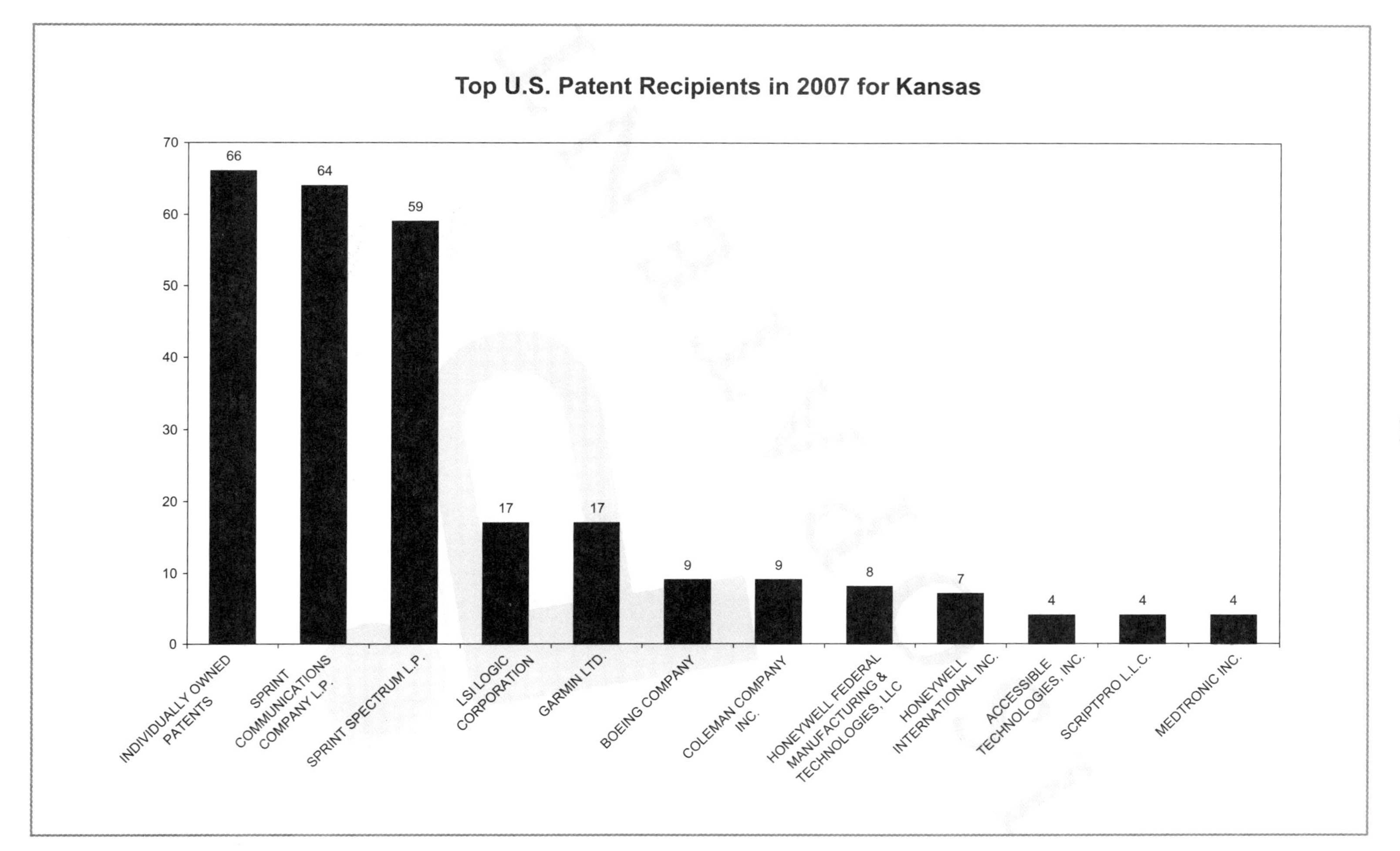

Top U.S. Patent Recipients in 2007 for Kansas
70
60
50
40
30
20
10
0
66
64
59
17
17
9
9
8
7
4
4
4
INDIVIDUALLY OWNED PATENTS
SPRINT COMMUNICATIONS COMPANY L.P.
SPRINT SPECTRUM L.P.
LSI LOGIC CORPORATION
GARMIN LTD.
BOEING COMPANY
COLEMAN COMPANY INC.
HONEYWELL FEDERAL MANUFACTURING & TECHNOLOGIES, LLC
HONEYWELL INTERNATIONAL INC.
ACCESSIBLE TECHNOLOGIES, INC.
SCRIPTPRO L.L.C.
MEDTRONIC INC.

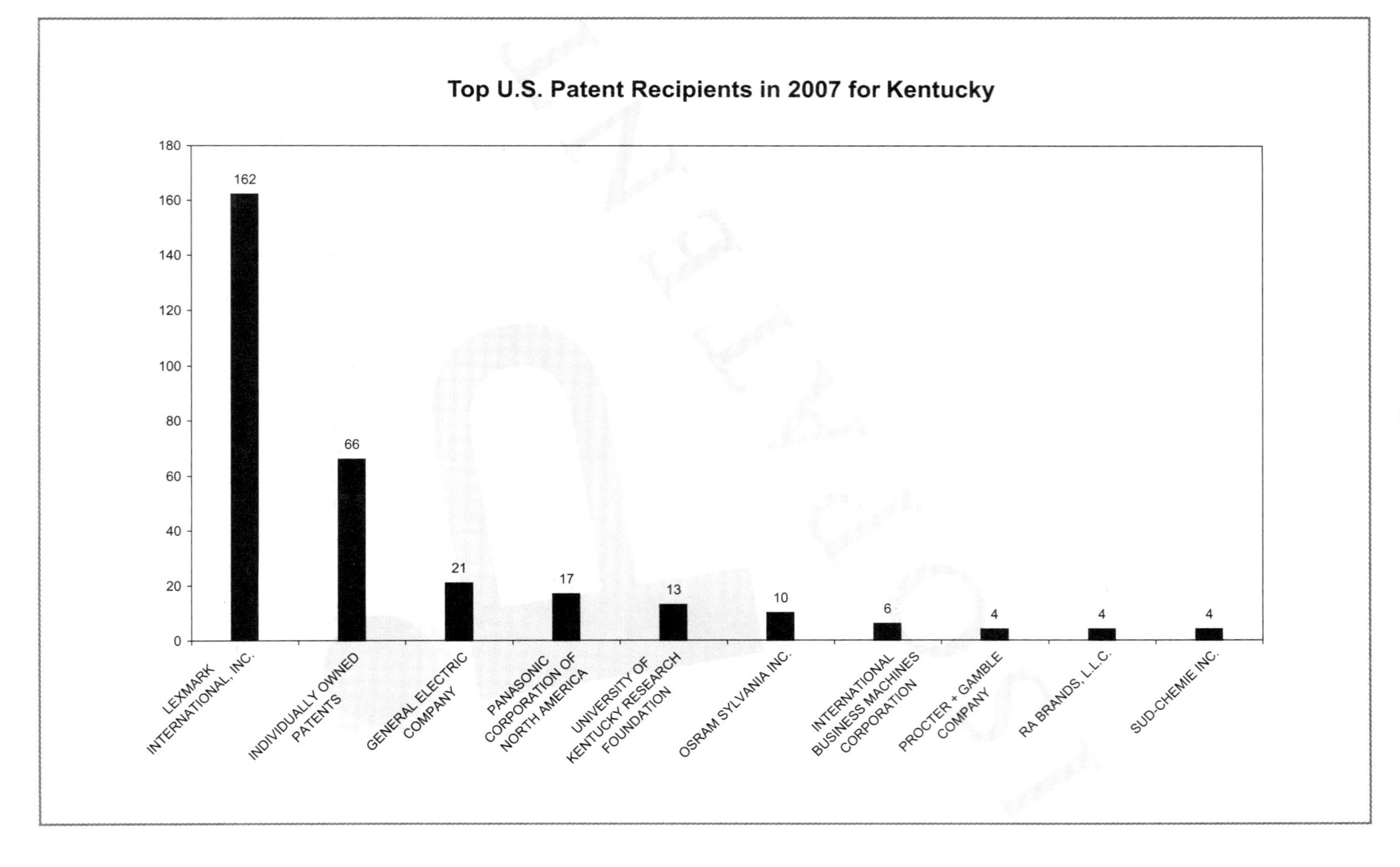

Top U.S. Patent Recipients in 2007 for Kentucky
180
160
140
120
100
80
60
40
20
0
162
66
21
17
13
10
6
4
4
4
LEXMARK INTERNATIONAL, INC.
INDIVIDUALLY OWNED PATENTS
GENERAL ELECTRIC COMPANY
PANASONIC CORPORATION OF NORTH AMERICA
UNIVERSITY OF KENTUCKY RESEARCH FOUNDATION
OSRAM SYLVANIA INC.
INTERNATIONAL BUSINESS MACHINES CORPORATION
PROCTER + GAMBLE COMPANY
RA BRANDS, L.L.C.
SUD-CHEMIE INC.

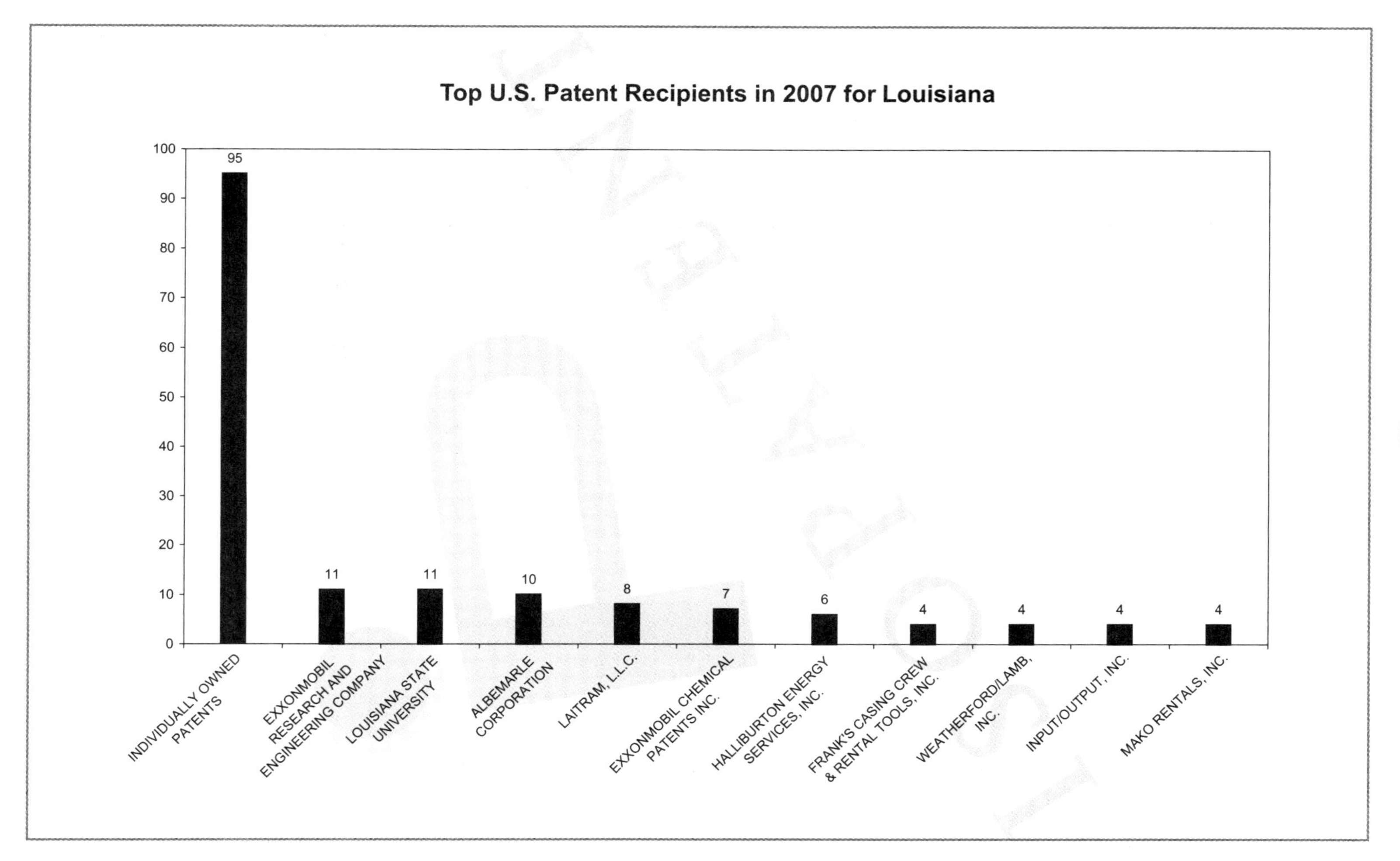

Top U.S. Patent Recipients in 2007 for Louisiana
100
90
80
70
60
50
40
30
20
10
0
95
11
11
10
8
7
6
4
4
4
4
INDIVIDUALLY OWNED PATENTS
EXXONMOBIL RESEARCH AND ENGINEERING COMPANY
LOUISIANA STATE UNIVERSITY
ALBEMARLE CORPORATION
LAITRAM, L.L.C.
EXXONMOBIL CHEMICAL PATENTS INC.
HALLIBURTON ENERGY SERVICES, INC.
FRANK'S CASING CREW & RENTAL TOOLS, INC.
WEATHERFORD/LAMB, INC.
INPUT/OUTPUT, INC.
MAKO RENTALS, INC.

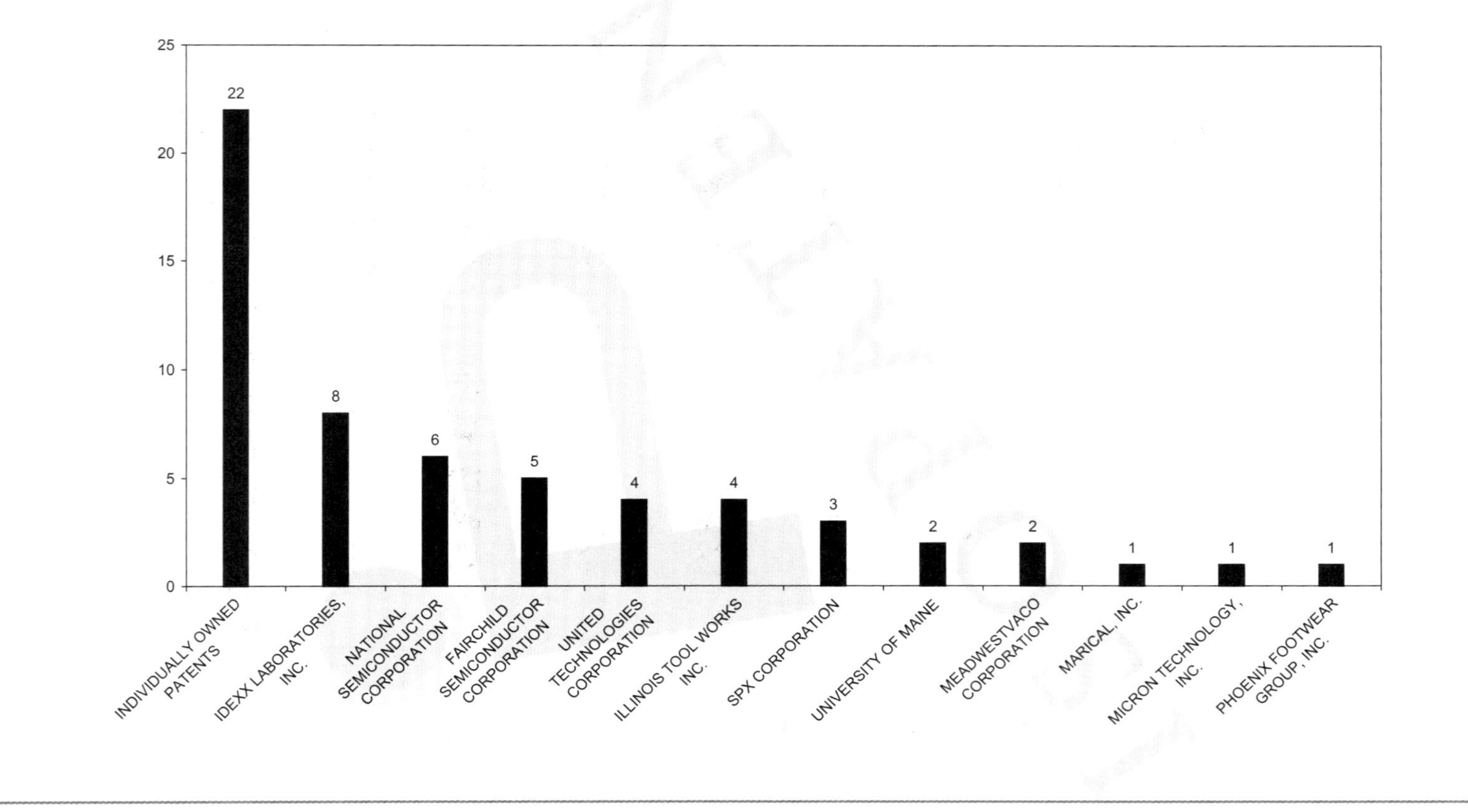

Top U.S. Patent Recipients in 2007 for Maine
25
20
15
10
5
0
22
8
6
5
4
4
3
2
2
1
1
1
INDIVIDUALLY OWNED PATENTS
IDEXX LABORATORIES, INC.
NATIONAL SEMICONDUCTOR CORPORATION
FAIRCHILD SEMICONDUCTOR CORPORATION
UNITED TECHNOLOGIES CORPORATION
ILLINOIS TOOL WORKS INC.
SPX CORPORATION
UNIVERSITY OF MAINE
MEADWESTVACO CORPORATION
MARICAL, INC.
MICRON TECHNOLOGY, INC.
PHOENIX FOOTWEAR GROUP, INC.

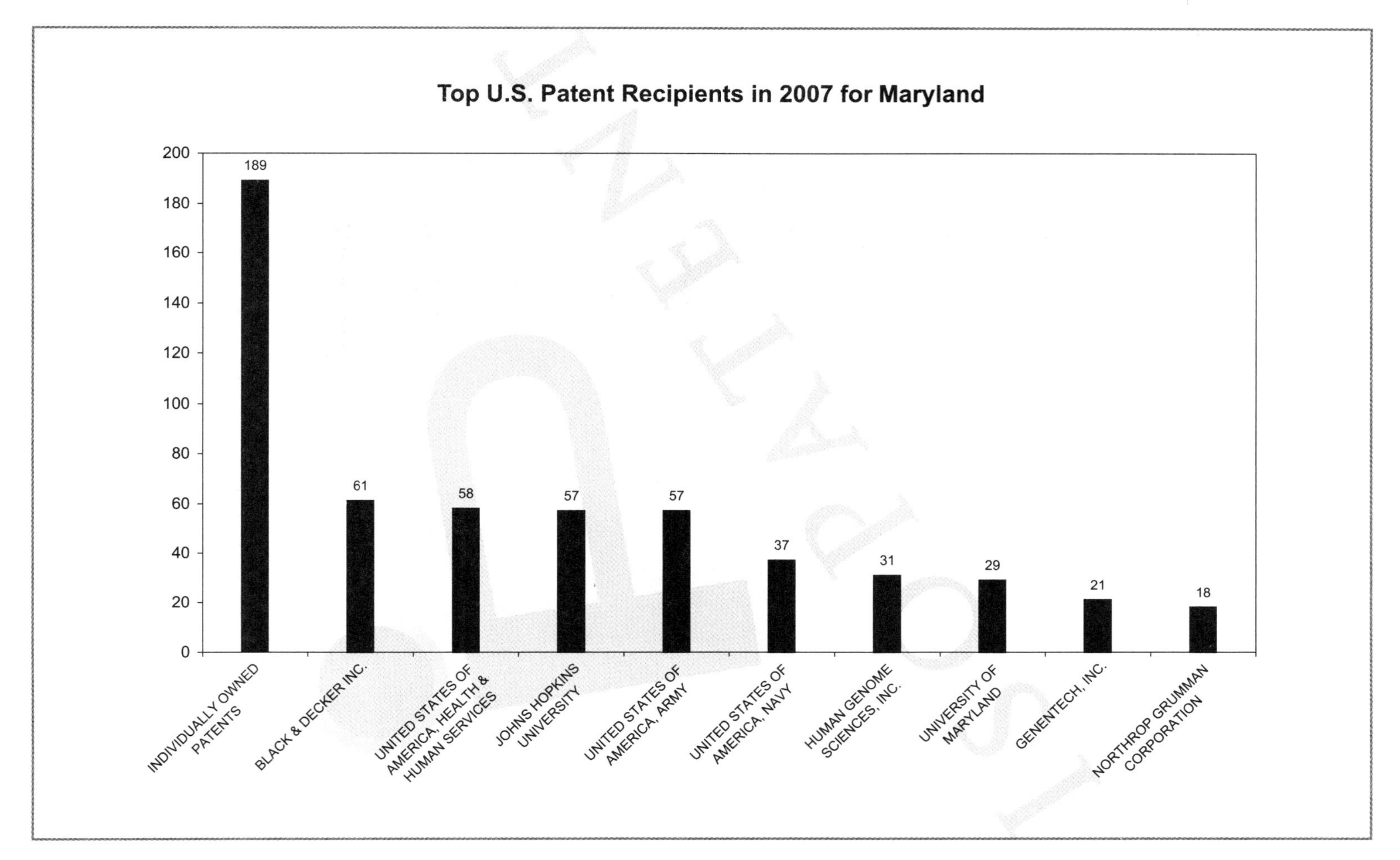
Top U.S. Patent Recipients in 2007 for Maryland
200
180
160
140
120
100
80
60
40
20
0
189
61
58
57
57
37
31
29
21
18
INDIVIDUALLY OWNED PATENTS
BLACK & DECKER INC.
UNITED STATES OF AMERICA, HEALTH & HUMAN SERVICES
JOHNS HOPKINS UNIVERSITY
UNITED STATES OF AMERICA, ARMY
UNITED STATES OF AMERICA, NAVY
HUMAN GENOME SCIENCES, INC.
UNIVERSITY OF MARYLAND
GENENTECH, INC.
NORTHROP GRUMMAN CORPORATION

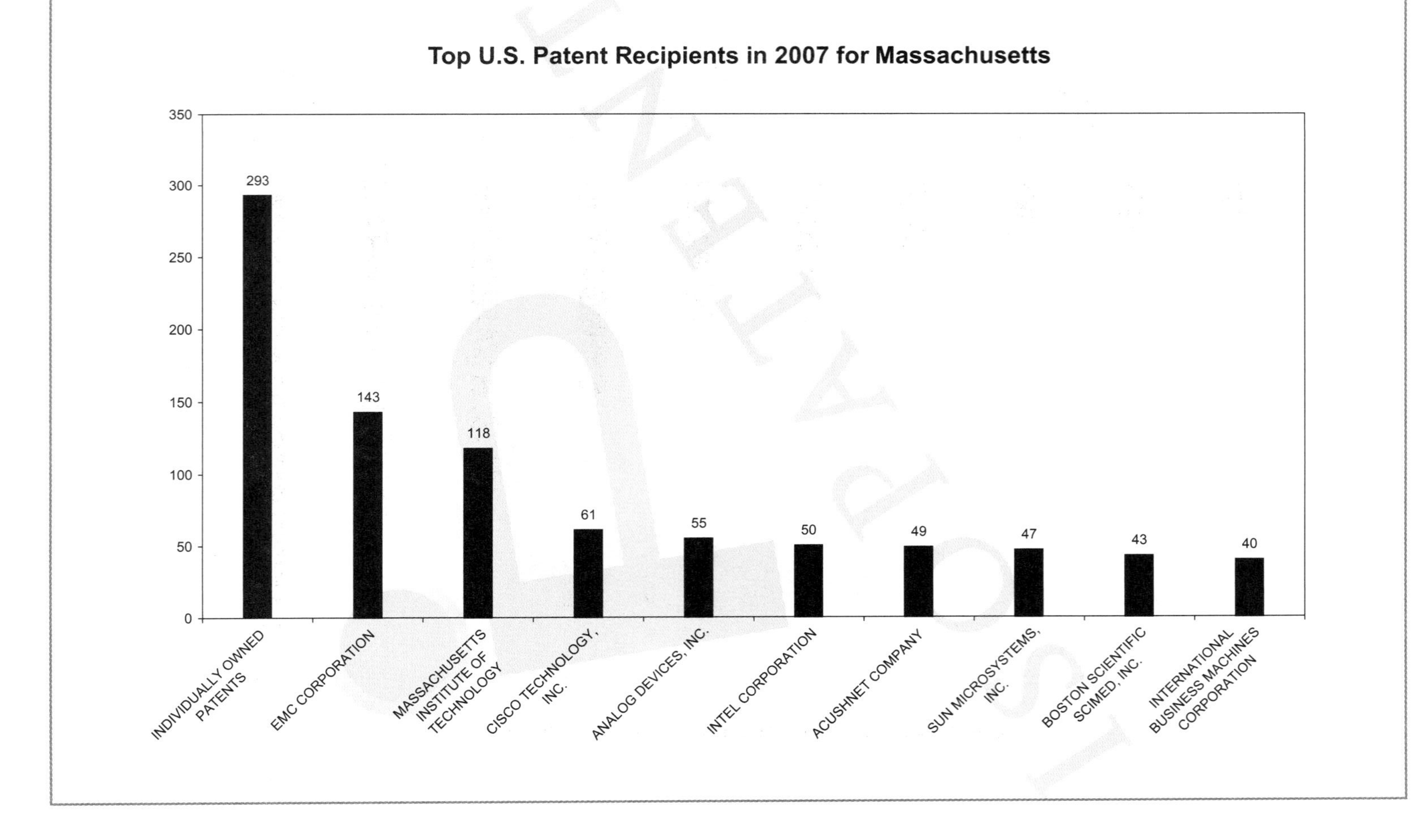

Top U.S. Patent Recipients in 2007 for Massachusetts
350
300
250
200
150
100
50
0
293
143
118
61
55
50
49
47
43
40
INDIVIDUALLY OWNED PATENTS
EMC CORPORATION
MASSACHUSETTS INSTITUTE OF TECHNOLOGY
CISCO TECHNOLOGY, INC.
ANALOG DEVICES, INC.
INTEL CORPORATION
ACUSHNET COMPANY
SUN MICROSYSTEMS, INC.
BOSTON SCIENTIFIC SCIMED, INC.
INTERNATIONAL BUSINESS MACHINES CORPORATION

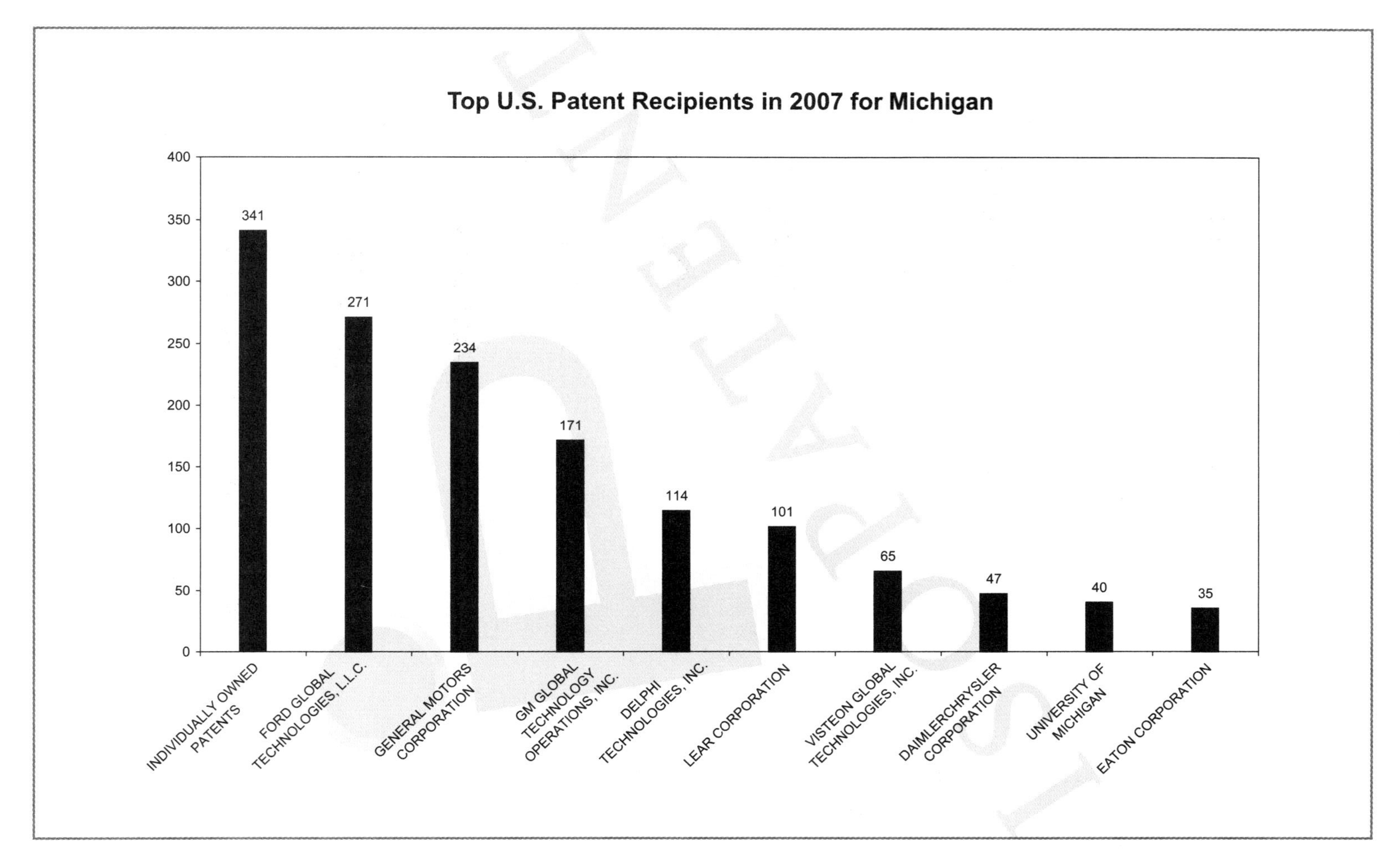
Top U.S. Patent Recipients in 2007 for Michigan
400
350
300
250
200
150
100
50
0
341
271
234
171
114
101
65
47
40
35
INDIVIDUALLY OWNED PATENTS
FORD GLOBAL TECHNOLOGIES, L.L.C.
GENERAL MOTORS CORPORATION
GM GLOBAL TECHNOLOGY OPERATIONS, INC.
DELPHI TECHNOLOGIES, INC.
LEAR CORPORATION
VISTEON GLOBAL TECHNOLOGIES, INC.
DAIMLERCHRYSLER CORPORATION
UNIVERSITY OF MICHIGAN
EATON CORPORATION

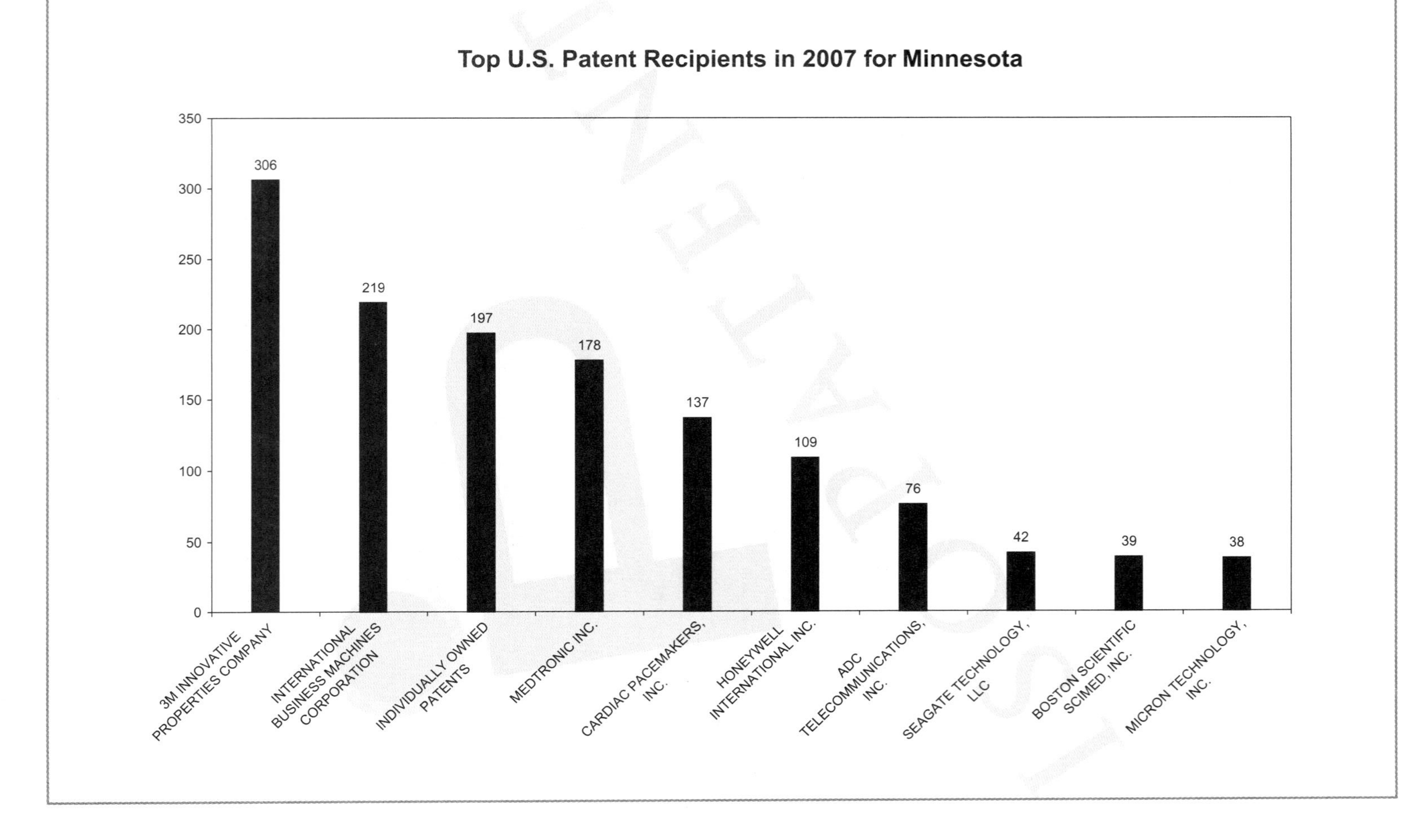
Top U.S. Patent Recipients in 2007 for Minnesota
350
300
250
200
150
100
50
0
306
219
197
178
137
109
76
42
39
38
3M INNOVATIVE PROPERTIES COMPANY
INTERNATIONAL BUSINESS MACHINES CORPORATION
INDIVIDUALLY OWNED PATENTS
MEDTRONIC INC.
CARDIAC PACEMAKERS, INC.
HONEYWELL INTERNATIONAL INC.
ADC TELECOMMUNICATIONS, INC.
SEAGATE TECHNOLOGY, LLC
BOSTON SCIENTIFIC SCIMED, INC.
MICRON TECHNOLOGY, INC.

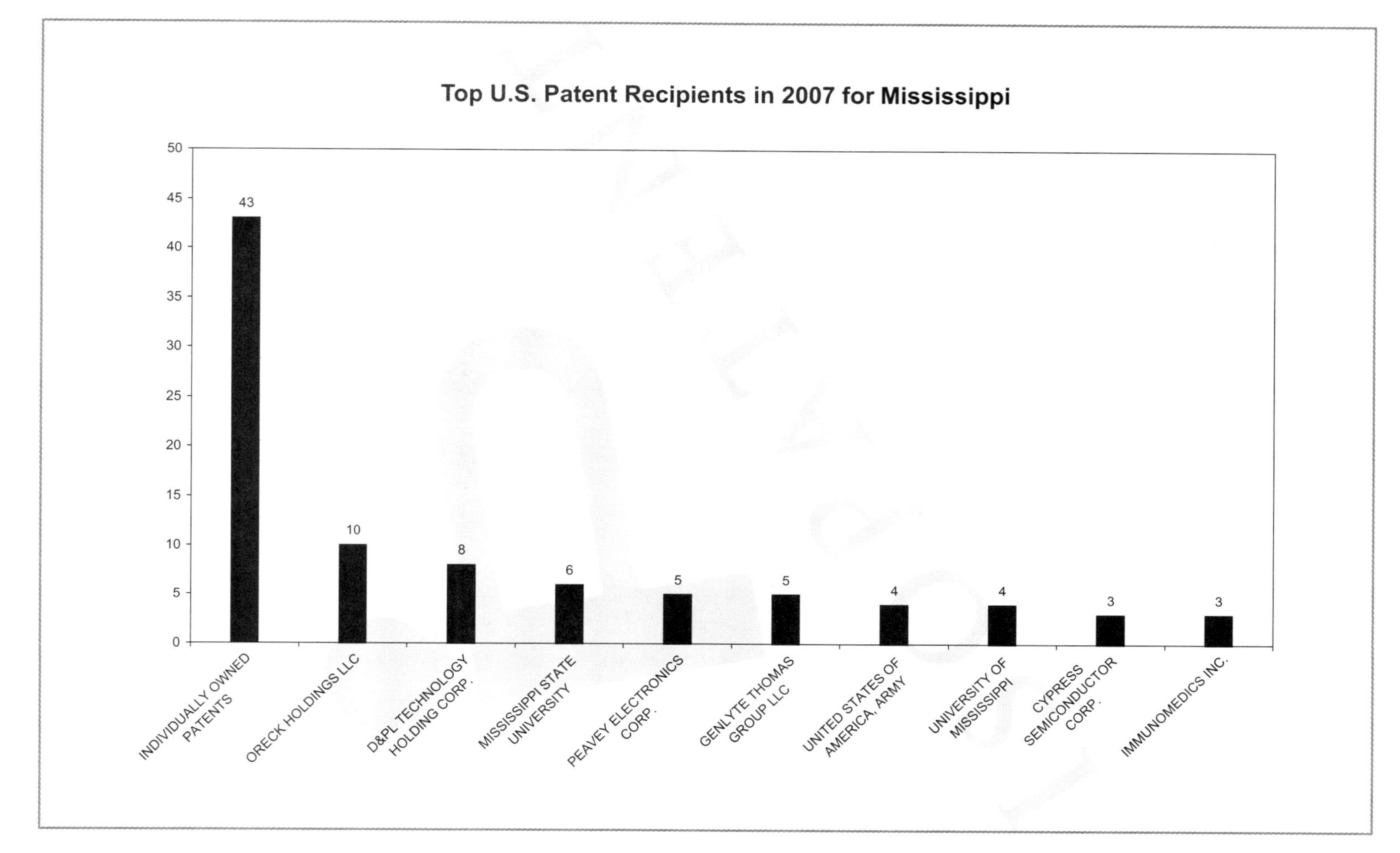
Top U.S. Patent Recipients in 2007 for Mississippi
50
45
40
35
30
25
20
15
10
5
0
43
10
8
6
5
5
4
4
3
3
INDIVIDUALLY OWNED PATENTS
ORECK HOLDINGS LLC
D&PL TECHNOLOGY HOLDING CORP.
MISSISSIPPI STATE UNIVERSITY
PEAVEY ELECTRONICS CORP.
GENLYTE THOMAS GROUP LLC
UNITED STATES OF AMERICA, ARMY
UNIVERSITY OF MISSISSIPPI
CYPRESS SEMICONDUCTOR CORP.
IMMUNOMEDICS INC.

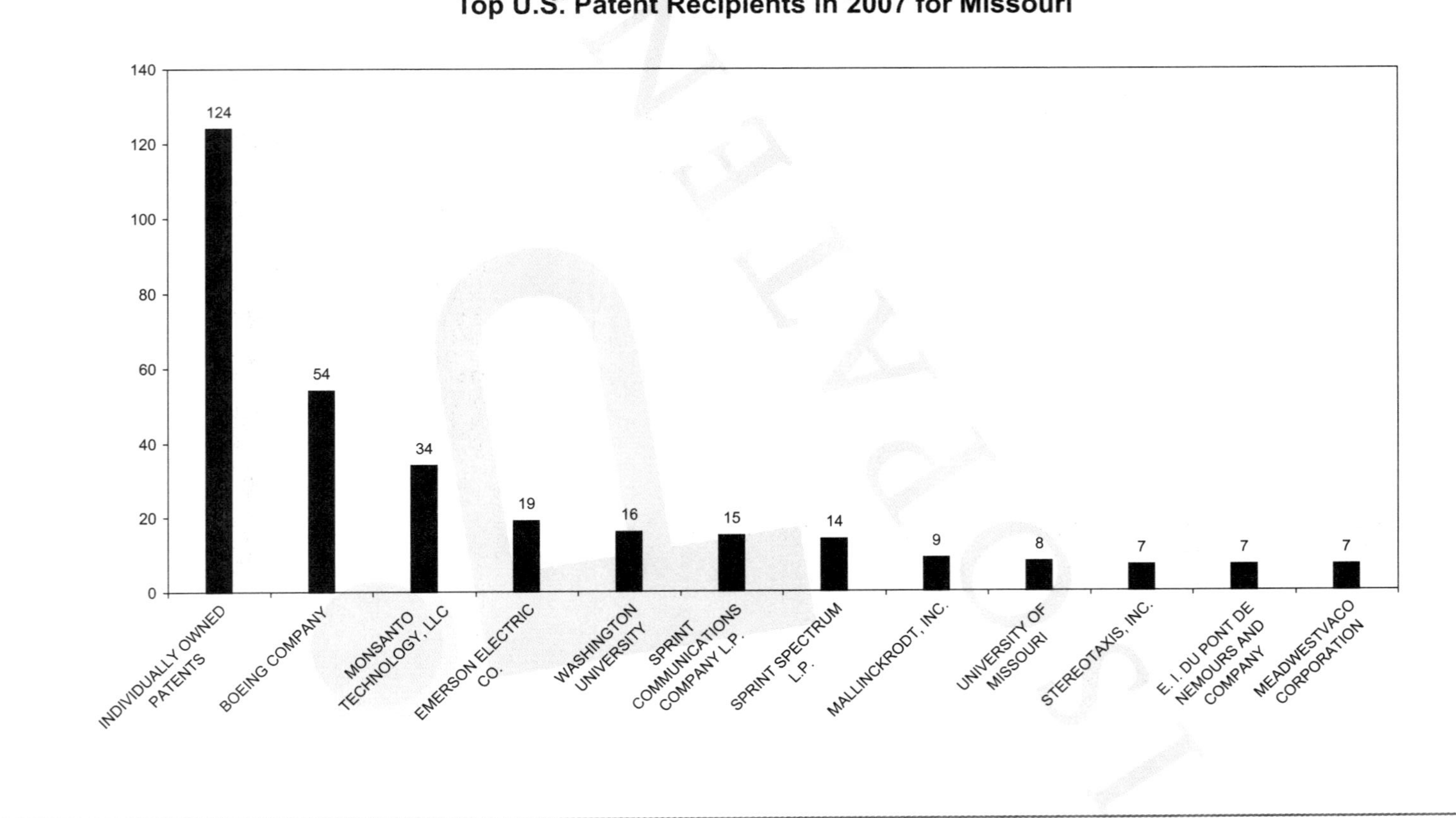
Top U.S. Patent Recipients in 2007 for Missouri
140
120
100
80
60
40
20
0
124
54
34
19
16
15
14
9
8
7
7
7
INDIVIDUALLY OWNED PATENTS
BOEING COMPANY
MONSANTO TECHNOLOGY, LLC
EMERSON ELECTRIC CO.
WASHINGTON UNIVERSITY
SPRINT COMMUNICATIONS COMPANY L.P.
SPRINT SPECTRUM L.P.
MALLINCKRODT, INC.
UNIVERSITY OF MISSOURI
STEREOTAXIS, INC.
E. I. DU PONT DE NEMOURS AND COMPANY
MEADWESTVACO CORPORATION

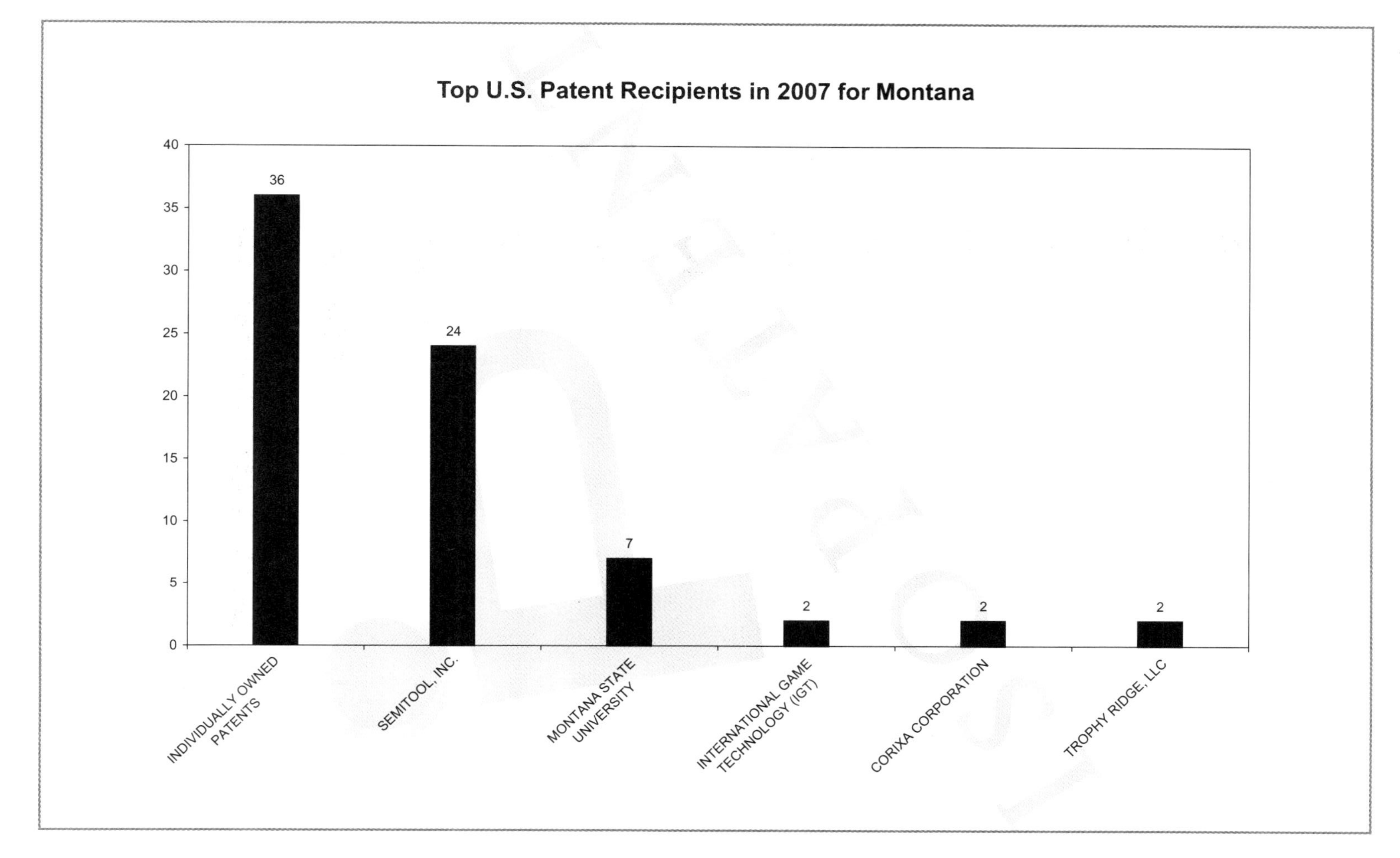

Top U.S. Patent Recipients in 2007 for Montana
40
35
30
25
20
15
10
5
0
36
24
7
2
2
2
INDIVIDUALLY OWNED PATENTS
SEMITOOL, INC.
MONTANA STATE UNIVERSITY
INTERNATIONAL GAME TECHNOLOGY (IGT)
CORIXA CORPORATION
TROPHY RIDGE, LLC

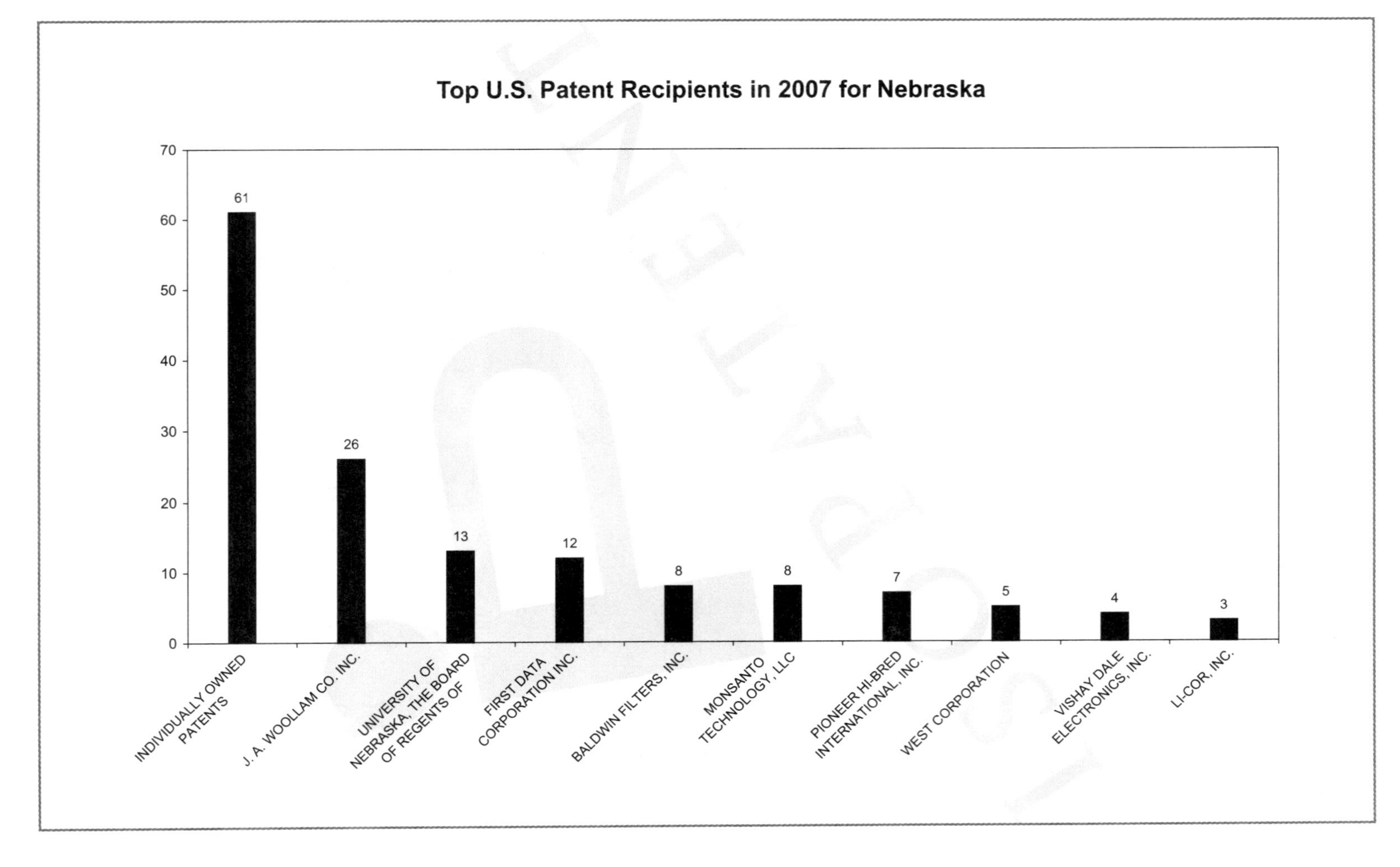

Top U.S. Patent Recipients in 2007 for Nebraska
70
60
50
40
30
20
10
0
61
26
13
12
8
8
7
5
4
3
INDIVIDUALLY OWNED PATENTS
J. A. WOOLLAM CO. INC.
UNIVERSITY OF NEBRASKA, THE BOARD OF REGENTS OF
FIRST DATA CORPORATION INC.
BALDWIN FILTERS, INC.
MONSANTO TECHNOLOGY, LLC
PIONEER HI-BRED INTERNATIONAL, INC.
WEST CORPORATION
VISHAY DALE ELECTRONICS, INC.
LI-COR, INC.

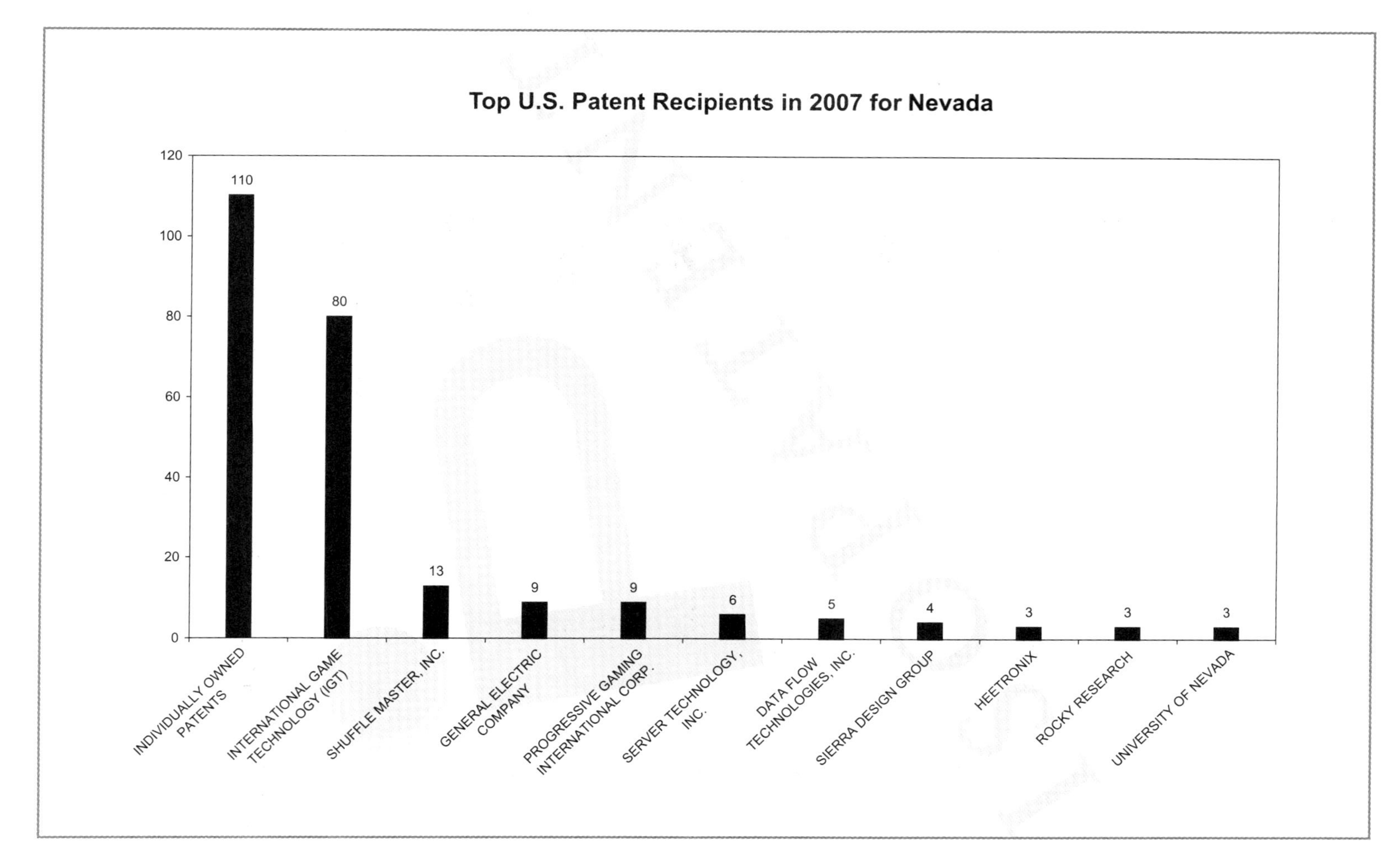
Top U.S. Patent Recipients in 2007 for Nevada
120
100
80
60
40
20
0
110
80
13
9
9
6
5
4
3
3
3
INDIVIDUALLY OWNED PATENTS
INTERNATIONAL GAME TECHNOLOGY (IGT)
SHUFFLE MASTER, INC.
GENERAL ELECTRIC COMPANY
PROGRESSIVE GAMING INTERNATIONAL CORP.
SERVER TECHNOLOGY, INC.
DATA FLOW TECHNOLOGIES, INC.
SIERRA DESIGN GROUP
HEETRONIX
ROCKY RESEARCH
UNIVERSITY OF NEVADA

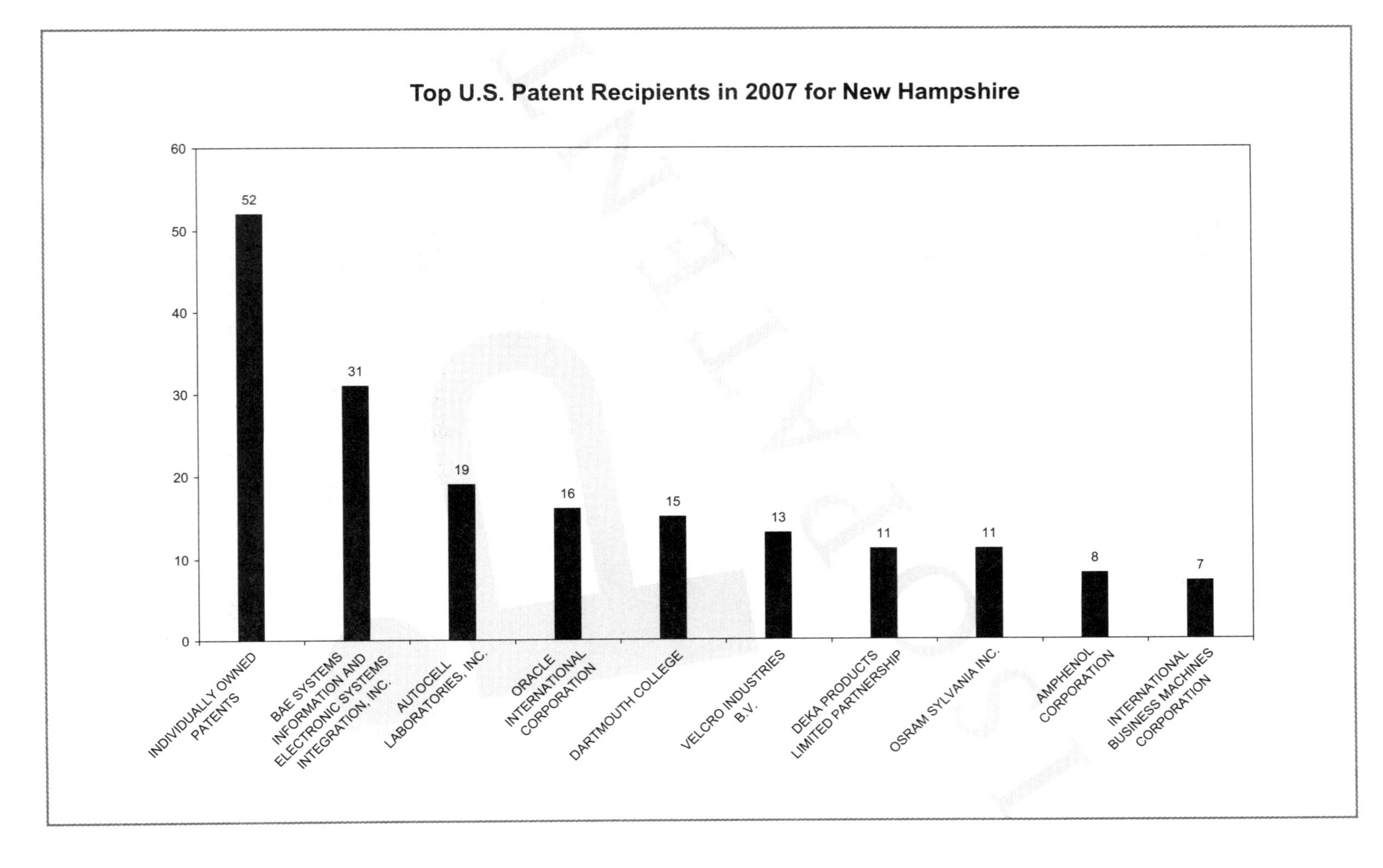

Top U.S. Patent Recipients in 2007 for New Hampshire
60
50
40
30
20
10
0
52
31
19
16
15
13
11
11
8
7
INDIVIDUALLY OWNED PATENTS
BAE SYSTEMS INFORMATION AND ELECTRONIC SYSTEMS INTEGRATION, INC.
AUTOCELL LABORATORIES, INC.
ORACLE INTERNATIONAL CORPORATION
DARTMOUTH COLLEGE
VELCRO INDUSTRIES B.V.
DEKA PRODUCTS LIMITED PARTNERSHIP
OSRAM SYLVANIA INC.
AMPHENOL CORPORATION
INTERNATIONAL BUSINESS MACHINES CORPORATION

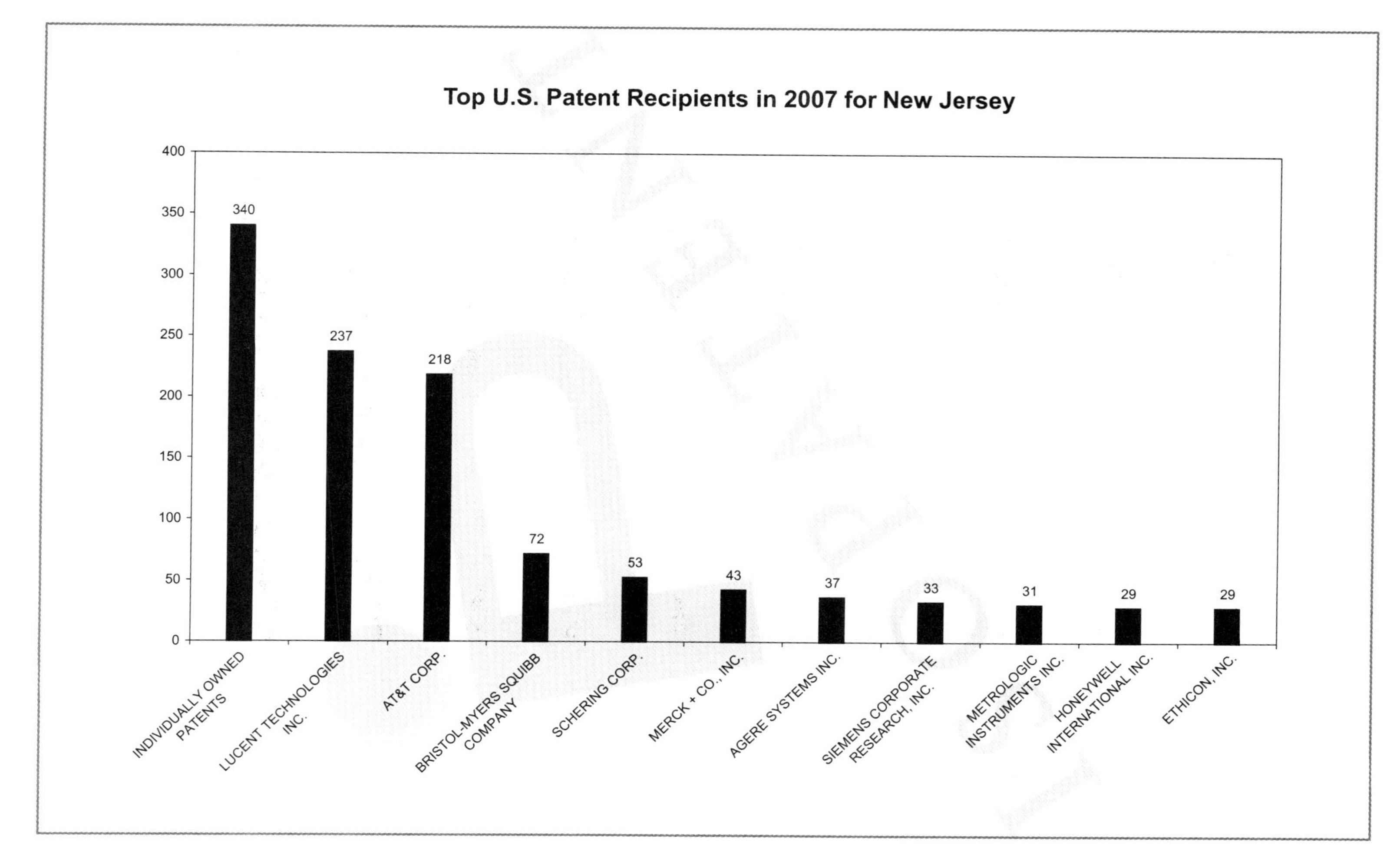
Top U.S. Patent Recipients in 2007 for New Jersey
400
350
300
250
200
150
100
50
0
340
237
218
72
53
43
37
33
31
29
29
INDIVIDUALLY OWNED PATENTS
LUCENT TECHNOLOGIES INC.
AT&T CORP.
BRISTOL-MYERS SQUIBB COMPANY
SCHERING CORP.
MERCK + CO., INC.
AGERE SYSTEMS INC.
SIEMENS CORPORATE RESEARCH, INC.
METROLOGIC INSTRUMENTS INC.
HONEYWELL INTERNATIONAL INC.
ETHICON, INC.

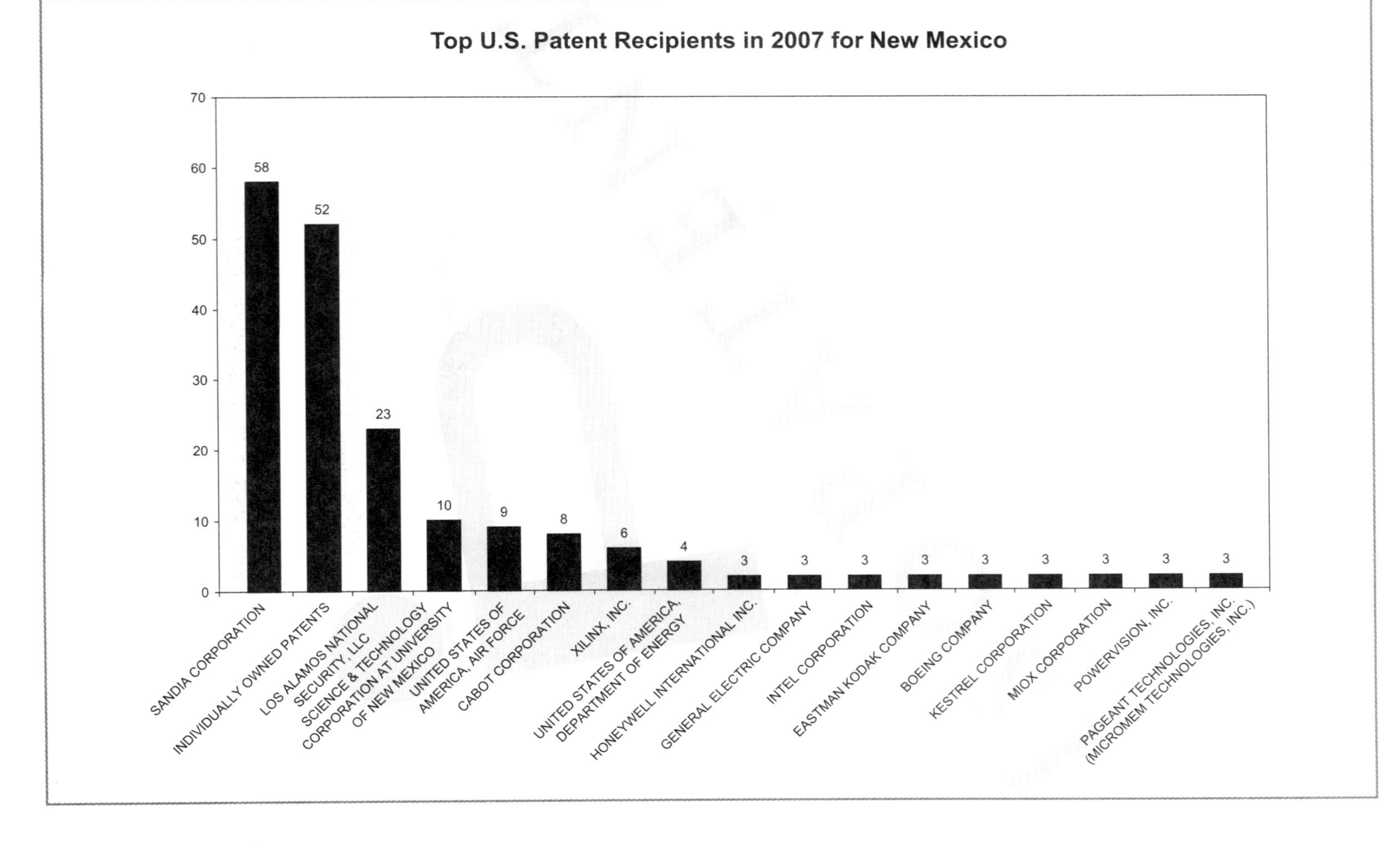

Top U.S. Patent Recipients in 2007 for New Mexico
70
60
50
40
30
20
10
0
58
52
23
10
9
8
6
4
3
3
3
3
3
3
3
3
3
SANDIA CORPORATION
INDIVIDUALLY OWNED PATENTS
LOS ALAMOS NATIONAL SECURITY, LLC
SCIENCE & TECHNOLOGY CORPORATION AT UNIVERSITY OF NEW MEXICO
UNITED STATES OF AMERICA, AIR FORCE
CABOT CORPORATION
XILINX, INC.
UNITED STATES OF AMERICA, DEPARTMENT OF ENERGY
HONEYWELL INTERNATIONAL INC.
GENERAL ELECTRIC COMPANY
INTEL CORPORATION
EASTMAN KODAK COMPANY
BOEING COMPANY
KESTREL CORPORATION
MIOX CORPORATION
POWERVISION, INC.
PAGEANT TECHNOLOGIES, INC. (MICROMEM TECHNOLOGIES, INC.)

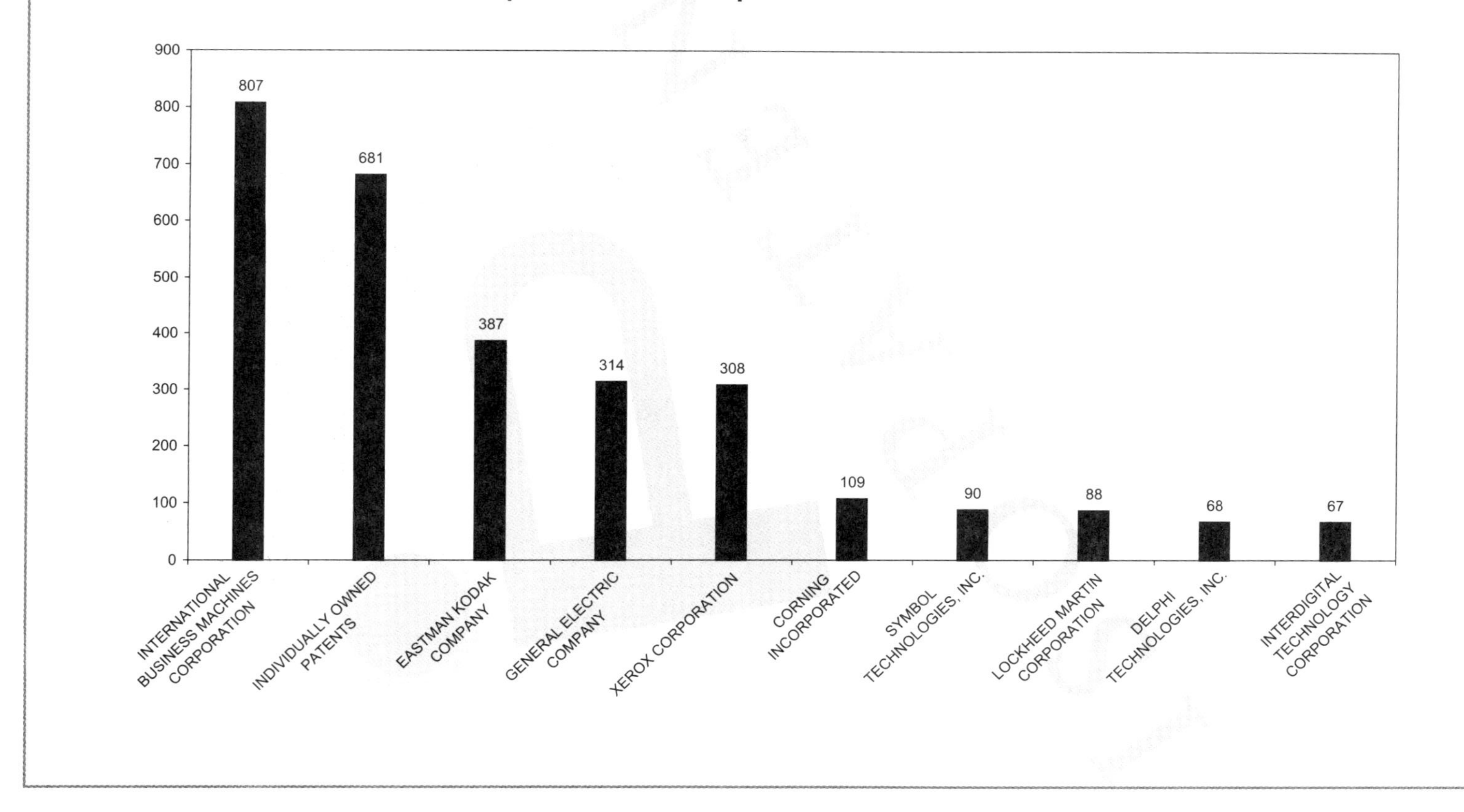
Top U.S. Patent Recipients in 2007 for New York
900
800
700
600
500
400
300
200
100
0
807
681
387
314
308
109
90
88
68
67
INTERNATIONAL BUSINESS MACHINES CORPORATION
INDIVIDUALLY OWNED PATENTS
EASTMAN KODAK COMPANY
GENERAL ELECTRIC COMPANY
XEROX CORPORATION
CORNING INCORPORATED
SYMBOL TECHNOLOGIES, INC.
LOCKHEED MARTIN CORPORATION
DELPHI TECHNOLOGIES, INC.
INTERDIGITAL TECHNOLOGY CORPORATION

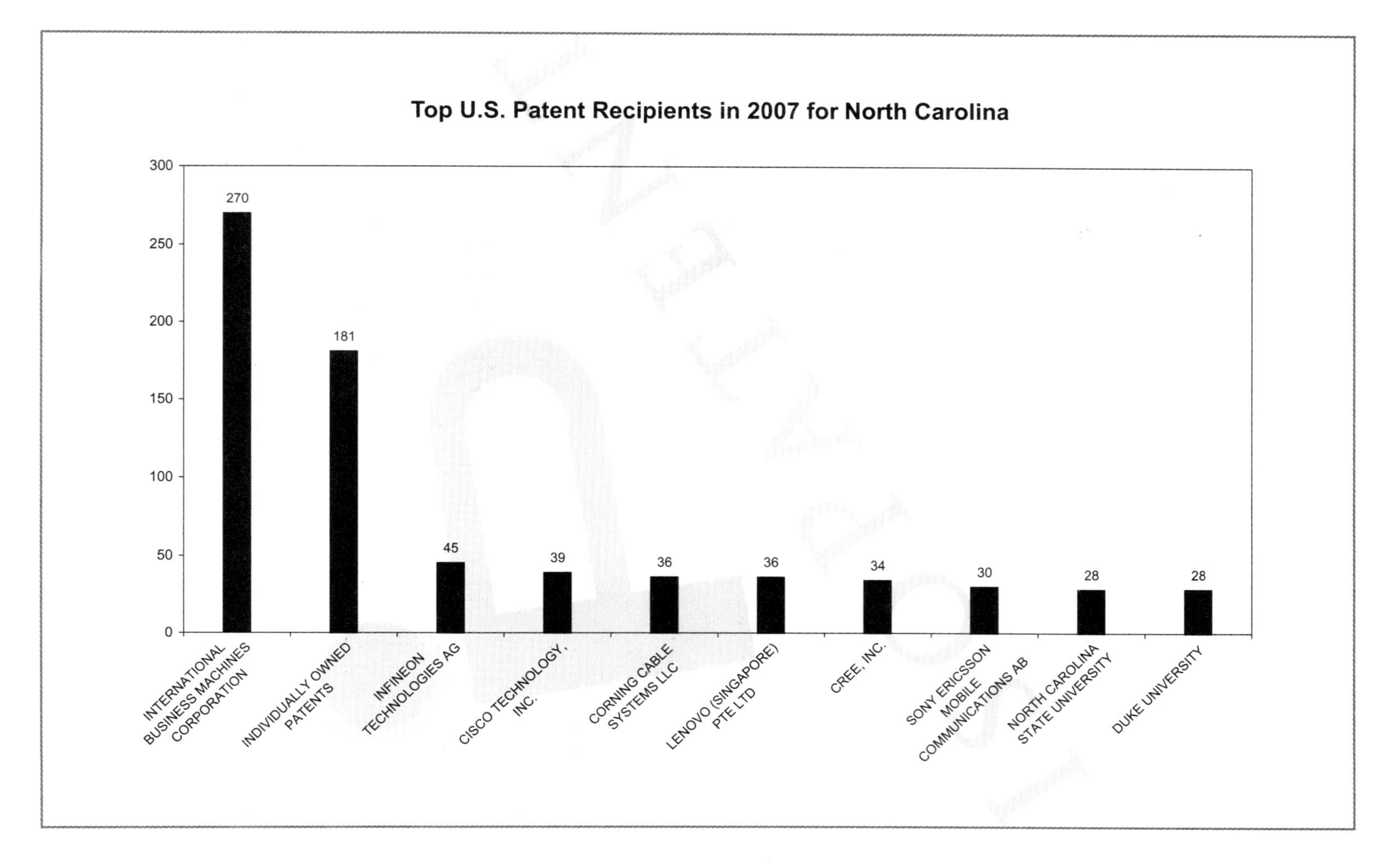
Top U.S. Patent Recipients in 2007 for North Carolina
300
250
200
150
100
50
0
270
181
45
39
36
36
34
30
28
28
INTERNATIONAL BUSINESS MACHINES CORPORATION
INDIVIDUALLY OWNED PATENTS
INFINEON TECHNOLOGIES AG
CISCO TECHNOLOGY, INC.
CORNING CABLE SYSTEMS LLC
LENOVO (SINGAPORE) PTE LTD
CREE, INC.
SONY ERICSSON MOBILE COMMUNICATIONS AB
NORTH CAROLINA STATE UNIVERSITY
DUKE UNIVERSITY

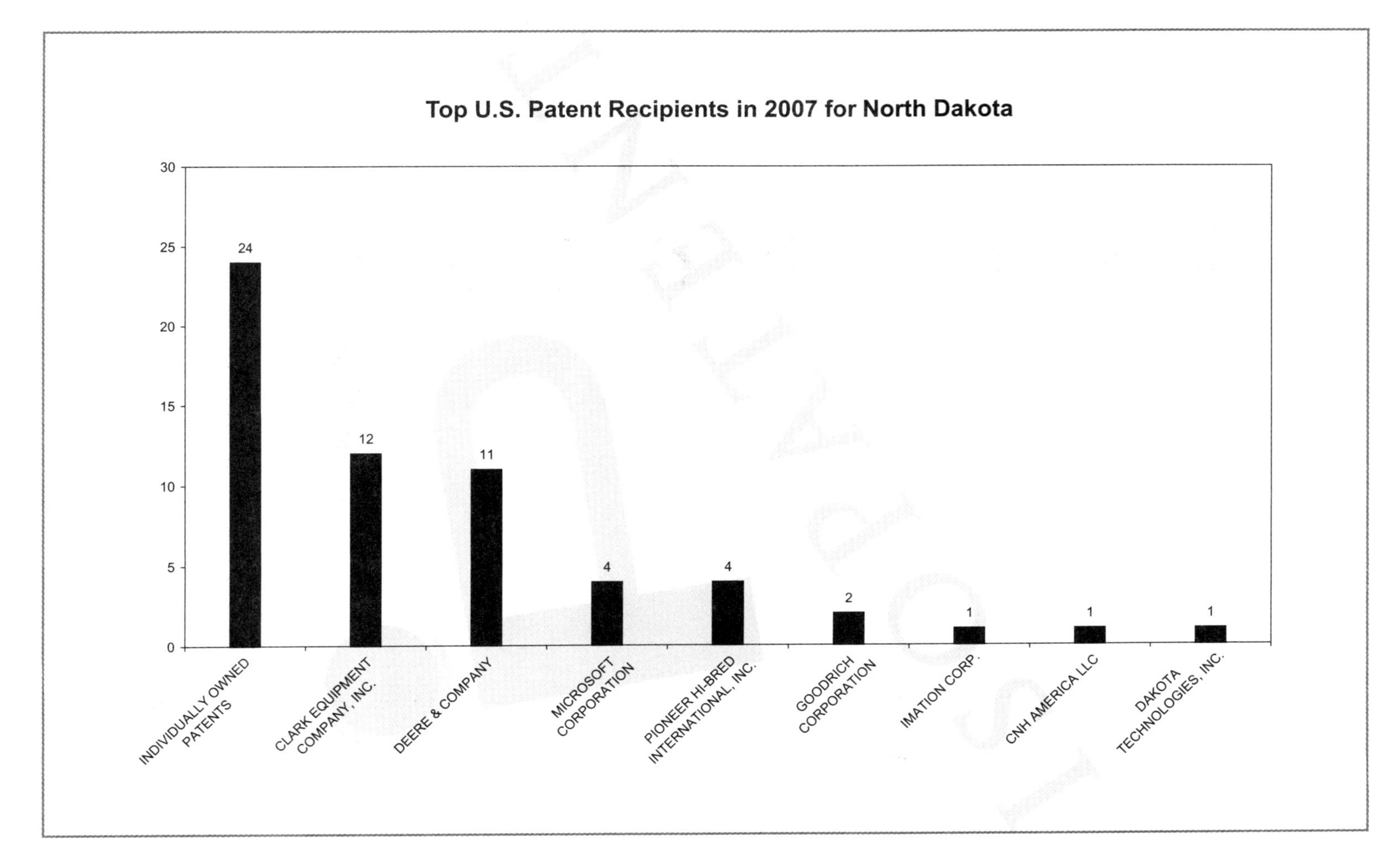
Top U.S. Patent Recipients in 2007 for North Dakota
30
25
20
15
10
5
0
24
12
11
4
4
2
1
1
1
INDIVIDUALLY OWNED PATENTS
CLARK EQUIPMENT COMPANY, INC.
DEERE & COMPANY
MICROSOFT CORPORATION
PIONEER HI-BRED INTERNATIONAL, INC.
GOODRICH CORPORATION
IMATION CORP.
CNH AMERICA LLC
DAKOTA TECHNOLOGIES, INC.

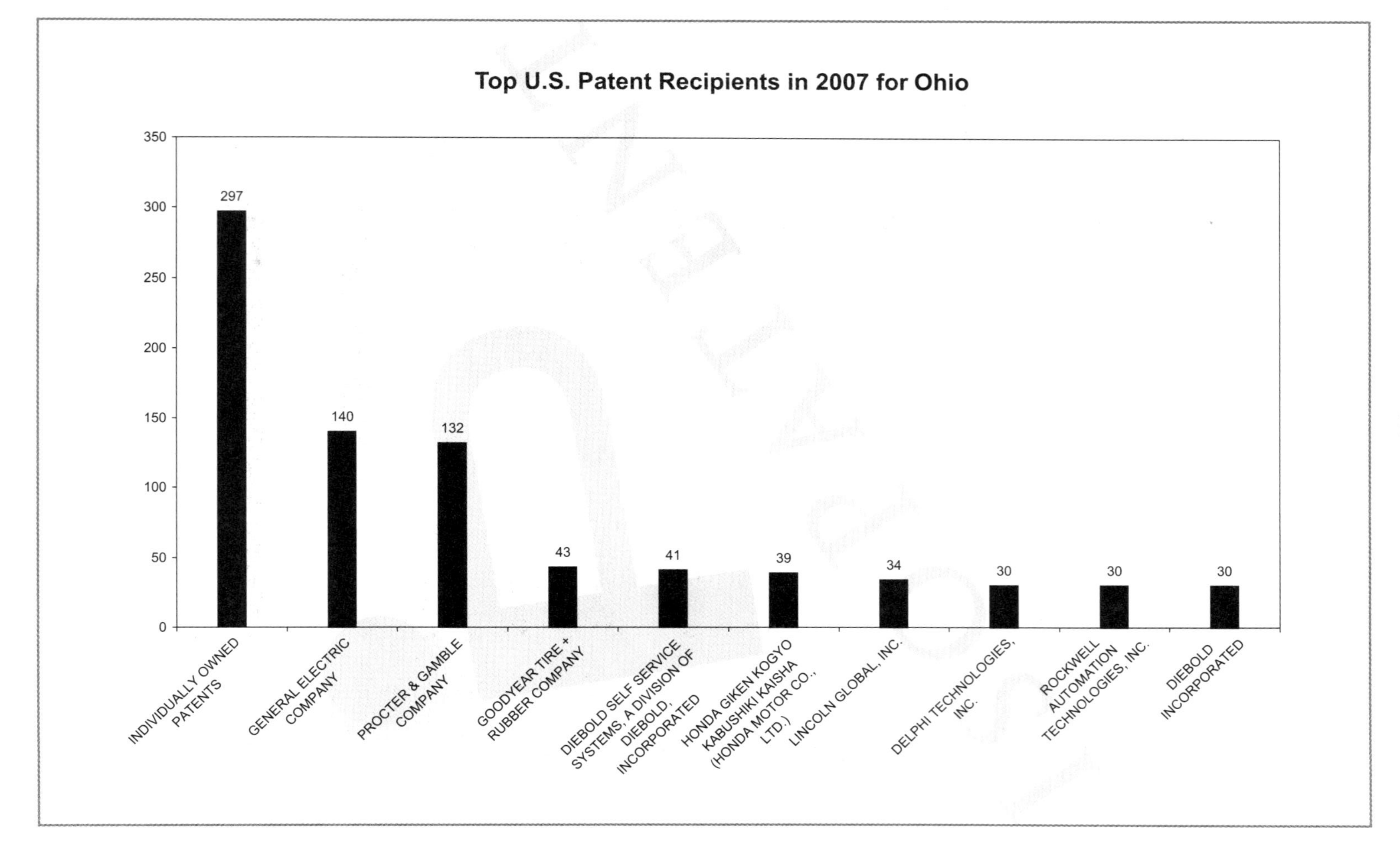

Top U.S. Patent Recipients in 2007 for Ohio
350
300
250
200
150
100
50
0
297
140
132
43
41
39
34
30
30
30
INDIVIDUALLY OWNED PATENTS
GENERAL ELECTRIC COMPANY
PROCTER & GAMBLE COMPANY
GOODYEAR TIRE + RUBBER COMPANY
DIEBOLD SELF SERVICE SYSTEMS, A DIVISION OF DIEBOLD, INCORPORATED
HONDA GIKEN KOGYO KABUSHIKI KAISHA (HONDA MOTOR CO., LTD.)
LINCOLN GLOBAL, INC.
DELPHI TECHNOLOGIES, INC.
ROCKWELL AUTOMATION TECHNOLOGIES, INC.
DIEBOLD INCORPORATED

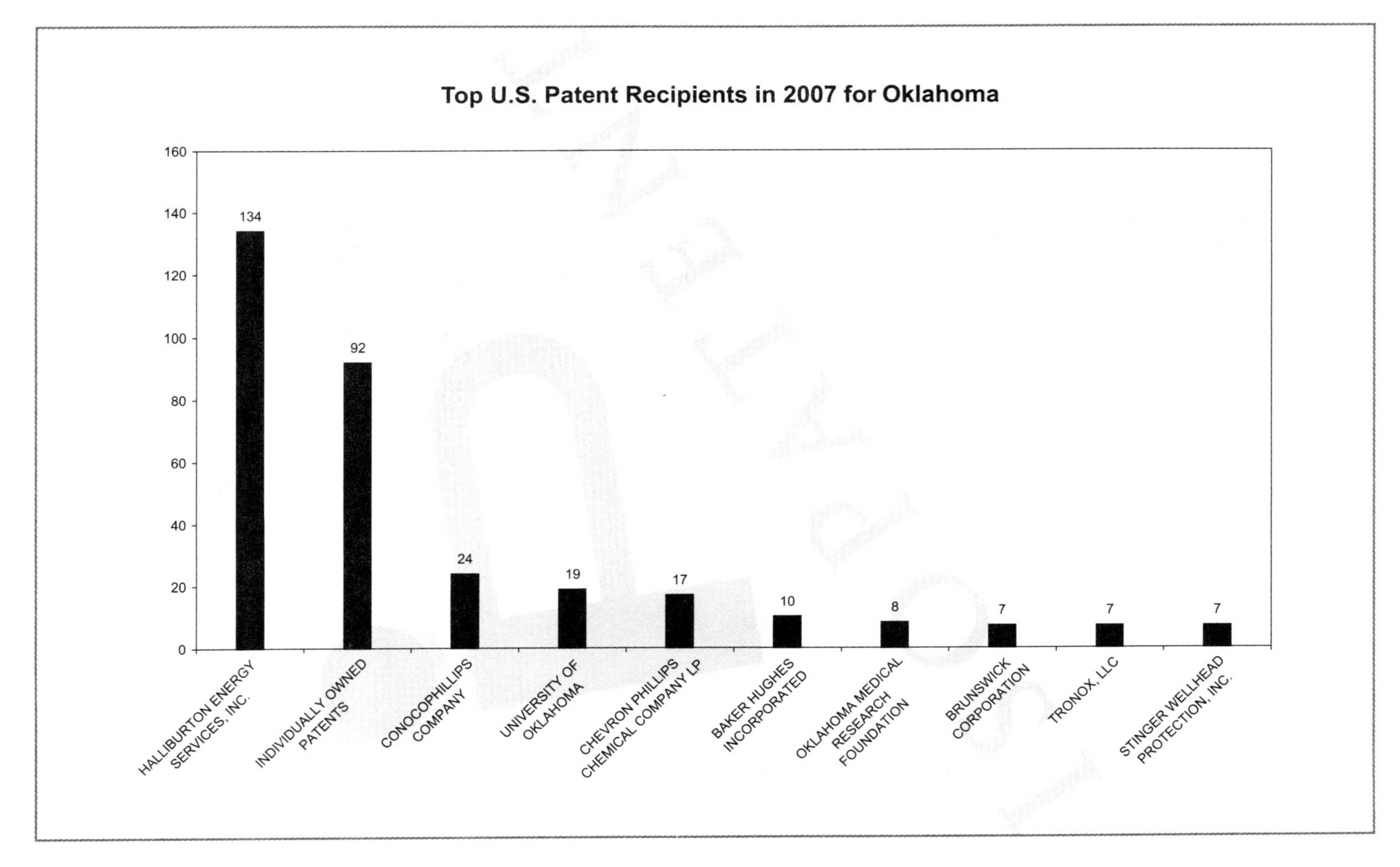

Top U.S. Patent Recipients in 2007 for Oklahoma
160
140
120
100
80
60
40
20
0
134
92
24
19
17
10
8
7
7
7
HALLIBURTON ENERGY SERVICES, INC.
INDIVIDUALLY OWNED PATENTS
CONOCOPHILLIPS COMPANY
UNIVERSITY OF OKLAHOMA
CHEVRON PHILLIPS CHEMICAL COMPANY LP
BAKER HUGHES INCORPORATED
OKLAHOMA MEDICAL RESEARCH FOUNDATION
BRUNSWICK CORPORATION
TRONOX, LLC
STINGER WELLHEAD PROTECTION, INC.

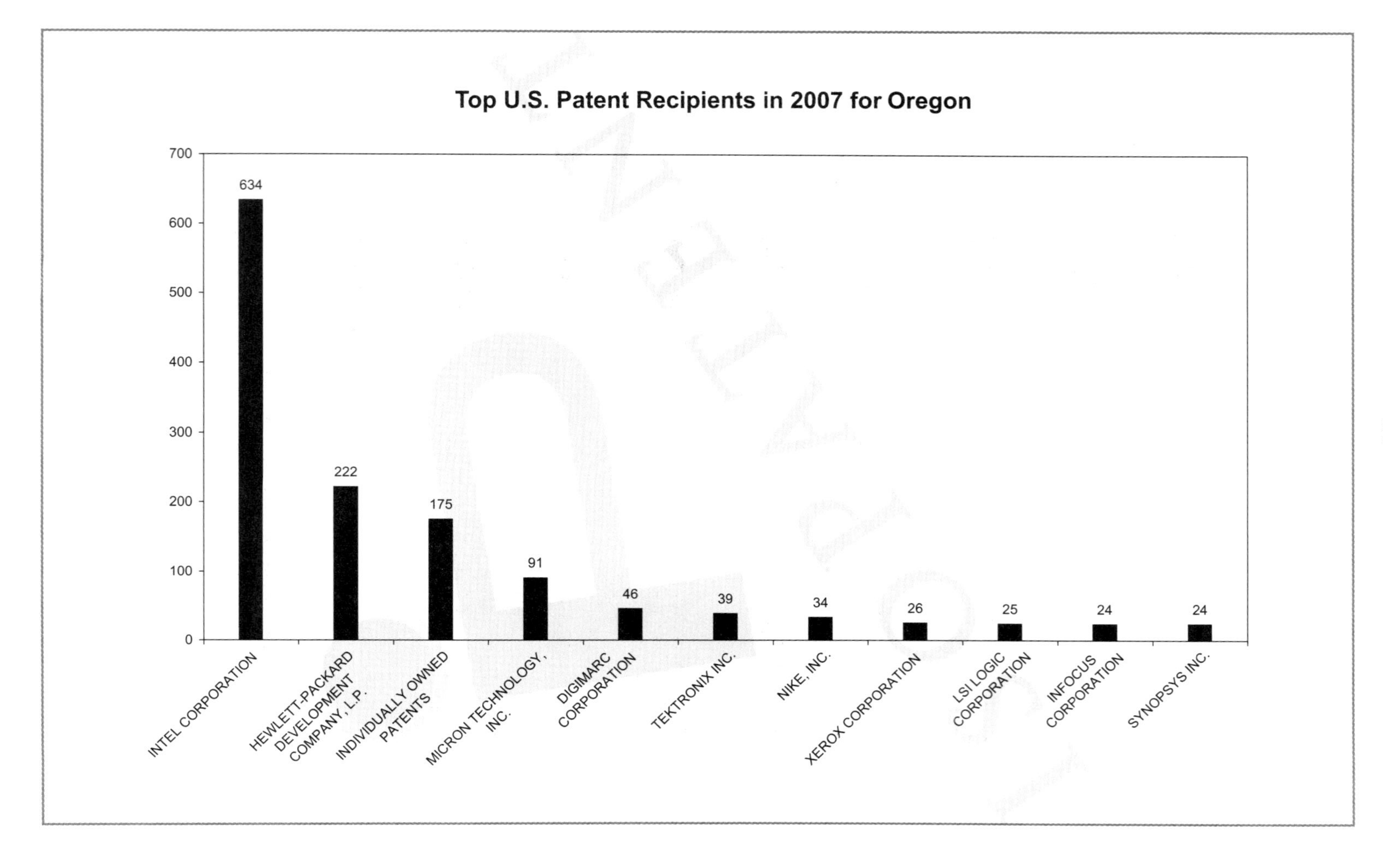

Top U.S. Patent Recipients in 2007 for Oregon
700
600
500
400
300
200
100
0
634
222
175
91
46
39
34
26
25
24
24
INTEL CORPORATION
HEWLETT-PACKARD DEVELOPMENT COMPANY, L.P.
INDIVIDUALLY OWNED PATENTS
MICRON TECHNOLOGY, INC.
DIGIMARC CORPORATION
TEKTRONIX INC.
NIKE, INC.
XEROX CORPORATION
LSI LOGIC CORPORATION
INFOCUS CORPORATION
SYNOPSYS INC.

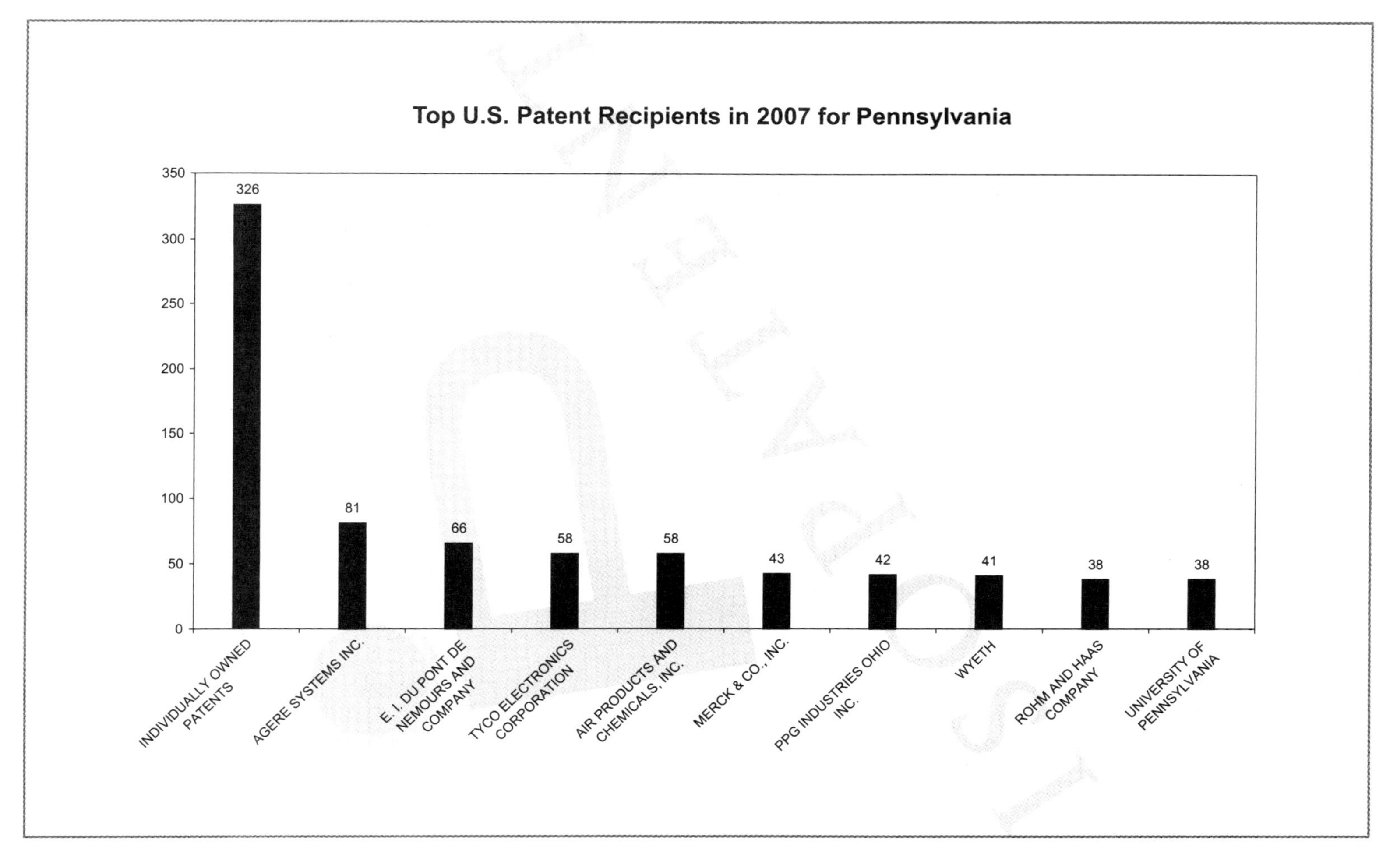

Top U.S. Patent Recipients in 2007 for Pennsylvania
350
300
250
200
150
100
50
0
326
81
66
58
58
43
42
41
38
38
INDIVIDUALLY OWNED PATENTS
AGERE SYSTEMS INC.
E. I. DU PONT DE NEMOURS AND COMPANY
TYCO ELECTRONICS CORPORATION
AIR PRODUCTS AND CHEMICALS, INC.
MERCK & CO., INC.
PPG INDUSTRIES OHIO INC.
WYETH
ROHM AND HAAS COMPANY
UNIVERSITY OF PENNSYLVANIA

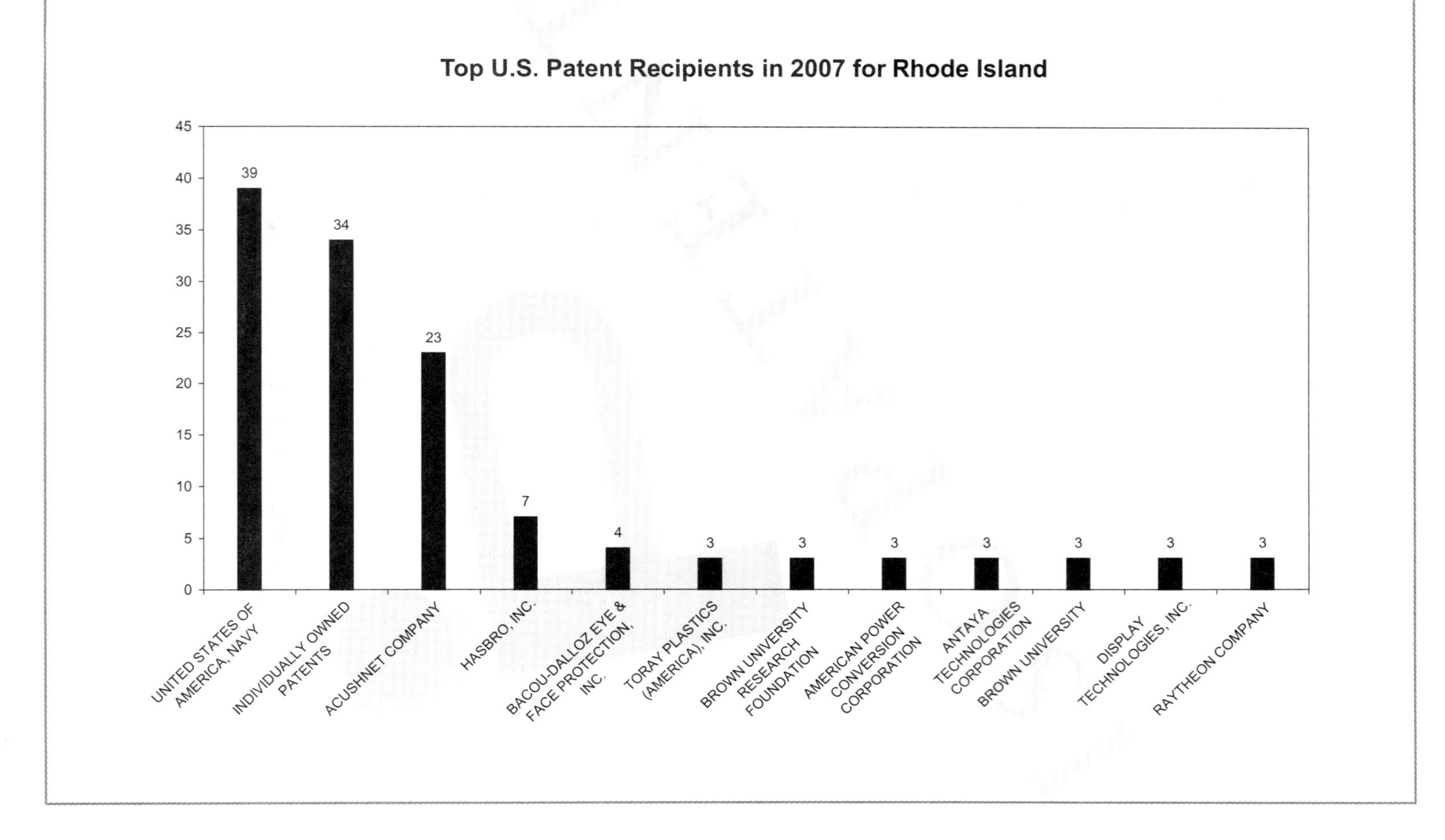
Top U.S. Patent Recipients in 2007 for Rhode Island
45
40
35
30
25
20
15
10
5
0
39
34
23
7
4
3
3
3
3
3
3
3
UNITED STATES OF AMERICA, NAVY
INDIVIDUALLY OWNED PATENTS
ACUSHNET COMPANY
HASBRO, INC.
BACOU-DALLOZ EYE & FACE PROTECTION, INC.
TORAY PLASTICS (AMERICA), INC.
BROWN UNIVERSITY RESEARCH FOUNDATION
AMERICAN POWER CONVERSION CORPORATION
ANTAYA TECHNOLOGIES CORPORATION
BROWN UNIVERSITY
DISPLAY TECHNOLOGIES, INC.
RAYTHEON COMPANY

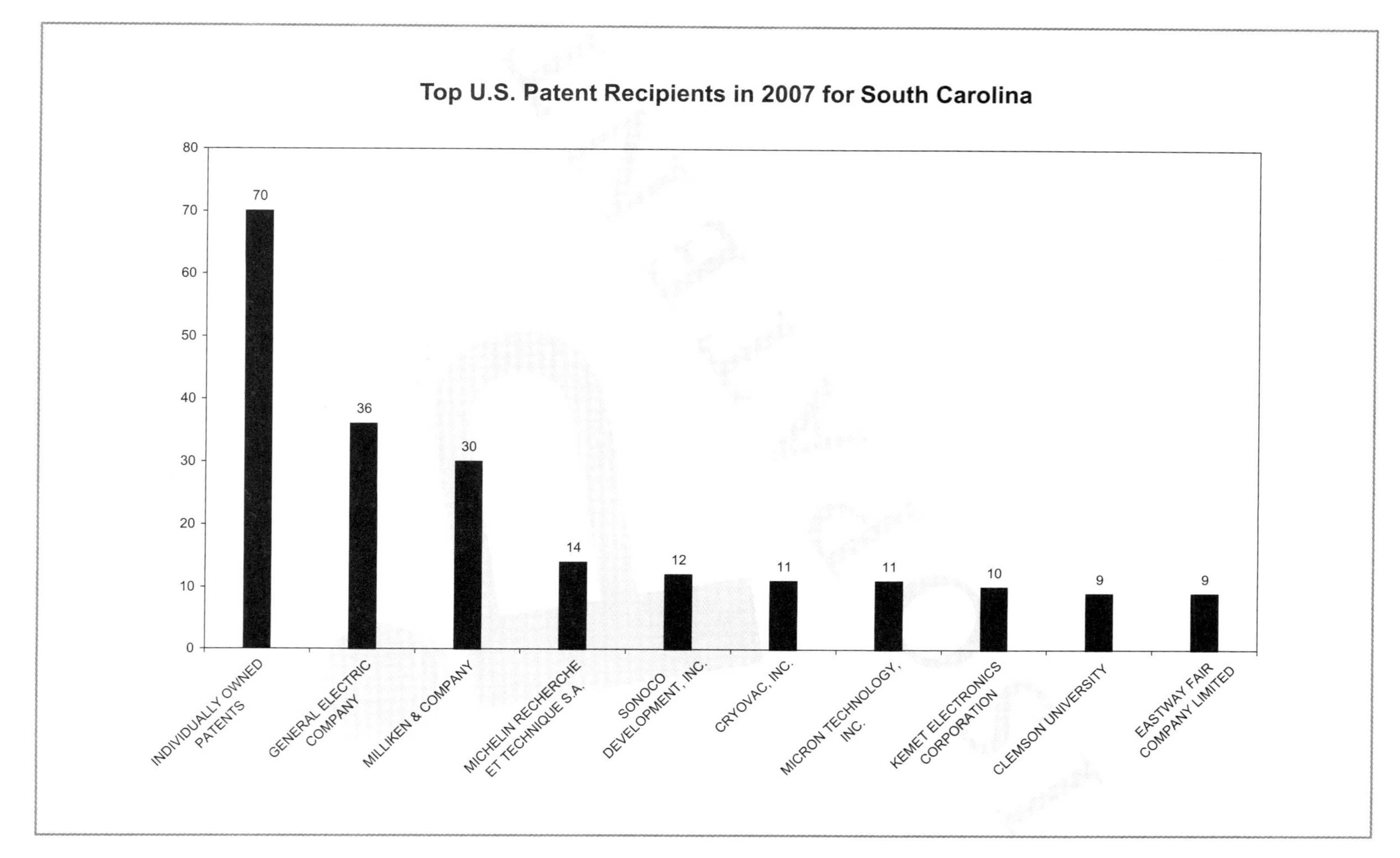
Top U.S. Patent Recipients in 2007 for South Carolina
80
70
60
50
40
30
20
10
0
70
36
30
14
12
11
11
10
9
9
INDIVIDUALLY OWNED PATENTS
GENERAL ELECTRIC COMPANY
MILLIKEN & COMPANY
MICHELIN RECHERCHE ET TECHNIQUE S.A.
SONOCO DEVELOPMENT, INC.
CRYOVAC, INC.
MICRON TECHNOLOGY, INC.
KEMET ELECTRONICS CORPORATION
CLEMSON UNIVERSITY
EASTWAY FAIR COMPANY LIMITED

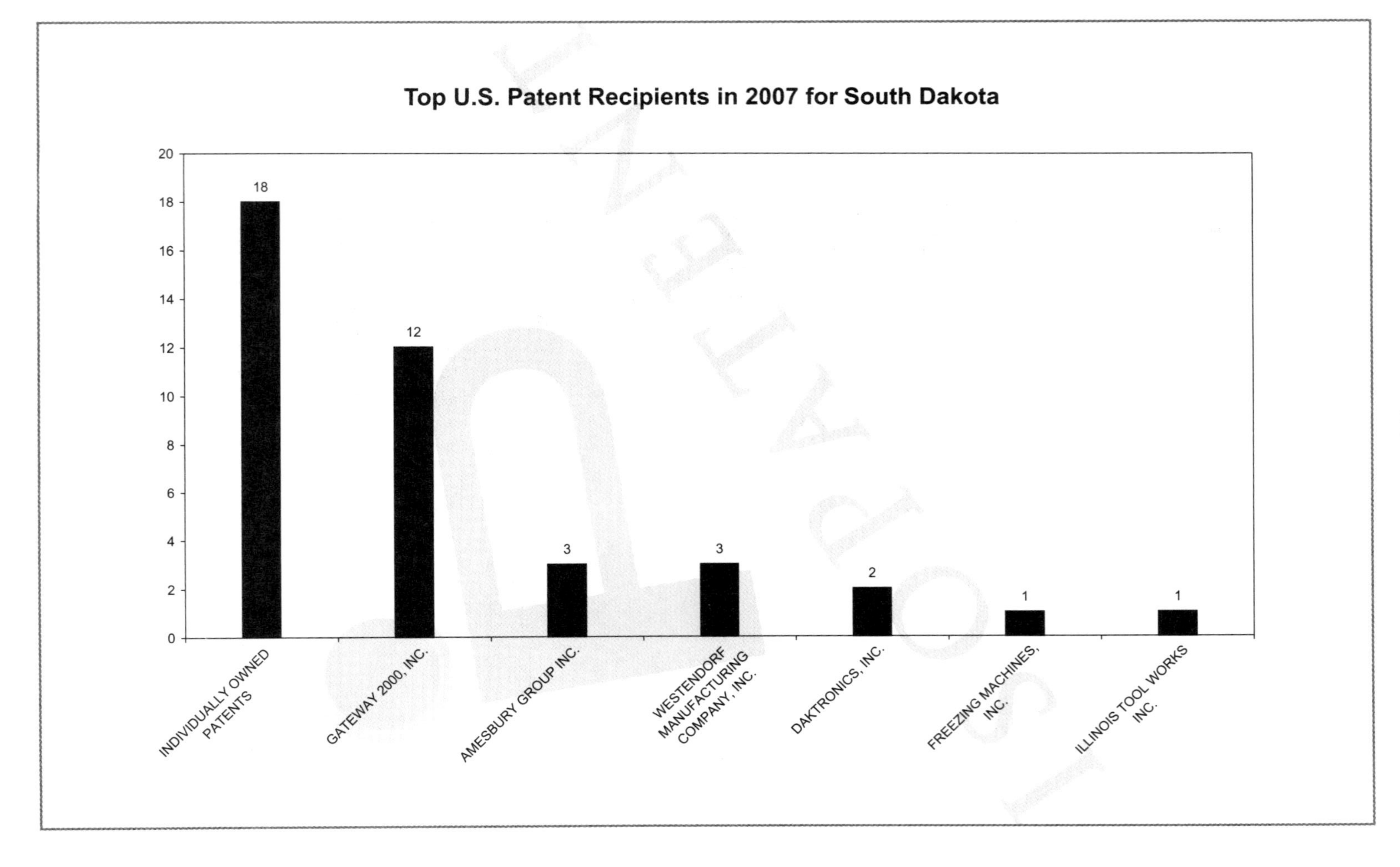

Top U.S. Patent Recipients in 2007 for South Dakota
20
18
16
14
12
10
8
6
4
2
0
18
12
3
3
2
1
1
INDIVIDUALLY OWNED PATENTS
GATEWAY 2000, INC.
AMESBURY GROUP INC.
WESTENDORF MANUFACTURING COMPANY, INC.
DAKTRONICS, INC.
FREEZING MACHINES, INC.
ILLINOIS TOOL WORKS INC.

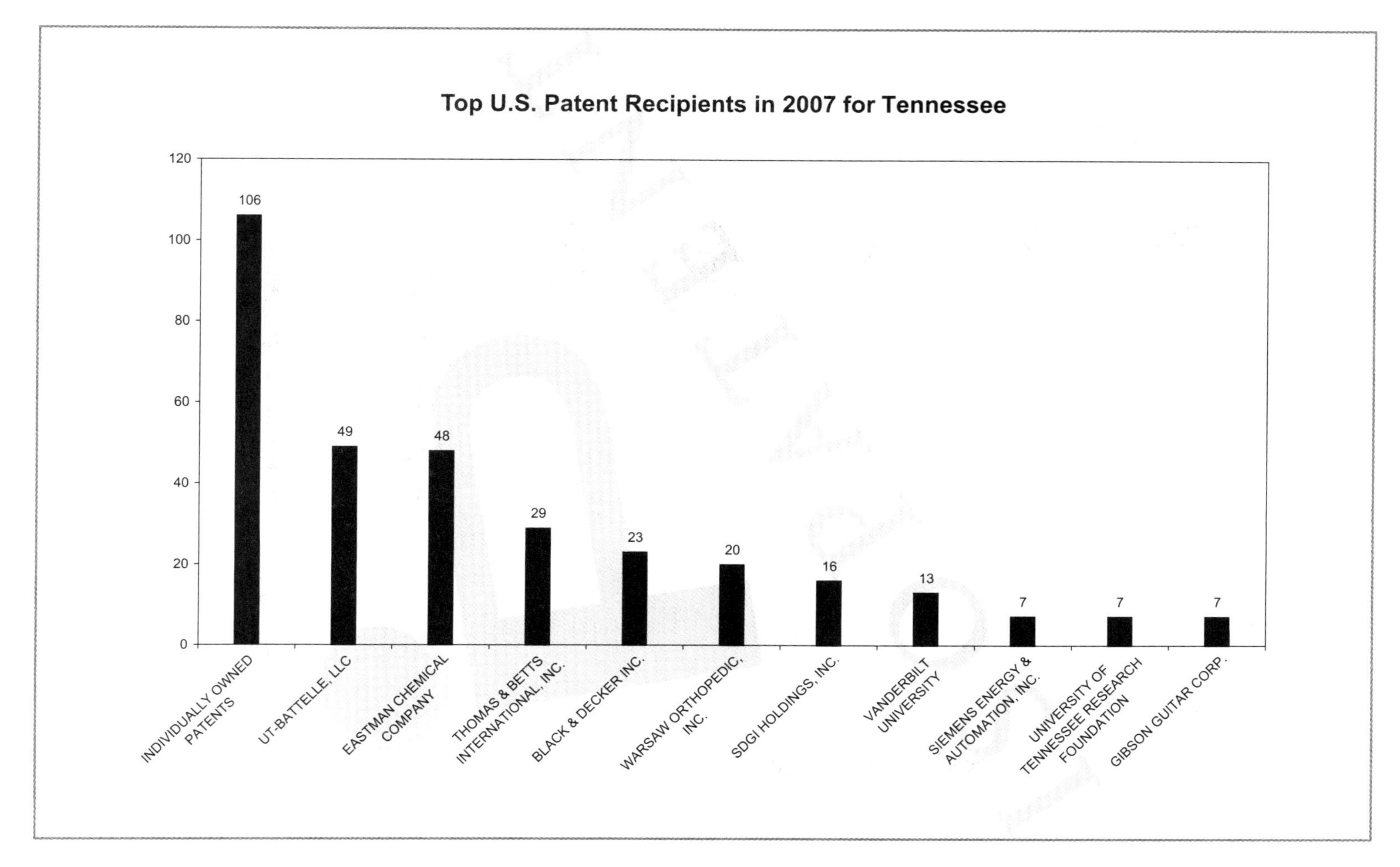

Top U.S. Patent Recipients in 2007 for Tennessee
120
100
80
60
40
20
0
106
49
48
29
23
20
16
13
7
7
7
INDIVIDUALLY OWNED PATENTS
UT-BATTELLE, LLC
EASTMAN CHEMICAL COMPANY
THOMAS & BETTS INTERNATIONAL, INC.
BLACK & DECKER INC.
WARSAW ORTHOPEDIC, INC.
SDGI HOLDINGS, INC.
VANDERBILT UNIVERSITY
SIEMENS ENERGY & AUTOMATION, INC.
UNIVERSITY OF TENNESSEE RESEARCH FOUNDATION
GIBSON GUITAR CORP.

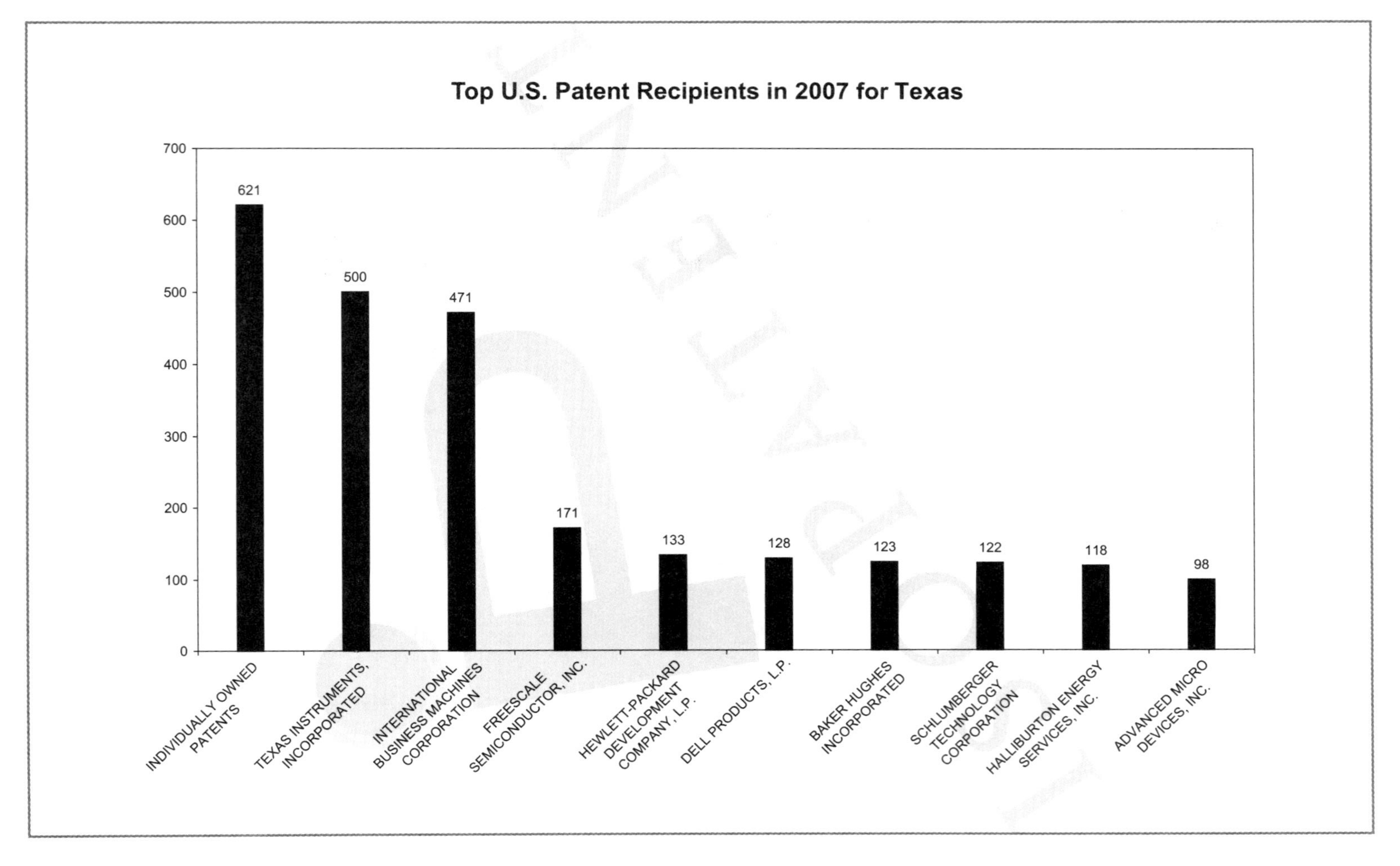

Top U.S. Patent Recipients in 2007 for Texas
700
600
500
400
300
200
100
0
621
500
471
171
133
128
123
122
118
98
INDIVIDUALLY OWNED PATENTS
TEXAS INSTRUMENTS, INCORPORATED
INTERNATIONAL BUSINESS MACHINES CORPORATION
FREESCALE SEMICONDUCTOR, INC.
HEWLETT-PACKARD DEVELOPMENT COMPANY, L.P.
DELL PRODUCTS, L.P.
BAKER HUGHES INCORPORATED
SCHLUMBERGER TECHNOLOGY CORPORATION
HALLIBURTON ENERGY SERVICES, INC.
ADVANCED MICRO DEVICES, INC.

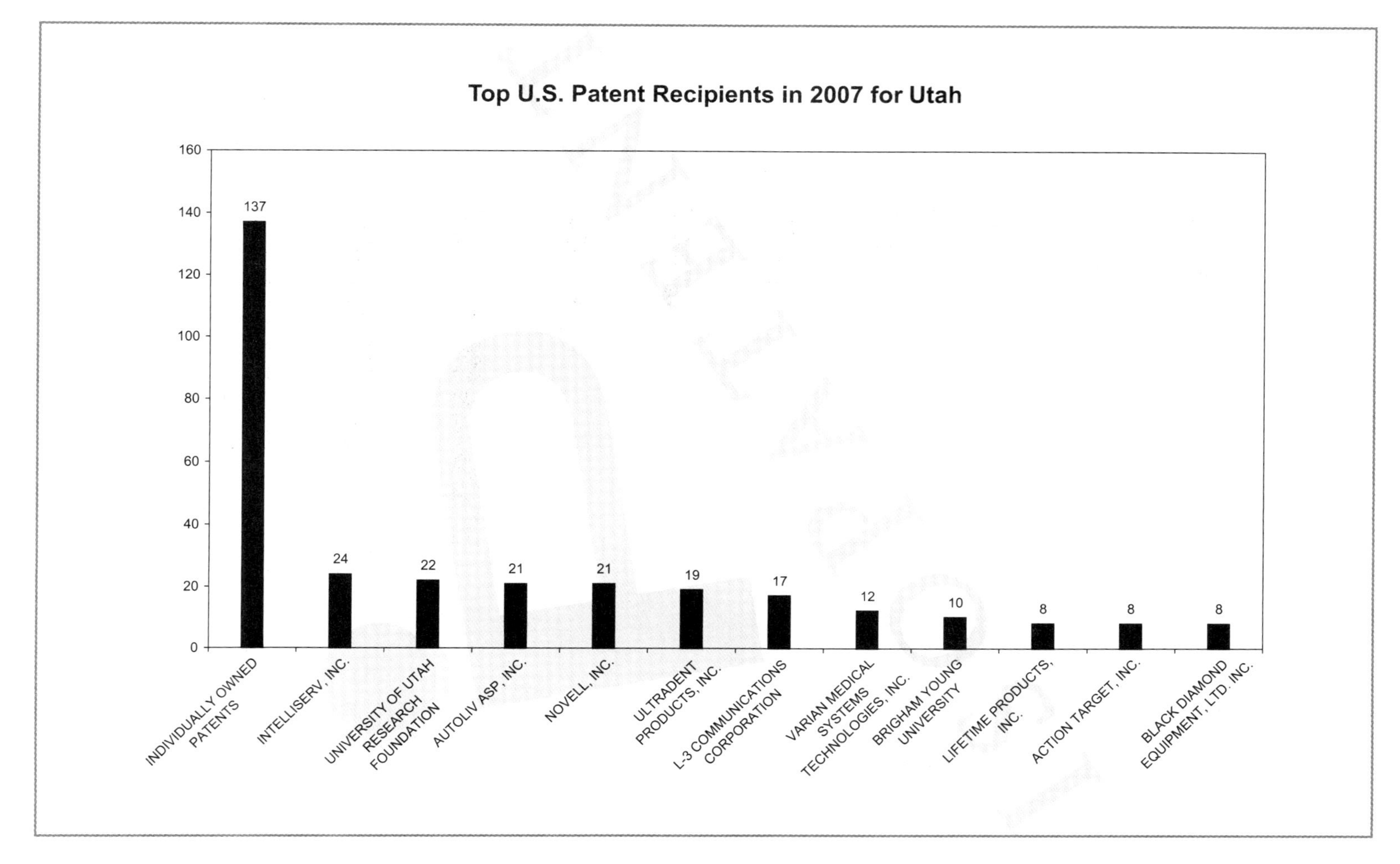

Top U.S. Patent Recipients in 2007 for Utah
160
140
120
100
80
60
40
20
0
137
24
22
21
21
19
17
12
10
8
8
8
INDIVIDUALLY OWNED PATENTS
INTELLISERV, INC.
UNIVERSITY OF UTAH RESEARCH FOUNDATION
AUTOLIV ASP, INC.
NOVELL, INC.
ULTRADENT PRODUCTS, INC.
L-3 COMMUNICATIONS CORPORATION
VARIAN MEDICAL SYSTEMS TECHNOLOGIES, INC.
BRIGHAM YOUNG UNIVERSITY
LIFETIME PRODUCTS, INC.
ACTION TARGET, INC.
BLACK DIAMOND EQUIPMENT, LTD. INC.

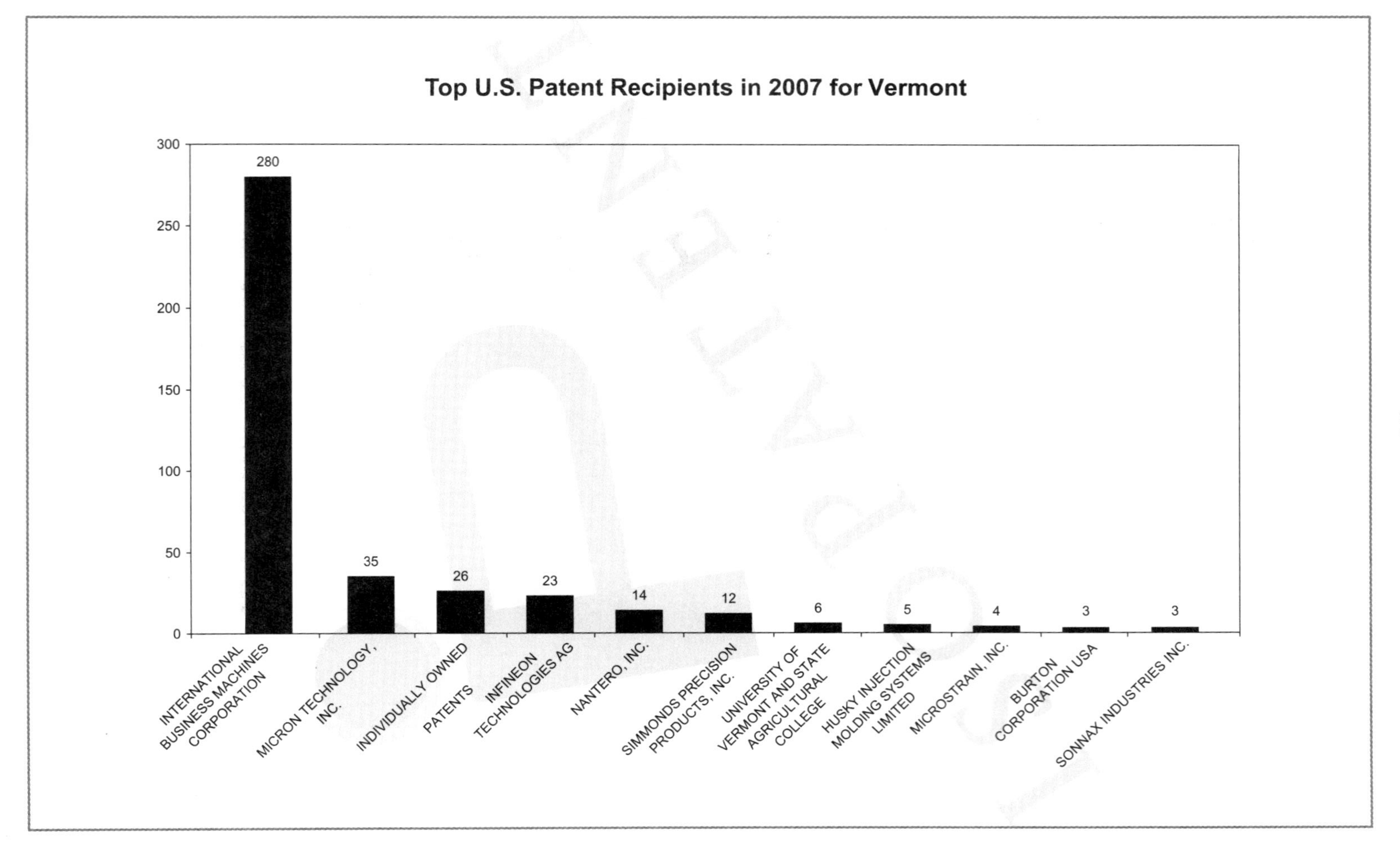

Top U.S. Patent Recipients in 2007 for Vermont
300
250
200
150
100
50
0
280
35
26
23
14
12
6
5
4
3
3
INTERNATIONAL BUSINESS MACHINES CORPORATION
MICRON TECHNOLOGY, INC.
INDIVIDUALLY OWNED PATENTS
INFINEON TECHNOLOGIES AG
NANTERO, INC.
SIMMONDS PRECISION PRODUCTS, INC.
UNIVERSITY OF VERMONT AND STATE AGRICULTURAL COLLEGE
HUSKY INJECTION MOLDING SYSTEMS LIMITED
MICROSTRAIN, INC.
BURTON CORPORATION USA
SONNAX INDUSTRIES INC.

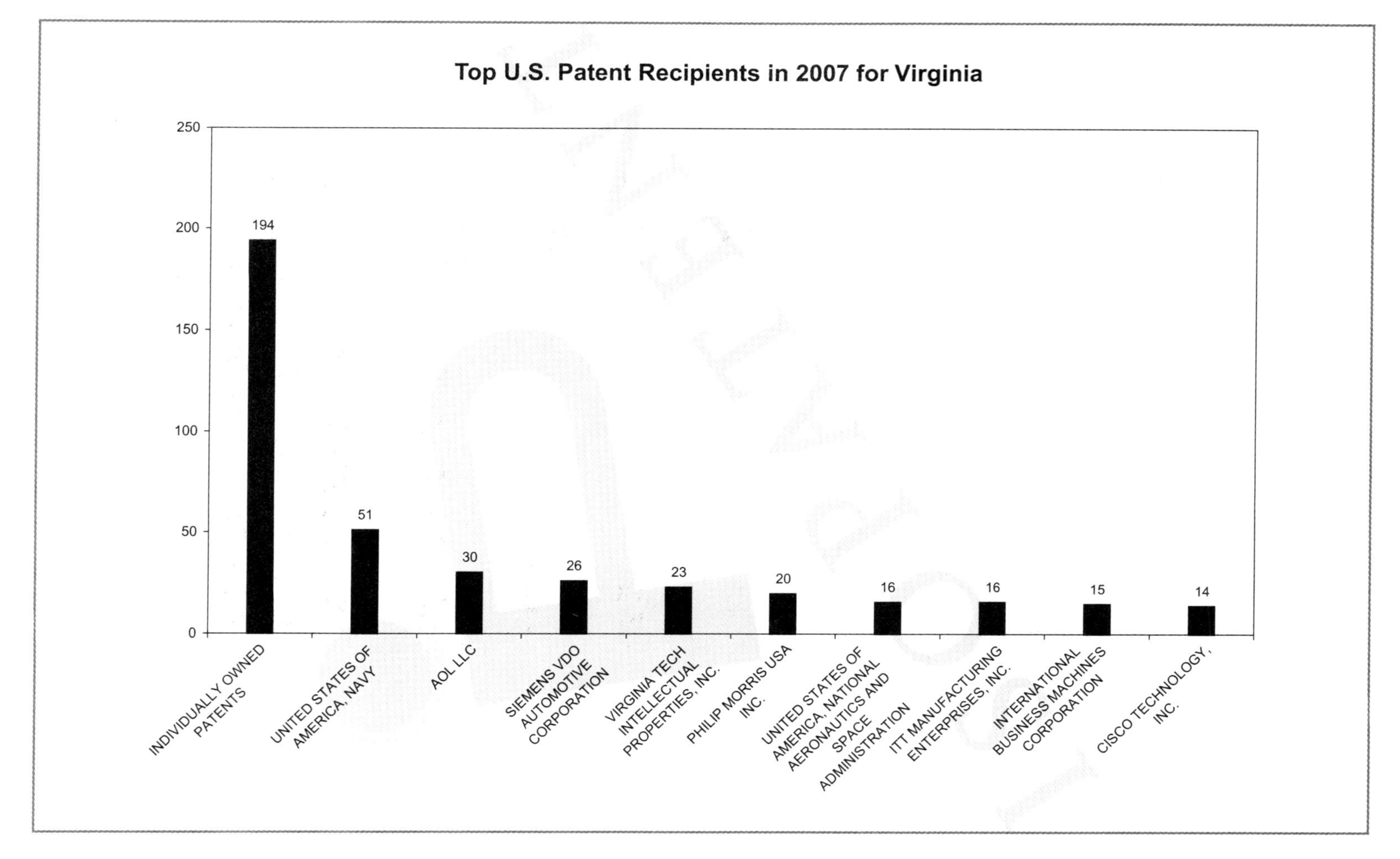

Top U.S. Patent Recipients in 2007 for Virginia
250
200
150
100
50
0
194
51
30
26
23
20
16
16
15
14
INDIVIDUALLY OWNED PATENTS
UNITED STATES OF AMERICA, NAVY
AOL LLC
SIEMENS VDO AUTOMOTIVE CORPORATION
VIRGINIA TECH INTELLECTUAL PROPERTIES, INC.
PHILIP MORRIS USA INC.
UNITED STATES OF AMERICA, NATIONAL AERONAUTICS AND SPACE ADMINISTRATION
ITT MANUFACTURING ENTERPRISES, INC.
INTERNATIONAL BUSINESS MACHINES CORPORATION
CISCO TECHNOLOGY, INC.

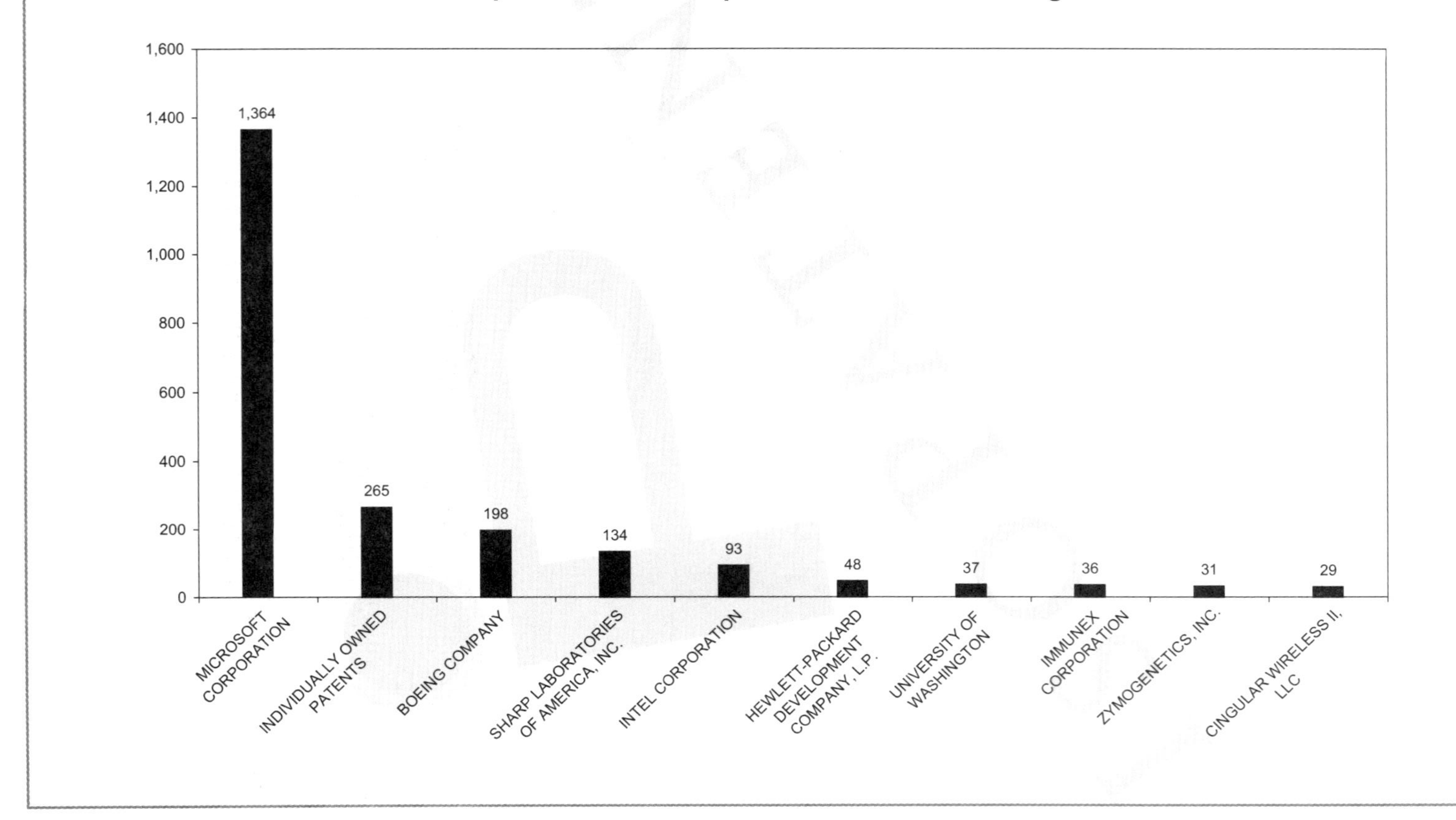

Top U.S. Patent Recipients in 2007 for Washington
1,600
1,400
1,200
1,000
800
600
400
200
0
1,364
265
198
134
93
48
37
36
31
29
MICROSOFT CORPORATION
INDIVIDUALLY OWNED PATENTS
BOEING COMPANY
SHARP LABORATORIES OF AMERICA, INC.
INTEL CORPORATION
HEWLETT-PACKARD DEVELOPMENT COMPANY, L.P.
UNIVERSITY OF WASHINGTON
IMMUNEX CORPORATION
ZYMOGENETICS, INC.
CINGULAR WIRELESS II, LLC

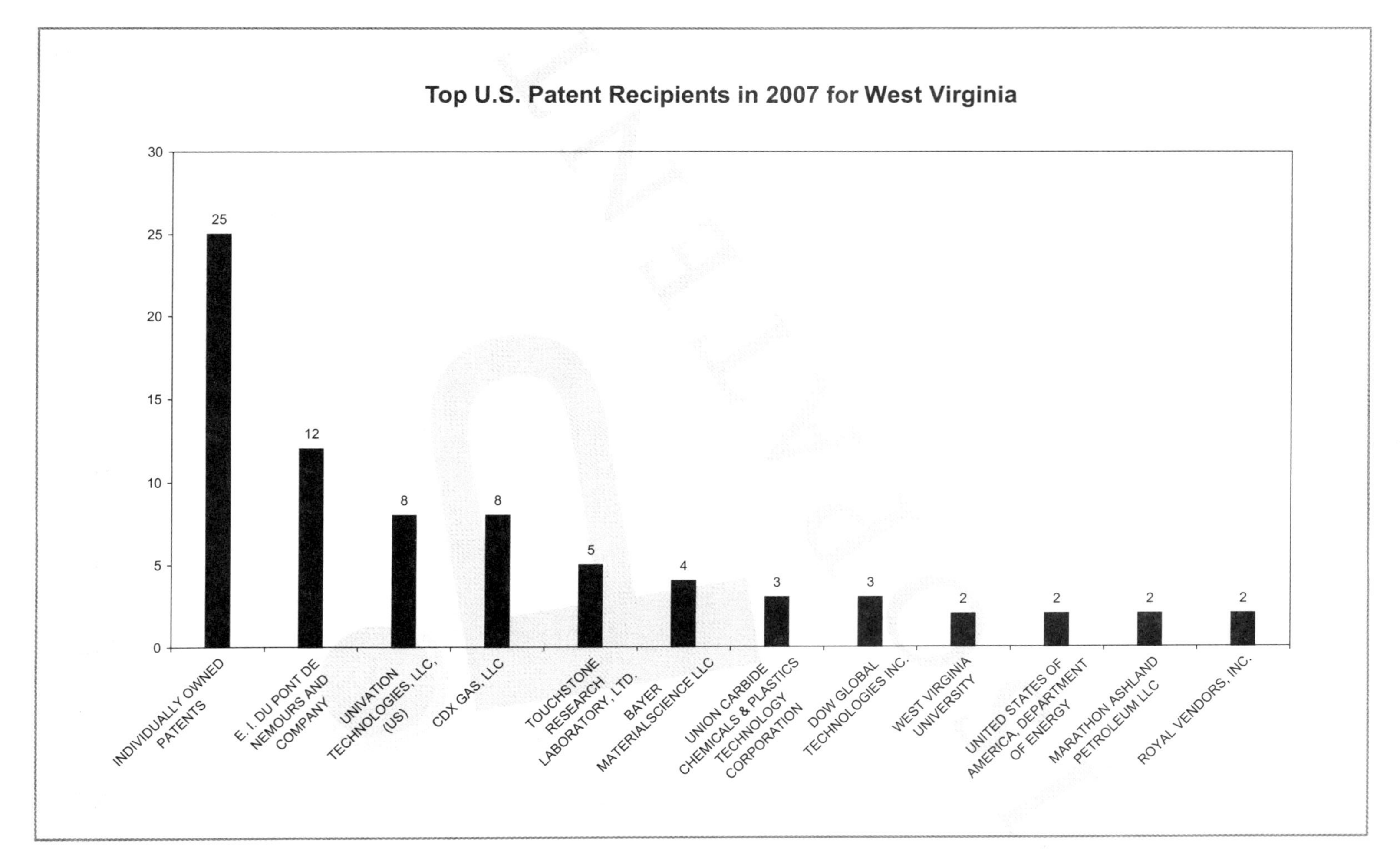
Top U.S. Patent Recipients in 2007 for West Virginia
30
25
20
15
10
5
0
25
12
8
8
5
4
3
3
2
2
2
2
INDIVIDUALLY OWNED PATENTS
E. I. DU PONT DE NEMOURS AND COMPANY
UNIVATION TECHNOLOGIES, LLC, (US)
CDX GAS, LLC
TOUCHSTONE RESEARCH LABORATORY, LTD.
BAYER MATERIALSCIENCE LLC
UNION CARBIDE CHEMICALS & PLASTICS TECHNOLOGY CORPORATION
DOW GLOBAL TECHNOLOGIES INC.
WEST VIRGINIA UNIVERSITY
UNITED STATES OF AMERICA, DEPARTMENT OF ENERGY
MARATHON ASHLAND PETROLEUM LLC
ROYAL VENDORS, INC.

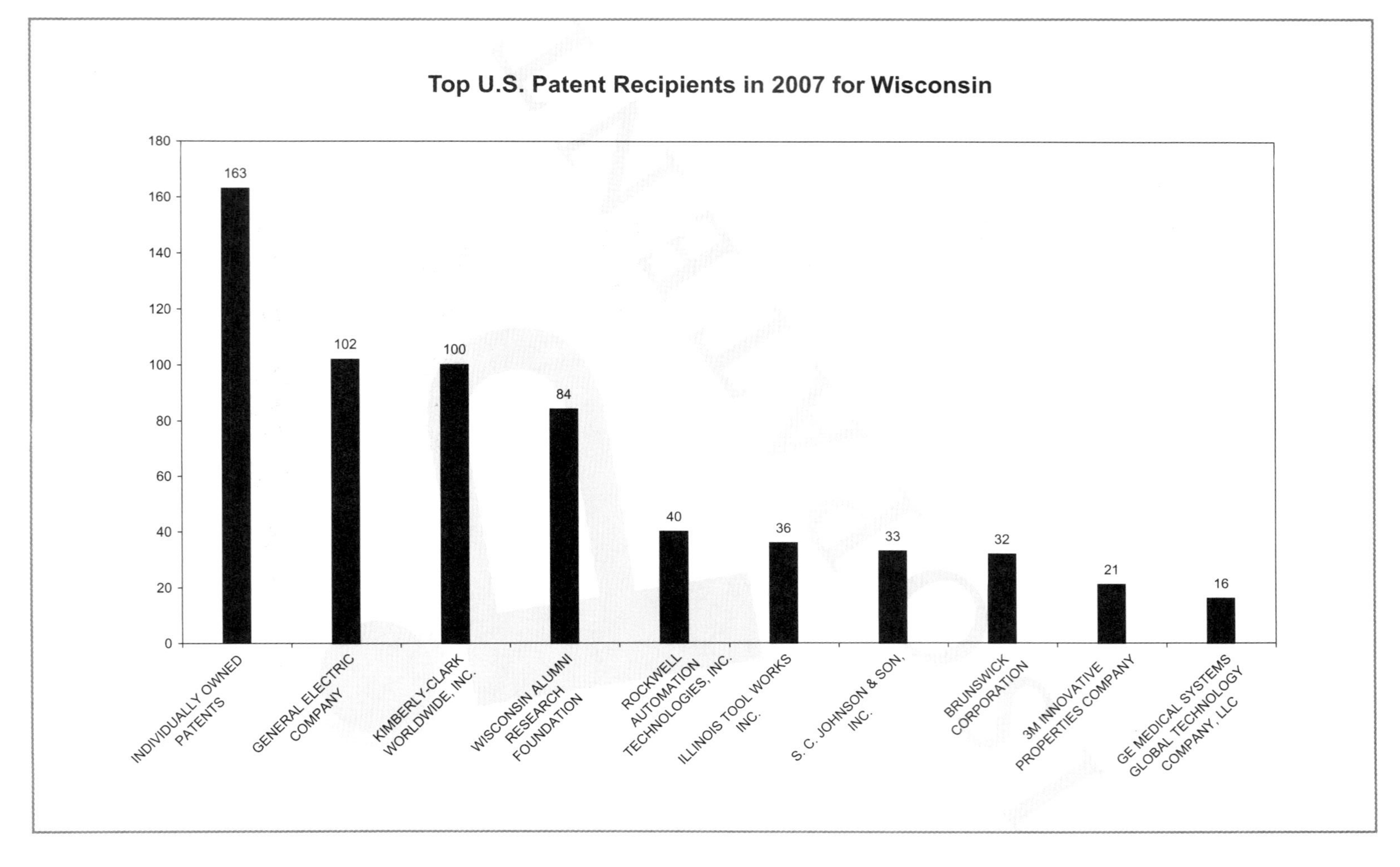
Top U.S. Patent Recipients in 2007 for Wisconsin
180
160
140
120
100
80
60
40
20
0
163
102
100
84
40
36
33
32
21
16
INDIVIDUALLY OWNED PATENTS
GENERAL ELECTRIC COMPANY
KIMBERLY-CLARK WORLDWIDE, INC.
WISCONSIN ALUMNI RESEARCH FOUNDATION
ROCKWELL AUTOMATION TECHNOLOGIES, INC.
ILLINOIS TOOL WORKS INC.
S. C. JOHNSON & SON, INC.
BRUNSWICK CORPORATION
3M INNOVATIVE PROPERTIES COMPANY
GE MEDICAL SYSTEMS GLOBAL TECHNOLOGY COMPANY, LLC

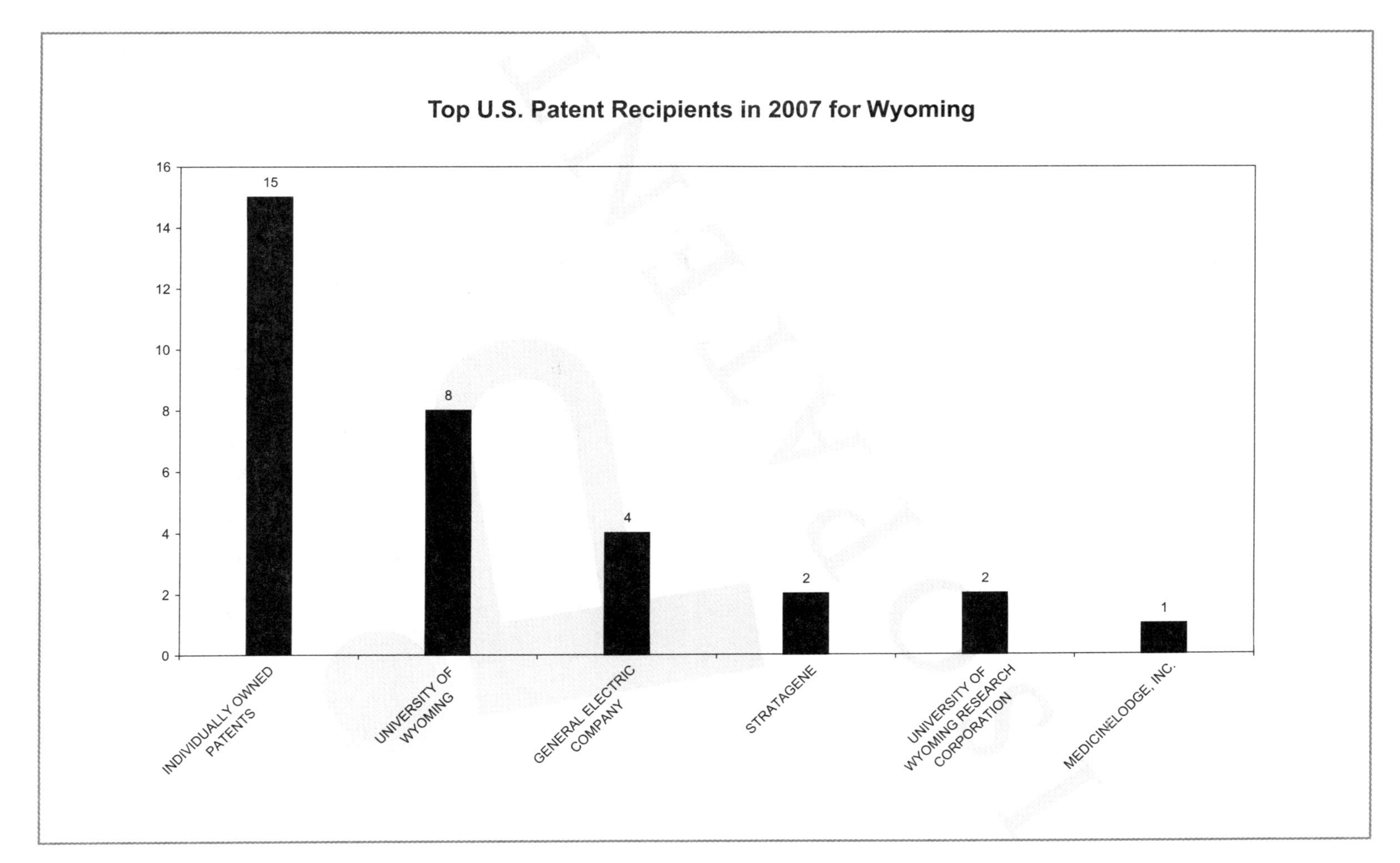

Top U.S. Patent Recipients in 2007 for Wyoming
0
2
4
6
8
10
12
14
16
15
8
4
2
2
1
INDIVIDUALLY OWNED PATENTS
UNIVERSITY OF WYOMING
GENERAL ELECTRIC COMPANY
STRATAGENE
UNIVERSITY OF WYOMING RESEARCH CORPORATION
MEDICINELODGE, INC.

Chapter 18
Exemplary Patents for Each of the Fifty United States

The following are first pages (not the entire patents) of exemplary utility patents for each of the fifty United States. The patent titles and abstracts are represented here with a surrounding border to highlight them for easier review. Additionally, the full names of the inventors and their cities and states of residence have been removed. You will see that patents are granted for a wide variety of innovations. The patents listed in this chapter include ones with the following titles:

- Apparatus for and Method of Processing Meat
- Intelligent Vehicle Identification System
- Adjustment of a Hearing Aid Using a Phone
- Method of Predicting a Change in an Economy
- Adjustable Table Lamp
- Footwear with Knit Upper and Method of Manufacturing the Footwear
- Direct Drive Wind Turbine
- Heated Pet Bed
- Composition and Methods for Inhibiting Pancreatic Cancer Mestastisis
- Action Figure Game Piece and Method of Playing Figure Game

Alabama

(12) **United States Patent**
Allen et al.

(10) **Patent No.:** **US 7,136,828 B1**
(45) **Date of Patent:** **Nov. 14, 2006**

(54) **INTELLIGENT VEHICLE IDENTIFICATION SYSTEM**

(75) Inventors:

(73) Assignee: **Jim Allen**, Wetumpka, AL (US)

(*) Notice: Subject to any disclaimer, the term of this patent is extended or adjusted under 35 U.S.C. 154(b) by 1036 days.

(21) Appl. No.: **09/977,937**

(22) Filed: **Oct. 17, 2001**

(51) **Int. Cl.**
G06F 17/00 (2006.01)

(52) **U.S. Cl.** **705/13**; 340/907; 340/928; 340/940; 340/941; 347/40; 235/384

(58) **Field of Classification Search** 705/13, 705/1; 340/907, 940, 941, 928; 347/40
See application file for complete search history.

(56) **References Cited**

U.S. PATENT DOCUMENTS

3,705,976 A * 12/1972 Platzman 235/384
3,927,389 A * 12/1975 Neeloff 340/940
5,173,692 A * 12/1992 Shapiro et al. 340/943
5,278,555 A 1/1994 Hoekman 340/941
5,339,081 A 8/1994 Jefferis et al. 342/28
5,614,894 A 3/1997 Stanczyk 340/933
5,617,086 A * 4/1997 Klashinsky et al. 340/907
5,717,390 A * 2/1998 Hasselbring 340/933
5,764,163 A * 6/1998 Waldman et al. 340/934
5,777,565 A * 7/1998 Hayashi et al. 340/928
5,900,825 A 5/1999 Pressel et al. 340/905
6,040,785 A 3/2000 Park et al. 340/928
6,121,898 A 9/2000 Moetteli 340/933
6,198,987 B1 3/2001 Park et al. 701/1
6,337,640 B1 1/2002 Lees et al.
6,342,845 B1 * 1/2002 Hilliard et al. 340/941
6,345,228 B1 2/2002 Lees et al.
6,490,443 B1 * 12/2002 Freeny, Jr. 455/406
6,744,383 B1 * 6/2004 Alfred et al. 340/988
6,764,163 B1 * 7/2004 Anderson et al. 347/40

FOREIGN PATENT DOCUMENTS

EP 577328 A2 * 1/1994
WO PCT/GB00/01206 10/2000
WO PCT/GB00/01221 10/2000

OTHER PUBLICATIONS

Hartje, Ronald L., "Tomorrow's Toll Road", Feb. 1991, Civil Engineering; 61, 2 ps.*
Research Document- Report No. FHWA-IP-90-002, Traffic Handbook, Jul. 1990.

* cited by examiner

Primary Examiner—John W. Hayes
Assistant Examiner—Freda Nelson
(74) *Attorney, Agent, or Firm*—Pillsbury Winthrop Shaw Pittman LLP

(57) **ABSTRACT**

An intelligent vehicle identification system that uses inductive loop technology to profile and classify a vehicle. In a tolling industry application, classification of the vehicle is made prior to the vehicle arriving at a payment point in a toll lane in which the vehicle travels. A predetermined fare associated with the classification is then solicited from an operator of the vehicle without efforts from a toll attendant. In a preferred embodiment, the system also includes an intelligent queue loop that verifies the vehicle at the payment point to prevent misclassification due to a second vehicle, e.g., a motorcycle, that changes from a different lane to the toll lane in question.

19 Claims, 26 Drawing Sheets

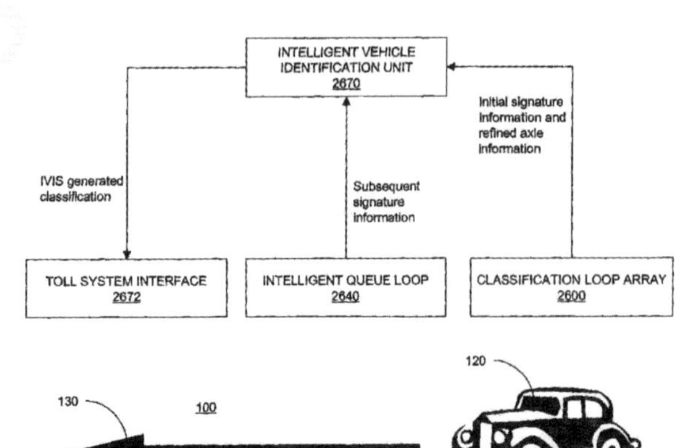

Alaska

US007018383B2

(12) **United States Patent**
McGuire

(10) **Patent No.:** **US 7,018,383 B2**
(45) **Date of Patent:** **Mar. 28, 2006**

(54) **SYSTEMS AND METHODS FOR PRODUCING OSTEOTOMIES**

(75) Inventor:

(73) Assignee: **David A. McGuire**, Anchorage, AK (US)

(*) Notice: Subject to any disclaimer, the term of this patent is extended or adjusted under 35 U.S.C. 154(b) by 68 days.

(21) Appl. No.: **10/402,826**

(22) Filed: **Mar. 28, 2003**

(65) **Prior Publication Data**

US 2003/0191475 A1 Oct. 9, 2003

Related U.S. Application Data

(60) Continuation of application No. 09/506,714, filed on Feb. 18, 2000, now Pat. No. 6,547,793, which is a division of application No. 08/985,568, filed on Dec. 5, 1997, now Pat. No. 6,027,504.

(60) Provisional application No. 60/063,195, filed on Oct. 21, 1997, provisional application No. 60/031,989, filed on Dec. 6, 1996.

(51) **Int. Cl.**
A61B 17/17 (2006.01)

(52) **U.S. Cl.** **606/102**; 606/96

(58) **Field of Classification Search** 606/53–55, 606/58, 59, 86, 87, 96, 102

See application file for complete search history.

(56) **References Cited**

U.S. PATENT DOCUMENTS

1,590,499	A *	6/1926	Cozad	600/595
4,335,715	A *	6/1982	Kirkley	606/87
4,919,119	A *	4/1990	Jonsson et al.	606/54
5,376,091	A *	12/1994	Hotchkiss et al.	606/55
5,437,667	A *	8/1995	Papierski et al.	606/55
5,645,548	A *	7/1997	Augsburger	606/87

* cited by examiner

Primary Examiner—David O. Reip
(74) *Attorney, Agent, or Firm*—Bromberg & Sunstein LLP

(57) **ABSTRACT**

Systems and methods for producing minimally invasive osteotomies to correct angular deformities of bones in and about the knee are disclosed. A method includes locating a plane in which the angle exhibited by the deformity is situated. An oblique cut is then made along a surface of the bone, such that the cut is transverse to the plane in which the angle is situated. Thereafter, the bone pieces are rotated about the cut relative to one another until a desired alignment between the bone pieces is achieved. To maintain the bone pieces in alignment, a device having an elongated body for extending into a tunnel between the bone pieces is provided. The system also includes a rigid member fixedly positioned at one end of the body. The rigid member is transverse to the body to engage one bone piece. The system further includes a locking mechanism at an opposite end of the body to engage the other bone piece. The system permits the bone pieces to be pulled against one another between the rigid member and the locking mechanism.

9 Claims, 26 Drawing Sheets

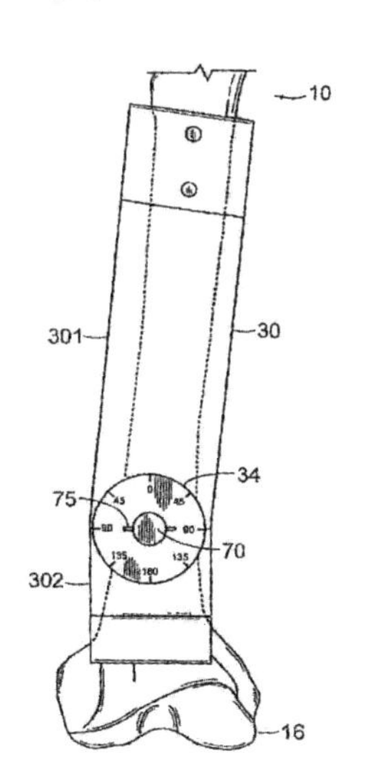

Arizona

US007019494B2

(12) **United States Patent**
Mikhaylik

(10) **Patent No.: US 7,019,494 B2**
(45) **Date of Patent: Mar. 28, 2006**

(54) **METHODS OF CHARGING LITHIUM SULFUR CELLS**

(75) Inventor:

(73) Assignee: **Moltech Corporation**, Tucson, AZ (US)

(*) Notice: Subject to any disclaimer, the term of this patent is extended or adjusted under 35 U.S.C. 154(b) by 167 days.

(21) Appl. No.: **10/753,123**

(22) Filed: **Jan. 6, 2004**

(65) **Prior Publication Data**

US 2005/0156575 A1 Jul. 21, 2005

(51) **Int. Cl.**
H02J 7/04 (2006.01)

(52) **U.S. Cl.** **320/148**

(58) **Field of Classification Search** 320/148, 320/122, 135; 429/231.95, 344

See application file for complete search history.

(56) **References Cited**

U.S. PATENT DOCUMENTS

4,238,721	A *	12/1980	DeLuca et al.	320/122
4,264,689	A	4/1981	Moses	429/338
4,410,609	A	10/1983	Peled et al.	429/105
4,816,358	A	3/1989	Holleck et al.	429/231.5
4,857,423	A	8/1989	Abraham et al.	429/329
5,021,308	A	6/1991	Armand et al.	429/336
5,352,967	A	10/1994	Nutz et al.	320/160
5,514,493	A	5/1996	Waddell et al.	429/199
5,529,860	A	6/1996	Skotheim et al.	429/213
5,601,947	A	2/1997	Skotheim et al.	429/213
5,686,201	A	11/1997	Chu	429/52
5,690,702	A	11/1997	Skotheim et al.	29/623.1
5,882,812	A	3/1999	Visco et al.	429/50
5,900,718	A	5/1999	Tsenter	320/151
5,919,587	A	7/1999	Mukherjee et al.	429/213
6,017,651	A	1/2000	Nimon et al.	429/101
6,025,094	A	2/2000	Visco et al.	429/231.95
6,030,720	A	2/2000	Chu et al.	429/105
6,060,184	A	5/2000	Gan et al.	429/3
6,117,590	A	9/2000	Skotheim et al.	429/213
6,136,477	A	10/2000	Gan et al.	429/307
6,153,337	A	11/2000	Carlson et al.	429/247
6,194,099	B1	2/2001	Geronov et al.	429/213
6,201,100	B1	3/2001	Gorkovenko et al.	528/388
6,210,831	B1	4/2001	Gorkovenko et al.	429/213
6,210,836	B1 *	4/2001	Takada et al.	429/231.95
6,210,839	B1	4/2001	Gan et al.	429/307
6,225,002	B1	5/2001	Nimon et al.	429/212
6,406,815	B1	6/2002	Sandberg et al.	429/231.95
6,436,583	B1	8/2002	Mikhaylik	429/340
6,632,573	B1 *	10/2003	Nimon et al.	429/344
6,882,130	B1 *	4/2005	Handa et al.	320/135

FOREIGN PATENT DOCUMENTS

WO WO 02/067344 8/2002

OTHER PUBLICATIONS

Peled et al., "The Electrochemical Behavior of Alkali and Alkaline Earth Metals in Nonaqueous Battery Systems", *J. Electrochem. Soc.*, vol. 126, pp. 2047-2051 (1979).

(Continued)

Primary Examiner—Pia Tibbits
(74) *Attorney, Agent, or Firm*—David E. Rogers; Squire, Sanders & Dempsey L.L.P.

(57) **ABSTRACT**

Disclosed is a method of charging a lithium-sulfur electrochemical cell wherein the lithium-sulfur cell comprises a cathode comprising an electroactive sulfur-containing material, an anode comprising lithium, and a nonaqueous electrolyte.

20 Claims, 5 Drawing Sheets

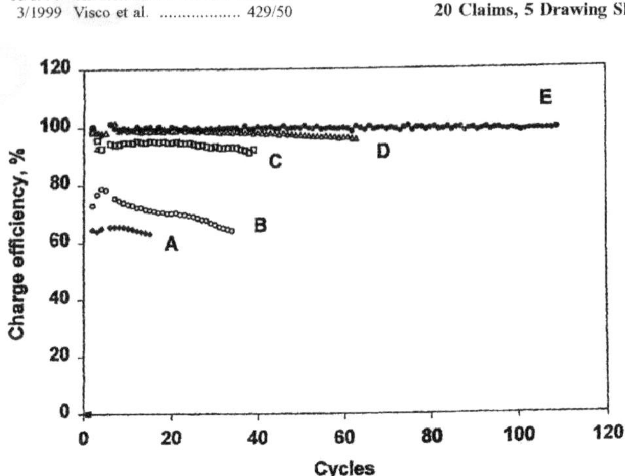

Arkansas

(12) **United States Patent**
Benson et al.

(10) **Patent No.: US 7,077,738 B2**
(45) **Date of Patent: Jul. 18, 2006**

(54) **APPARATUS FOR AND METHOD OF PROCESSING MEAT**

(75) Inventors:

(73) Assignee: **Tyson Fresh Meats, Inc.**, Springdale, AR (US)

(*) Notice: Subject to any disclaimer, the term of this patent is extended or adjusted under 35 U.S.C. 154(b) by 156 days.

(21) Appl. No.: **10/249,275**

(22) Filed: **Mar. 27, 2003**

(65) **Prior Publication Data**

US 2003/0186638 A1 Oct. 2, 2003

Related U.S. Application Data

(60) Provisional application No. 60/368,945, filed on Mar. 29, 2002.

(51) **Int. Cl.**
A22C 18/00 (2006.01)

(52) **U.S. Cl.** **452/149**

(58) **Field of Classification Search** 452/149, 452/173, 131, 123
See application file for complete search history.

(56) **References Cited**

U.S. PATENT DOCUMENTS

3,603,102 A * 9/1971 Banas 62/64
3,740,795 A * 6/1973 Cox 452/5
3,829,933 A * 8/1974 Lambert 452/19
4,020,528 A * 5/1977 Lindbladh et al. 452/131
4,193,373 A * 3/1980 Hanson et al. 118/17
4,217,679 A * 8/1980 Gordon 452/140
5,195,921 A * 3/1993 Ledet 452/4
5,902,177 A * 5/1999 Tessier et al. 452/156
6,019,033 A * 2/2000 Wilson et al. 99/470
6,383,068 B1 * 5/2002 Tollett et al. 452/170
6,601,499 B1 * 8/2003 Bifulco 100/73

OTHER PUBLICATIONS

Hobart Corporation, Instruction Manual Diagram for the Model ABR-1U Automatic Bone Dust Remover, p. 14.
Bettcher Industries, Inc. Foodservice Group, Automatic Batter-Breading System, website printout dated Nov. 16, 2001.

* cited by examiner

Primary Examiner—Thomas Price
(74) *Attorney, Agent, or Firm*—Blackwell Sanders Peper Martin LLP; Mark E. Stallion

(57) **ABSTRACT**

An apparatus for and method of cleaning cutting residue from one or both sides of cut meat. Meat, after being cut, for example with a band saw or the like, is passed through a curtain of fluid to impinge fluid on to at least one side of the meat for removing a substantial portion of the deleterious residue. The cleaned meat surface may then be passed through a drying area for removing cleaning fluid if the cleaning fluid contains liquid. After passing through a drying zone, brine or other ingredients may be applied to one or both sides of the cut meat.

18 Claims, 4 Drawing Sheets

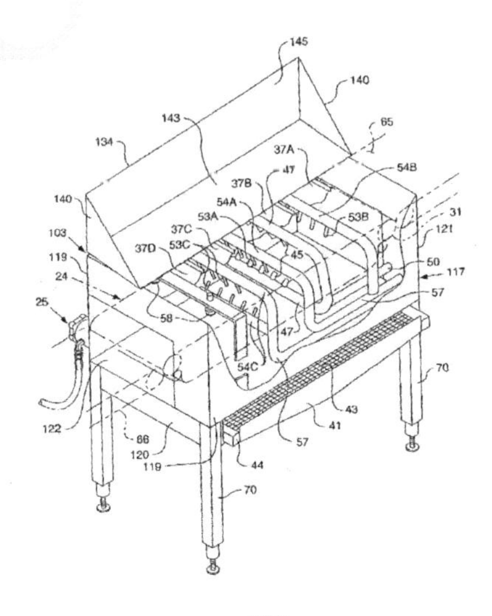

California

(12) **United States Patent**
Ellis et al.

(10) **Patent No.:** **US 7,025,734 B1**
(45) **Date of Patent:** **Apr. 11, 2006**

(54) **GUIDEWIRE WITH CHEMICAL SENSING CAPABILITIES**

(75) Inventors:

(73) Assignee: **Advanced Cardiovascular Systmes, Inc.**, Santa Clara, CA (US)

(*) Notice: Subject to any disclaimer, the term of this patent is extended or adjusted under 35 U.S.C. 154(b) by 0 days.

(21) Appl. No.: **09/967,186**

(22) Filed: **Sep. 28, 2001**

(51) **Int. Cl.**
A61B 5/05 (2006.01)
A61B 5/00 (2006.01)
A61M 25/00 (2006.01)

(52) **U.S. Cl.** **600/585**; 600/345

(58) **Field of Classification Search** 600/585, 600/433, 434, 435, 345; 604/164.13, 170.01, 604/171, 158

See application file for complete search history.

(56) **References Cited**

U.S. PATENT DOCUMENTS

4,801,538 A		1/1989	Hanada et al.	435/25
4,966,148 A		10/1990	Millar	128/637
5,124,130 A		6/1992	Costello et al.	422/82.06
5,176,882 A		1/1993	Gray et al.	422/82.07
5,345,932 A	*	9/1994	Yafuso et al.	600/368
5,434,085 A		7/1995	Capomacchia et al.	436/116
5,582,170 A	*	12/1996	Soller	600/322
5,603,820 A		2/1997	Malinski et al.	205/781
5,617,870 A		4/1997	Hastings et al.	128/692
5,776,100 A		7/1998	Forman	604/102
5,788,647 A		8/1998	Eggers	600/526
5,806,517 A		9/1998	Gerhardt et al.	128/635
5,852,058 A		12/1998	Cooke et al.	514/564
5,860,938 A		1/1999	Lafontaine et al.	600/585
5,885,842 A		3/1999	Lai	436/116
5,935,075 A		8/1999	Casscells et al.	600/474
5,945,452 A	*	8/1999	Cooke et al.	514/564
5,980,705 A		11/1999	Allen et al.	204/291
6,002,817 A		12/1999	Kopelman et al.	385/12
6,100,096 A		8/2000	Bollinger et al.	436/116
6,112,598 A	*	9/2000	Tenerz et al.	73/756
6,336,906 B1	*	1/2002	Hammarstrom et al.	600/585
6,498,941 B1	*	12/2002	Jackson	600/310
6,615,067 B1	*	9/2003	Hoek et al.	600/381
2002/0072680 A1	*	6/2002	Schock et al	600/486
2003/0013985 A1	*	1/2003	Saadat	600/549
2003/0028128 A1	*	2/2003	Tenerz	600/585

FOREIGN PATENT DOCUMENTS

WO WO 94/02845 * 3/1994

OTHER PUBLICATIONS

http://dictionary.reference.com/search?=wire.*
Bennett et al., *Conductive Polymeric Porphyrin Films: Application in the Electrocatalytic Oxidation of Hydrazine*, Chem. Mater. 1991, 3, pp. 490-495.

(Continued)

Primary Examiner—Max F. Hindenburg
Assistant Examiner—Jonathan Foreman
(74) *Attorney, Agent, or Firm*—Squire, Sanders & Dempsey L.L.P.

(57) **ABSTRACT**

A guidewire with a sensor which can detect NO and/or superoxide levels is disclosed. This invention can be useful for in vivo analysis of vascular health.

17 Claims, 7 Drawing Sheets

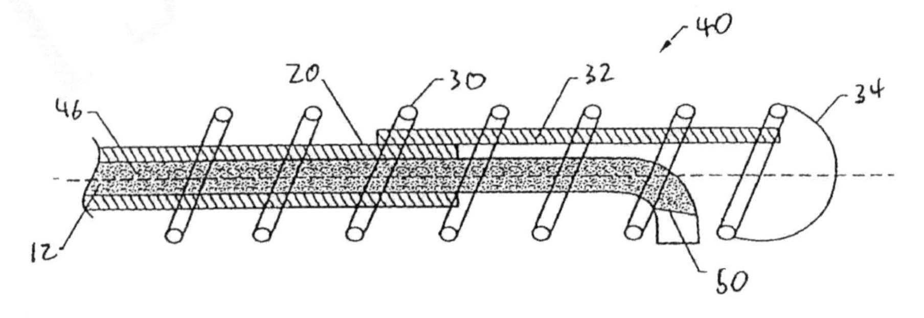

Colorado

(12) **United States Patent**
Gold et al.

(10) **Patent No.:** **US 7,311,357 B2**
(45) **Date of Patent:** **Dec. 25, 2007**

(54) **INFANT SURROUND SUPPORT**

(75) Inventors:

(73) Assignee: **Gold Bug, Inc.**, Aurora, CO (US)

(*) Notice: Subject to any disclaimer, the term of this patent is extended or adjusted under 35 U.S.C. 154(b) by 14 days.

(21) Appl. No.: **11/211,886**

(22) Filed: **Aug. 25, 2005**

(65) **Prior Publication Data**

US 2007/0138845 A1 Jun. 21, 2007

(51) **Int. Cl.**
A47C 31/10 (2006.01)

(52) **U.S. Cl.** **297/219.12**

(58) **Field of Classification Search** 297/250.1, 297/219.12, 230.13, 256.15
See application file for complete search history.

(56) **References Cited**

U.S. PATENT DOCUMENTS

4,383,713 A 5/1983 Roston
D300,694 S 4/1989 Krok
5,127,120 A 7/1992 Mason
D328,683 S 8/1992 Kalozdi
5,228,745 A * 7/1993 Hazel .... 297/219.12 X
5,310,245 A 5/1994 Lyszczasz
5,586,351 A 12/1996 Ive
D389,359 S 1/1998 Nowak
5,826,287 A 10/1998 Tandrup
5,829,830 A 11/1998 Maloney
5,842,739 A 12/1998 Noble
5,916,089 A 6/1999 Ive
5,918,933 A 7/1999 Hutchinson et al.
5,934,749 A 8/1999 Pond et al.
5,937,461 A 8/1999 Dombrowski et al.
6,036,263 A * 3/2000 Gold 297/219.12
6,341,818 B1 1/2002 Verbovszky et al.
6,363,558 B1 * 4/2002 Dunne .... 297/219.12 X
6,386,639 B1 * 5/2002 McMichael .... 297/219.12 X
6,454,352 B1 9/2002 Konovalov et al.
6,467,840 B1 * 10/2002 Verbovszky et al. ... 297/219.12
6,473,923 B1 11/2002 Straub
6,814,405 B2 * 11/2004 Norman 297/219.12
6,918,631 B2 * 7/2005 Verbovszky 297/219.12
6,926,359 B2 * 8/2005 Runk 297/219.12
6,966,089 B2 * 11/2005 Gold et al. .... 297/219.12 X
7,097,243 B2 * 8/2006 Verbovszky 297/219.12
2002/0014793 A1 2/2002 Santha

* cited by examiner

Primary Examiner—Rodney B. White
Assistant Examiner—Stephen Vu
(74) *Attorney, Agent, or Firm*—Ellen Reilly; John E. Reilly; The Reilly Intellectual Property Law Firm, P.C.

(57) **ABSTRACT**

An infant surround support having a head support member and opposite side support members and foot rest members forming a relatively continuous peripheral cushion support for an infant.

11 Claims, 3 Drawing Sheets

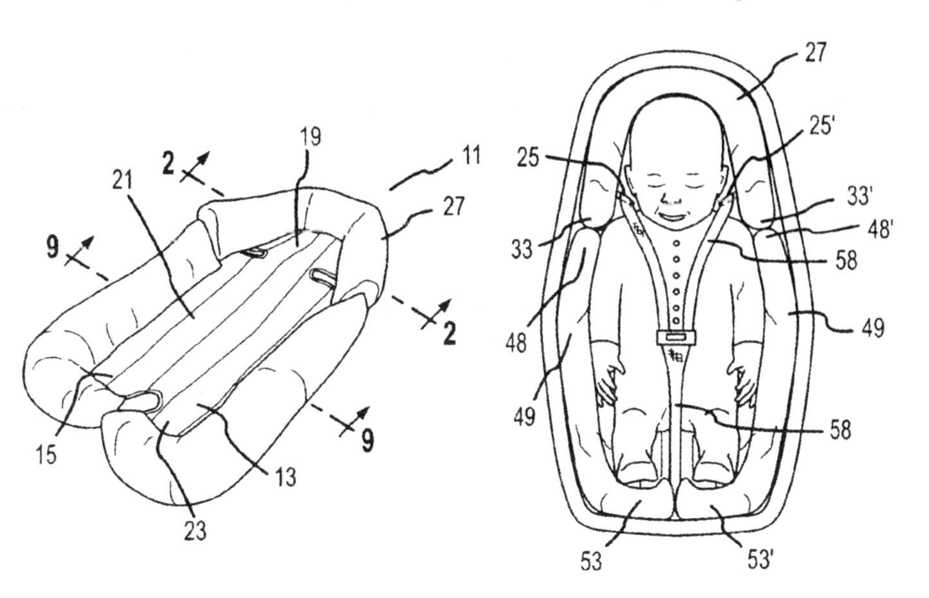

Connecticut

(12) **United States Patent**
Schlinker et al.

(10) **Patent No.:** **US 7,114,323 B2**
(45) **Date of Patent:** **Oct. 3, 2006**

(54) **JET EXHAUST NOISE REDUCTION SYSTEM AND METHOD**

(75) Inventors:

(73) Assignee: **United Technologies Corporation**, Hartford, CT (US)

(*) Notice: Subject to any disclaimer, the term of this patent is extended or adjusted under 35 U.S.C. 154(b) by 82 days.

(21) Appl. No.: **10/794,968**

(22) Filed: **Mar. 5, 2004**

(65) **Prior Publication Data**

US 2005/0193716 A1 Sep. 8, 2005

(51) **Int. Cl.**

B63H 11/00	(2006.01)
B64G 9/00	(2006.01)
F02K 9/00	(2006.01)
F03H 9/00	(2006.01)
F23R 9/00	(2006.01)

(52) **U.S. Cl.** **60/204**; 60/770; 60/262; 60/725; 239/265.19; 181/213

(58) **Field of Classification Search** 60/204, 60/228, 230, 232, 263, 770, 771, 725, 262, 60/264, 39.5, 226.1, 767; 239/265.19, 265.33, 239/265.39, 265.17, 265.13, 265.15, 265.25, 239/423; 181/213, 220, 206, 175, 196, 296, 181/400

See application file for complete search history.

(56) **References Cited**

U.S. PATENT DOCUMENTS

3,508,403	A *	4/1970	Neitzel	60/226.1
3,568,782	A	3/1971	Cox	
3,750,402	A *	8/1973	Vdoviak et al.	60/762
4,045,957	A *	9/1977	DiSabato	60/262
4,077,206	A *	3/1978	Ayyagari	60/262
4,117,671	A *	10/1978	Neal et al.	60/262
4,135,363	A *	1/1979	Packman	60/262
4,149,375	A *	4/1979	Wynosky et al.	60/262
4,175,384	A *	11/1979	Wagenknecht et al.	60/226.3
4,226,297	A *	10/1980	Cicon	181/213
4,302,934	A *	12/1981	Wynosky et al.	60/262

(Continued)

FOREIGN PATENT DOCUMENTS

GB 2207468 A * 2/1989

(Continued)

Primary Examiner—William H. Rodriguez
(74) *Attorney, Agent, or Firm*—Wall Marjama & Bilinski LLP

(57) **ABSTRACT**

A system for reducing jet noise emission from an internally mixed gas turbine engine exhaust, comprising a fan/core flow mixer having a plurality of mixer lobes and a common flow nozzle having an equal number of tabs located along a circumferential edge of an aft end of the nozzle. There is a predetermined clocking relationship between the plurality of mixer lobes and the plurality of nozzle tabs that results in reduced exhaust noise emission, most evident in the lower frequency range. A method for reducing jet noise emission from an internally mixed gas turbine engine exhaust comprises selectively aligning a circumferential distribution of a mixed flow vorticity field produced by a fan/core mixer with a circumferentially distributed exhaust flow vorticity field produced by a modified common flow nozzle at an exit plane of the engine exhaust.

26 Claims, 13 Drawing Sheets

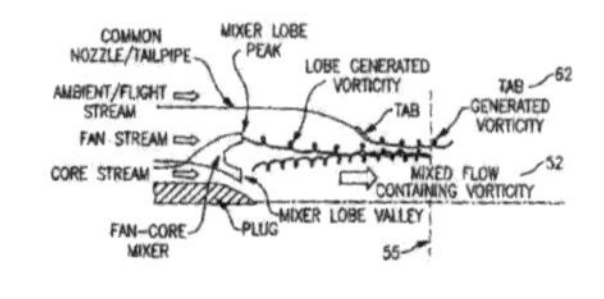

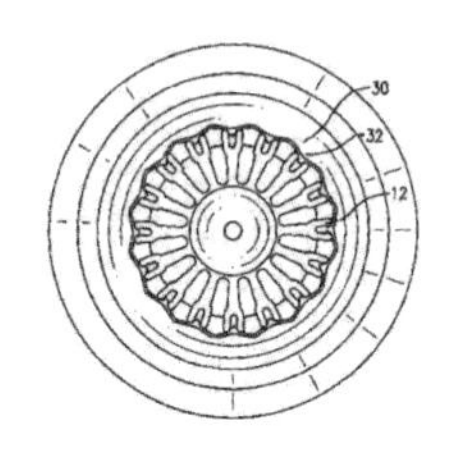

Delaware

US007138483B2

(12) **United States Patent**
Wang et al.

(10) **Patent No.: US 7,138,483 B2**
(45) **Date of Patent: Nov. 21, 2006**

(54) **MONOMERS, CONJUGATED POLYMERS AND ELECTRONIC DEVICES USING SUCH POLYMERS**

(75) Inventors:

(73) Assignees: **E.I. du Pont de Nemours and Company**, Wilmington, DE (US); **DuPont Displays, Inc.**, Santa Barbara, CA (US)

(*) Notice: Subject to any disclaimer, the term of this patent is extended or adjusted under 35 U.S.C. 154(b) by 0 days.

(21) Appl. No.: **10/771,014**

(22) Filed: **Feb. 3, 2004**

(65) **Prior Publication Data**

US 2004/0192871 A1 Sep. 30, 2004

Related U.S. Application Data

(60) Provisional application No. 60/446,823, filed on Feb. 12, 2003.

(51) **Int. Cl.**
C08G 69/00 (2006.01)
C08G 75/00 (2006.01)
C08G 73/00 (2006.01)
C07D 291/00 (2006.01)
C07D 277/00 (2006.01)

(52) **U.S. Cl.** **528/327**; 528/373; 528/377; 528/380; 528/394; 528/396; 528/397; 528/401; 528/422; 528/423; 528/341; 528/345; 528/347; 528/364; 528/424; 548/100; 548/122; 548/126; 548/146; 548/147

(58) **Field of Classification Search** 528/327, 528/397, 394
See application file for complete search history.

(56) **References Cited**

U.S. PATENT DOCUMENTS

4,486,576	A	12/1984	Colon et al.
4,508,639	A	4/1985	Camps et al.
5,597,890	A	1/1997	Jenekhe
5,621,131	A	4/1997	Kreuder et al.
5,708,130	A	1/1998	Woo et al.
5,777,070	A	7/1998	Inbasekaran et al.
5,814,244	A	9/1998	Kreuder et al.
5,821,002	A	10/1998	Ohnishi et al.
5,856,434	A	1/1999	Stern et al.
5,900,327	A	5/1999	Pei et al.
5,962,631	A	10/1999	Woo et al.
5,998,045	A	12/1999	Chen et al.
6,107,452	A	8/2000	Miller et al.
6,124,046	A	9/2000	Jin et al.
6,169,163	B1	1/2001	Woo et al.
6,204,515	B1	3/2001	Bernius et al.
6,255,449	B1	7/2001	Woo et al.
6,309,763	B1	10/2001	Woo et al.
6,353,083	B1	3/2002	Inbasekaran et al.
6,355,756	B1	3/2002	Hawker et al.
6,541,602	B1	4/2003	Spreitzer et al.
6,605,373	B1	8/2003	Woo et al.
6,653,438	B1	11/2003	Spreitzer et al.
2001/0024738	A1	9/2001	Hawker et al.
2002/0013451	A1	1/2002	Huang et al.
2002/0028347	A1	3/2002	Marrocco, III et al.
2002/0045719	A1	4/2002	Hawker et al.
2002/0051895	A1	5/2002	Cho et al.
2002/0103332	A1	8/2002	Leclerc et al.
2004/0192871	A1*	9/2004	Wang et al. 528/4
2004/0204557	A1*	10/2004	Uckert et al. 528/4

FOREIGN PATENT DOCUMENTS

EP	0 259 229	B1	9/1987
EP	0 956 312	B1	1/1998
EP	1 213 336	A2	6/2002
EP	1 281 745	A1	2/2003
EP	1 205 526	B1	9/2004
WO	WO 99/54943	A1	10/1999
WO	WO 00/53656	A1	9/2000
WO	WO 00/55927	A1	9/2000
WO	WO 01/07502	A2	2/2001
WO	WO 01/77203	A2	10/2001
WO	WO 02/090415	A1	11/2002
WO	WO 02/092724	A1	11/2002
WO	WO 03/050086	A1	6/2003
WO	WO 2004/015025	A1	2/2004

OTHER PUBLICATIONS

Macromolecules 2002, 35, pp. 3474-3483, Jianfu Ding et al "Synthesis and Characterization of Alternating Copolymers of Fluorene and Oxadiazole".*

(Continued)

Primary Examiner—P. Hampton Hightower
(74) *Attorney, Agent, or Firm*—John H. Lamming

(57) **ABSTRACT**

The energy levels (HOMO, LUMO) of the conjugated polymer are tuned independently, so that an energy match on both sides of the device can be accomplished while keeping the emission color in the blue region. Such polymers can be formed by polymerization of a mixture of monomers. The mixture of the monomers contains at least one monomer having an electron-deficient group sandwiched by two aromatic hydrocarbon groups ("Monomer (I)") and at least one hole transporting ("HT") monomer. The mixture of monomers may also contain a solubility enhancement ("SE") monomer and/or a Branching Monomer. Such polymers can be used in fabricating light emitting diodes to achieve some of the best device performance to date including high efficiency and blue color purity.

7 Claims, 9 Drawing Sheets

Florida

(12) **United States Patent**
Lynch

(10) **Patent No.:** **US 7,311,645 B1**
(45) **Date of Patent:** **Dec. 25, 2007**

(54) **ABDOMINAL EXERCISE MACHINE**

(75) Inventor:

(73) Assignee: **AB-Stractor LLC**, Boca Raton, FL (US)

(*) Notice: Subject to any disclaimer, the term of this patent is extended or adjusted under 35 U.S.C. 154(b) by 0 days.

(21) Appl. No.: **11/540,329**

(22) Filed: **Sep. 29, 2006**

(51) **Int. Cl.**
A63B 26/00 (2006.01)

(52) **U.S. Cl.** **482/142**; 482/121

(58) **Field of Classification Search** 482/121–131, 482/142; D21/676, 686, 690
See application file for complete search history.

(56) **References Cited**

U.S. PATENT DOCUMENTS

4,517,966 A 5/1985 von Othegraven
D329,672 S 9/1992 Shieh
5,387,171 A 2/1995 Casey et al.
5,423,731 A * 6/1995 Chen 482/130
6,248,047 B1 6/2001 Abdo
6,309,329 B2 10/2001 Conner
6,602,171 B1 8/2003 Tsen et al.
2004/0043877 A1 3/2004 Brown
2005/0101460 A1 5/2005 Lobban
2006/0160683 A1 * 7/2006 Schenk 482/907

* cited by examiner

Primary Examiner—Lori Amerson
(74) *Attorney, Agent, or Firm*—Laurence A. Greenberg; Werner H. Stemer; Ralph E. Locher

(57) **ABSTRACT**

An exercise machine has a main frame for placement on a support surface and a seat supported on the main frame. The seat is movable between a horizontal position and a vertical position. A post having an integrated resilient member is supported by the main frame, the integrated resilient member is disposed at least one foot above the seat. A handle bar is attached to and extends from the post. A downward movement of the handle bar is resisted by the integrated resilient member.

19 Claims, 5 Drawing Sheets

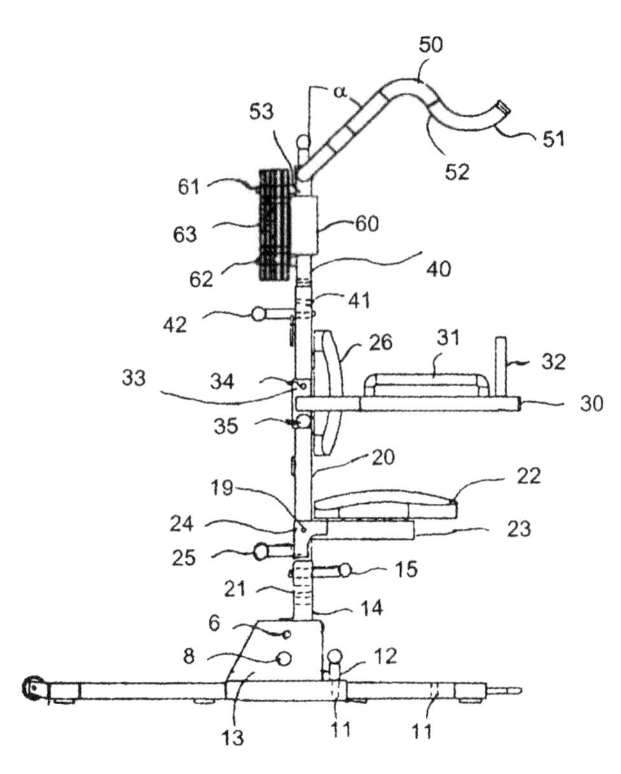

Georgia

US007028088B1

(12) **United States Patent**
Koperda et al.

(10) **Patent No.: US 7,028,088 B1**
(45) **Date of Patent: Apr. 11, 2006**

(54) SYSTEM AND METHOD FOR PROVIDING STATISTICS FOR FLEXIBLE BILLING IN A CABLE ENVIRONMENT

(75) Inventors:

(73) Assignee: **Scientific-Atlanta, Inc.**, Lawrenceville, GA (US)

(*) Notice: Subject to any disclaimer, the term of this patent is extended or adjusted under 35 U.S.C. 154(b) by 318 days.

(21) Appl. No.: **09/588,211**

(22) Filed: **Jun. 6, 2000**

Related U.S. Application Data

(63) Continuation of application No. 08/818,037, filed on Mar. 14, 1997, now Pat. No. 6,230,203, which is a continuation-in-part of application No. 08/627,062, filed on Apr. 3, 1996, now Pat. No. 5,790,806, and a continuation-in-part of application No. 08/732,668, filed on Oct. 16, 1996, now Pat. No. 5,996,163.

(51) **Int. Cl.**
G06F 15/16 (2006.01)

(52) **U.S. Cl.** **709/229**; 709/224

(58) **Field of Classification Search** 709/229, 709/224, 217, 219, 227, 228; 725/1, 2, 4, 725/5, 8

See application file for complete search history.

(56) **References Cited**

U.S. PATENT DOCUMENTS

3,733,430 A * 5/1973 Thompson et al. 725/1
3,985,962 A 10/1976 Jones et al.
4,186,380 A 1/1980 Edwin et al.
4,207,431 A 6/1980 McVoy
4,528,589 A 7/1985 Block et al.
4,601,028 A 7/1986 Huffman et al.
4,633,462 A 12/1986 Stifle et al.
4,641,304 A 2/1987 Raychaudhuri
4,757,460 A 7/1988 Bione et al.
4,858,224 A 8/1989 Nakano et al.
4,907,224 A 3/1990 Scoles et al.
4,912,721 A 3/1990 Pidgeon, Jr. et al.
5,012,469 A 4/1991 Sardana
5,014,125 A 5/1991 Pocock et al.
5,047,928 A 9/1991 Wiedemer
5,131,041 A 7/1992 Brunner et al.
5,136,690 A 8/1992 Becker et al.
5,157,657 A 10/1992 Potter et al.
5,159,592 A 10/1992 Perkins
5,166,931 A 11/1992 Riddle

(Continued)

OTHER PUBLICATIONS

PPP Bridging Control Protocol (BCP); F. Baker et al.; Network Working Group Request for Comments, Jun. 1994; pps. 1-28.

(Continued)

Primary Examiner—Jason D Cardone

(57) **ABSTRACT**

A system for providing flexible billing in a cable environment can establish billing practices based on tier level of service, quality of service or the amount of network resources consumed. A plurality of tiers or levels of service can be defined by parameters including at least a maximum or peak bit rate or bandwidth for providing services over a shared channel. A plurality of levels or tiers of services are defined by maximum bandwidth or bit rate and a subscriber receives service at that subscribe-to level or at a slower data rate depending on availability of shared bandwidth. Quality of service is maintaining the specified bandwidth, jitter or delay. The amount of network resources consumed may be expressed in terms of the amount of data transmitted or the connect time of a network access device to the network.

20 Claims, 14 Drawing Sheets

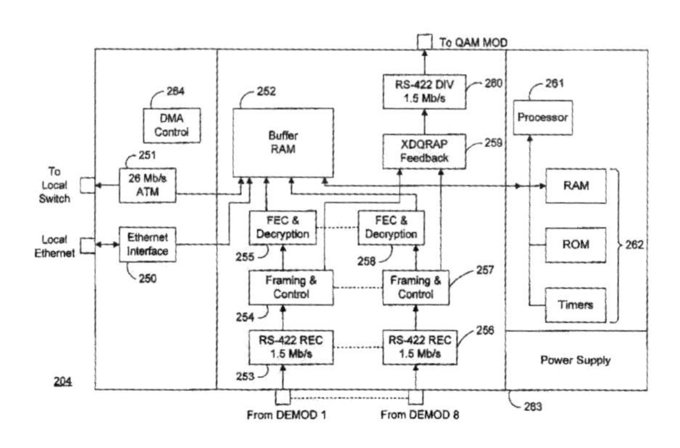

Hawaii

(12) **United States Patent**
Loui et al.

(10) **Patent No.: US 7,311,059 B2**
(45) **Date of Patent: Dec. 25, 2007**

(54) **WATERCRAFT HULL WITH ENTRAPMENT TUNNEL**

(75) Inventors:

(73) Assignee: **Navatek, Ltd.**, Honolulu, HI (US)

(*) Notice: Subject to any disclaimer, the term of this patent is extended or adjusted under 35 U.S.C. 154(b) by 17 days.

(21) Appl. No.: **11/315,304**

(22) Filed: **Dec. 23, 2005**

(65) **Prior Publication Data**

US 2006/0137592 A1 Jun. 29, 2006

Related U.S. Application Data

(60) Provisional application No. 60/639,856, filed on Dec. 27, 2004.

(51) **Int. Cl.**
B63B 1/32 (2006.01)

(52) **U.S. Cl.** **114/288**; 114/289

(58) **Field of Classification Search** 114/288, 114/289

See application file for complete search history.

(56) **References Cited**

U.S. PATENT DOCUMENTS

1,265,035 A 5/1918 Bazaine
1,665,149 A 4/1928 Van Wienen
2,753,135 A 7/1956 Gouge 244/106
3,146,752 A 9/1964 Ford 114/67
3,528,380 A 9/1970 Yost 114/66.5
3,561,389 A 2/1971 Hunt 114/66.5
3,662,700 A 5/1972 Roumejon 114/67
4,067,286 A 1/1978 Stout et al. 114/283
4,371,350 A 2/1983 Kruppa et al. 440/69
4,649,847 A 3/1987 Tinkler et al. 114/39
4,682,560 A 7/1987 Lieb et al. 114/343
4,689,026 A 8/1987 Small 440/66
4,713,028 A 12/1987 Duff 440/61
4,821,663 A 4/1989 Schad 114/43
4,915,668 A 4/1990 Hardy 440/69
4,924,792 A 5/1990 Sapp et al. 114/61
4,926,771 A * 5/1990 Hull 114/289
4,977,845 A 12/1990 Rundquist 114/289
5,111,767 A 5/1992 Haines 114/288
5,415,120 A 5/1995 Burg 114/67
5,588,389 A 12/1996 Carter, Jr. 114/271
6,006,689 A 12/1999 Olofsson 114/285
6,125,781 A 10/2000 White 114/288
6,406,341 B1 6/2002 Morejohn 440/69
6,425,341 B1 7/2002 Devin 114/288
6,604,478 B2 8/2003 Barsumian 114/67

* cited by examiner

Primary Examiner—Stephen Avila
(74) *Attorney, Agent, or Firm*—Fitzpatrick, Cella, Harper & Scinto

(57) **ABSTRACT**

A watercraft hull has at least one bent entrapment tunnel with an aft portion whose tunnel ceiling is at a higher elevation than portions of the tunnel ceiling immediately in front of it to provide a negative angle of attack with the free surface at design planing speeds.

37 Claims, 7 Drawing Sheets

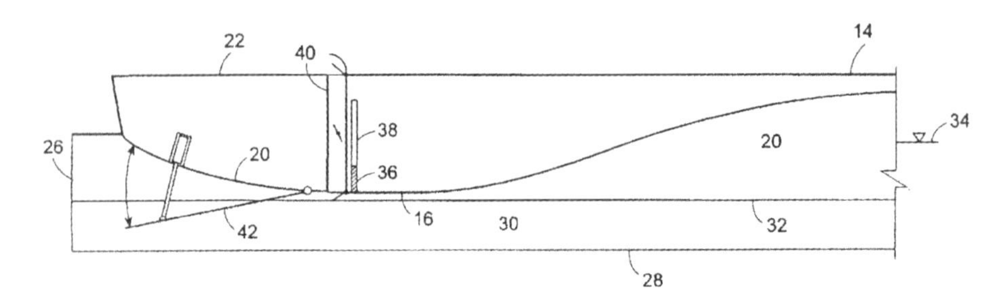

Idaho

US006983354B2

(12) **United States Patent**
Jeddeloh

(10) **Patent No.: US 6,983,354 B2**
(45) **Date of Patent: Jan. 3, 2006**

(54) **MEMORY DEVICE SEQUENCER AND METHOD SUPPORTING MULTIPLE MEMORY DEVICE CLOCK SPEEDS**

(75) Inventor:

(73) Assignee: **Micron Technology, Inc.**, Boise, ID (US)

(*) Notice: Subject to any disclaimer, the term of this patent is extended or adjusted under 35 U.S.C. 154(b) by 238 days.

(21) Appl. No.: **10/155,668**

(22) Filed: **May 24, 2002**

(65) **Prior Publication Data**

US 2003/0221078 A1 Nov. 27, 2003

(51) **Int. Cl.**
G06F 12/00 (2006.01)
G06F 13/00 (2006.01)

(52) **U.S. Cl.** **711/167**; 711/169; 713/400; 713/500; 713/501; 713/600

(58) **Field of Classification Search** 711/167; 713/500, 501, 400, 600
See application file for complete search history.

(56) **References Cited**

U.S. PATENT DOCUMENTS

4,245,303 A 1/1981 Durvasula et al. 364/200
5,155,809 A 10/1992 Baker et al. 395/200
5,448,715 A * 9/1995 Lelm et al. 713/600
5,471,587 A * 11/1995 Fernando 710/307
5,487,092 A 1/1996 Finney et al. 375/354
5,600,824 A * 2/1997 Williams et al. 713/400
5,915,107 A 6/1999 Maley et al. 395/551
5,923,193 A 7/1999 Bloch et al. 327/141
5,923,858 A 7/1999 Kanekal 395/287
6,000,022 A 12/1999 Manning 711/167
6,006,340 A 12/1999 O'Connell 713/600
6,016,549 A 1/2000 Matsushiba et al. 713/324
6,112,307 A 8/2000 Ajanovic et al. 713/400
6,128,749 A 10/2000 McDonnell et al. 713/600
6,134,638 A 10/2000 Olarig et al. 711/167
6,202,119 B1 3/2001 Manning 711/5
6,279,077 B1 8/2001 Nasserbakht et al. 711/118
6,279,090 B1 8/2001 Manning 711/167
6,363,076 B1 3/2002 Allison et al. 370/419
6,370,600 B1 4/2002 Hughes et al. 710/29
6,389,529 B1 * 5/2002 Arimilli et al. 712/225
6,414,903 B1 7/2002 Keeth et al. 365/233

(Continued)

OTHER PUBLICATIONS

Microsoft Dictionary, 1999, Microsoft Press, Fourth Edition, pp. 159 and 399.*

Primary Examiner—Mano Padmanabhan
Assistant Examiner—Midys Rojas
(74) *Attorney, Agent, or Firm*—Dorsey & Whitney, LLP.

(57) **ABSTRACT**

A sequence state matrix has a plurality of time slots for storing a plurality of memory device signals. The memory device signals are loaded into the matrix by a sequencer load unit, which loads the memory device signals at locations in the matrix corresponding to the times that the signals will be coupled to a memory device. The sequencer load unit loads the signals into the matrix at a rate corresponding to a frequency of a system clock signal controlling the operation of the electronic system. A first in, first out ("FIFO") buffer receives the memory device signals from the sequence state matrix at a rate corresponding to the frequency of the system clock signal. A command selector transfers the memory device signals from the FIFO buffer to the memory device at a rate corresponding to the frequency of a memory clock signal controlling the operation of the memory device.

35 Claims, 4 Drawing Sheets

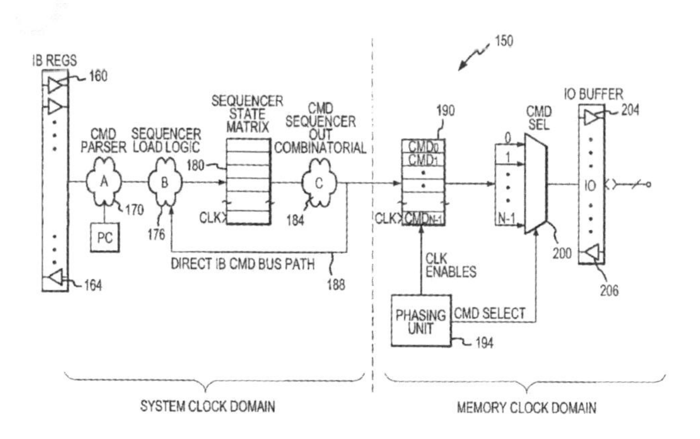

Illinois

(12) **United States Patent**
Shafer et al.

(10) **Patent No.: US 7,272,532 B2**
(45) **Date of Patent: *Sep. 18, 2007**

(54) **METHOD FOR PREDICTING THE QUALITY OF A PRODUCT**

(75) Inventors:

(73) Assignee: **Caterpillar Inc.**, Peoria, IL (US)

(*) Notice: Subject to any disclaimer, the term of this patent is extended or adjusted under 35 U.S.C. 154(b) by 402 days.

This patent is subject to a terminal disclaimer.

(21) Appl. No.: **10/952,309**

(22) Filed: **Sep. 28, 2004**

(65) **Prior Publication Data**

US 2005/0065752 A1 Mar. 24, 2005

Related U.S. Application Data

(63) Continuation of application No. 10/320,802, filed on Dec. 17, 2002, now Pat. No. 6,823,287.

(51) **Int. Cl.**
G06F 11/00 (2006.01)

(52) **U.S. Cl.** **702/183**; 702/80; 702/84; 700/108; 700/109; 700/110; 714/37

(58) **Field of Classification Search** 702/81-84, 702/182, 183; 700/75, 108, 109, 110; 714/37
See application file for complete search history.

(56) **References Cited**

U.S. PATENT DOCUMENTS

6,600,961 B2 * 7/2003 Liang et al. 700/48
6,823,287 B2 * 11/2004 Shafer et al. 702/183
7,130,748 B2 * 10/2006 Ueda et al. 702/42

OTHER PUBLICATIONS

Ho et al. 'Neural Network Modeling with Confidence Bounds: A Case Study on the Solder Paste Deposition Process', Oct. 2001, IEEE Publicaiton, vol. 24, No. 4, pp. 323332.*

Chi et al., 'A Fuzzy Basis Function Neural Network for Predicting Multiple Quality Characterstics of Plasma Arc Welding', 2001, IEEE Publication, pp. 2807-2812.*

* cited by examiner

Primary Examiner—Elisio Ramos-Feliciano
Assistant Examiner—Elias Desta
(74) *Attorney, Agent, or Firm*—W. Bryan McPherson; Finegan, Henderson, Farabow, Garrett & Dunne

(57) **ABSTRACT**

A method is provided for predicting a quality characteristic of a product to be manufactured. The method may integrate one or more of feature and tolerance information associated with the product, manufacturing characteristic information associated with the manufacture of the product, measurement capability characteristic information associated with the manufacture of the product, assembly characteristic information associated with an assembly of the product, and desired quality characteristic information associated with the product. Based on the integrated information, the quality characteristic of the product may be predicted.

23 Claims, 10 Drawing Sheets

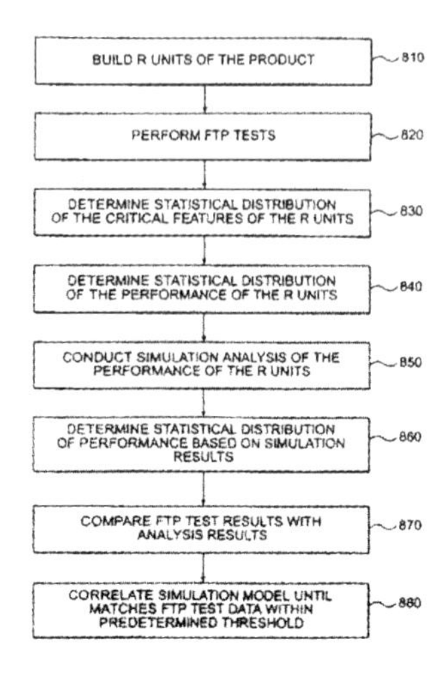

Indiana

(12) **United States Patent**
Gallops, Jr.

(10) **Patent No.: US 7,311,098 B2**
(45) **Date of Patent: *Dec. 25, 2007**

(54) **ZERO CENTER OF MASS ARCHERY CAM**

(75) Inventor:

(73) Assignee: **Bear Archery, Inc.**, Evansville, IN (US)

(*) Notice: Subject to any disclaimer, the term of this patent is extended or adjusted under 35 U.S.C. 154(b) by 206 days.

This patent is subject to a terminal disclaimer.

(21) Appl. No.: **11/221,117**

(22) Filed: **Sep. 7, 2005**

(65) **Prior Publication Data**

US 2006/0000463 A1 Jan. 5, 2006

Related U.S. Application Data

(63) Continuation of application No. 10/971,272, filed on Oct. 22, 2004, now Pat. No. 6,976,484.

(60) Provisional application No. 60/585,764, filed on Jul. 6, 2004, provisional application No. 60/576,664, filed on Jun. 3, 2004.

(51) **Int. Cl.**
F41B 5/10 (2006.01)

(52) **U.S. Cl.** **124/25.6**; 124/900

(58) **Field of Classification Search** 124/25.6, 124/900

See application file for complete search history.

(56) **References Cited**

U.S. PATENT DOCUMENTS

5,368,006 A 11/1994 McPherson
5,749,351 A 5/1998 Allshouse et al.
5,809,982 A 9/1998 McPherson
5,901,692 A 5/1999 Allshouse et al.
5,921,227 A 7/1999 Allshouse et al.
5,996,567 A 12/1999 McPherson
6,035,841 A 3/2000 Martin et al.
6,039,035 A 3/2000 McPherson
6,105,565 A 8/2000 Martin et al.
6,257,219 B1 7/2001 McPherson
6,267,108 B1 7/2001 McPherson et al.
6,382,201 B1 5/2002 McPherson et al.
6,976,484 B1 * 12/2005 Gallops, Jr. 124/25.6

* cited by examiner

Primary Examiner—John A. Ricci
(74) *Attorney, Agent, or Firm*—Woodard, Emhardt, Moriarty, McNett & Henry LLP

(57) **ABSTRACT**

One preferred embodiment of the present invention provides a cam having an axle location for mounting the cam to an archery bow, where the center of mass of the cam is substantially coaxial with the axle location. Preferable the cam has an eccentric geometric rotation profile with regard to a rotation axis, typically an irregular geometry with a non-centered axle location, or a circular profile with an axle location offset from the center of the circular profile. The mass of the cam is balanced to have an effectively equal mass distribution around the axle location. In an alternate preferred embodiment, the cam has a balanced center of mass aligned with the axle location in an X-Y orientation, and may also have a balanced center of mass through the thickness of the cam in an X-Z or Y-Z orientation.

24 Claims, 13 Drawing Sheets

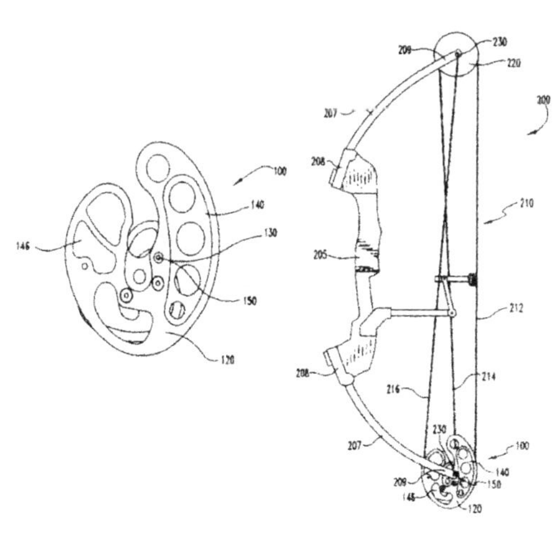

Iowa

US007109399B2

(12) **United States Patent**
Eby et al.

(10) **Patent No.: US 7,109,399 B2**
(45) **Date of Patent: Sep. 19, 2006**

(54) SOYBEAN CULTIVAR 0491735

(75) Inventors:

(73) Assignees: **Stine Seed Farm, Inc.**, Adel, IA (US); **Monsanto Technology LLC**, St. Louis, MO (US)

(*) Notice: Subject to any disclaimer, the term of this patent is extended or adjusted under 35 U.S.C. 154(b) by 400 days.

(21) Appl. No.: **10/625,055**

(22) Filed: **Jul. 22, 2003**

(65) **Prior Publication Data**

US 2005/0022272 A1 Jan. 27, 2005

(51) **Int. Cl.**
A01H 1/00 (2006.01)
A01H 4/00 (2006.01)
A01H 5/00 (2006.01)
A01H 5/10 (2006.01)
C12N 15/82 (2006.01)

(52) **U.S. Cl.** **800/312**; 800/260; 800/263; 800/264; 800/265; 800/279; 800/281; 800/284; 800/286; 800/300; 800/301; 800/302; 435/415; 435/426; 435/430

(58) **Field of Classification Search** 800/260, 800/312, 263, 264, 265, 279, 281, 284, 286, 800/300, 301, 302; 435/415, 426, 430
See application file for complete search history.

(56) **References Cited**

U.S. PATENT DOCUMENTS

5,304,719 A 4/1994 Segebart
5,367,109 A 11/1994 Segebart
5,523,520 A 6/1996 Hunsperger et al.
5,850,009 A 12/1998 Kevern
5,866,771 A * 2/1999 Holmes 800/260
5,968,830 A 10/1999 Dan et al.

OTHER PUBLICATIONS

Eshed, et al., 1996. Less-than-additive epistatic interactions of quantitative trait loci in tomato. Genetics 143:1807-1817.

Kraft, et al., 2000. Linkage disequilibrium and fingerprinting in sugar beet. Theor. App. Genet. 101:323-326.

Willmot, et al., 1989. Genetic analysis of brown stem rot resistance in soybean. Crop Sci. 29:672-674.

Poehlman, J.M. and Sleper, D.A., Methods in Plant Breeding, *In* Breeding Field Crops, 4th ed. (1995), Iowa State University Press, pp. 172-174.

* cited by examiner

Primary Examiner—Cynthia Collins
(74) *Attorney, Agent, or Firm*—Jondle & Associates P.C.

(57) **ABSTRACT**

A novel soybean cultivar, designated 0491735, is disclosed. The invention relates to the seeds of soybean cultivar 0491735, to the plants of soybean 0491735 and to methods for producing a soybean plant produced by crossing the cultivar 0491735 with itself or another soybean variety. The invention further relates to hybrid soybean seeds and plants produced by crossing the cultivar 0491735 with another soybean cultivar.

22 Claims, No Drawings

Kansas

US007058689B2

(12) **United States Patent**
Parker et al.

(10) **Patent No.: US 7,058,689 B2**
(45) **Date of Patent: Jun. 6, 2006**

(54) **SHARING OF STILL IMAGES WITHIN A VIDEO TELEPHONY CALL**

(75) Inventors:

(73) Assignee: **Sprint Communications Company L.P.**, Overland Park, KS (US)

(*) Notice: Subject to any disclaimer, the term of this patent is extended or adjusted under 35 U.S.C. 154(b) by 718 days.

(21) Appl. No.: **10/085,859**

(22) Filed: **Feb. 28, 2002**

(65) **Prior Publication Data**

US 2003/0074404 A1 Apr. 17, 2003

Related U.S. Application Data

(63) Continuation-in-part of application No. 10/033,813, filed on Oct. 20, 2001, which is a continuation-in-part of application No. 09/978,616, filed on Oct. 16, 2001, now Pat. No. 6,545,697.

(51) **Int. Cl.**
G06F 15/16 (2006.01)

(52) **U.S. Cl.** **709/206**; 709/204

(58) **Field of Classification Search** 709/204–206, 709/231

See application file for complete search history.

(56) **References Cited**

U.S. PATENT DOCUMENTS

5,689,553 A 11/1997 Ahuja et al.
5,764,916 A 6/1998 Busey et al.
5,872,923 A * 2/1999 Schwartz et al. 709/205
5,896,500 A * 4/1999 Ludwig et al. 709/204
5,915,091 A * 6/1999 Ludwig et al. 709/204
5,949,763 A 9/1999 Lund
5,978,835 A * 11/1999 Ludwig et al. 709/204
6,097,793 A 8/2000 Jandel
6,212,547 B1 * 4/2001 Ludwig et al. 709/204
6,237,025 B1 * 5/2001 Ludwig et al. 709/204
6,301,607 B1 * 10/2001 Barraclough et al. 709/204
6,337,858 B1 1/2002 Petty et al.
6,343,314 B1 * 1/2002 Ludwig et al. 709/204
6,351,762 B1 * 2/2002 Ludwig et al. 709/204
6,353,610 B1 * 3/2002 Bhattacharya et al. 370/352
6,370,137 B1 4/2002 Lund
6,381,220 B1 * 4/2002 Kung et al. 370/250
6,425,131 B1 * 7/2002 Crandall et al. 725/106
6,539,077 B1 3/2003 Ranalli et al.
6,674,745 B1 * 1/2004 Schuster et al. 370/352
6,677,979 B1 * 1/2004 Westfield 348/14.12
6,704,294 B1 3/2004 Cruickshank
6,714,536 B1 * 3/2004 Dowling 370/356
6,798,767 B1 * 9/2004 Alexander et al. 370/352

(Continued)

FOREIGN PATENT DOCUMENTS

EP 0 721266 7/1996

(Continued)

Primary Examiner—David Wiley
Assistant Examiner—Yemane M. Gerezgiher

(57) **ABSTRACT**

A data call between at least two internetworked computers is established using a central server and call clients in the two computers. Once the data call is in place, a video telephony conversation is initiated. In addition, still images such as a slideshow of personal photographs are shared between the two computers while the video telephony call fully continues. Image viewer subclients utilize the same network session as is used by the call clients. Either user can pause, navigate through, or change the display a parameters of the slideshow as it is being viewed by both users.

8 Claims, 7 Drawing Sheets

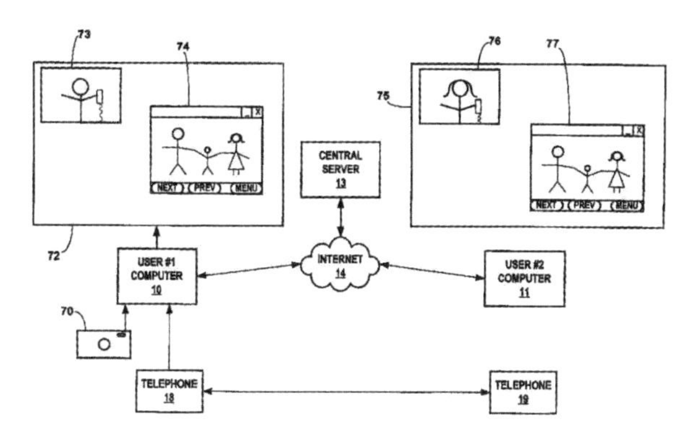

Kentucky

US007116452B2

(12) **United States Patent**
Han et al.

(10) **Patent No.:** **US 7,116,452 B2**
(45) **Date of Patent:** **Oct. 3, 2006**

(54) **METHOD OF CALIBRATING A SCANNER TO A PRINTER**

(75) Inventors:

(73) Assignee: **Lexmark International, Inc.**, Lexington, KY (US)

(*) Notice: Subject to any disclaimer, the term of this patent is extended or adjusted under 35 U.S.C. 154(b) by 969 days.

(21) Appl. No.: **10/122,748**

(22) Filed: **Apr. 15, 2002**

(65) **Prior Publication Data**

US 2003/0193678 A1 Oct. 16, 2003

(51) **Int. Cl.**
H04N 1/46 (2006.01)
G06K 1/00 (2006.01)

(52) **U.S. Cl.** **358/530**; 358/504; 358/1.9

(58) **Field of Classification Search** 382/162, 382/167; 358/500, 504, 515, 518, 520, 525, 358/530, 2.1, 1
See application file for complete search history.

(56) **References Cited**

U.S. PATENT DOCUMENTS

4,500,919 A 2/1985 Schreiber
5,331,439 A 7/1994 Bachar
5,420,704 A 5/1995 Winkelman
5,483,360 A 1/1996 Rolleston et al.
5,579,031 A * 11/1996 Liang 345/604
5,581,376 A 12/1996 Harrington
5,774,238 A 6/1998 Tsukada
5,787,193 A 7/1998 Balasubramanian
5,987,168 A 11/1999 Decker et al.
6,061,501 A 5/2000 Decker et al.
6,072,901 A 6/2000 Balonon-Rosen et al.
6,104,829 A 8/2000 Nakajima
6,137,594 A 10/2000 Decker et al.
6,137,596 A 10/2000 Decker et al.
6,151,135 A 11/2000 Tanaka et al.
6,297,826 B1 10/2001 Semba et al.
6,301,025 B1 10/2001 DeLean
6,327,052 B1 12/2001 Falk
6,765,691 B1 * 7/2004 Kubo et al. 358/1.9
2003/0006951 A1 * 1/2003 Ito 345/89
2003/0142377 A1 * 7/2003 Yamada et al. 358/521
2003/0193678 A1 * 10/2003 Han et al. 358/1.9

FOREIGN PATENT DOCUMENTS

GB 2213674 * 8/1989

* cited by examiner

Primary Examiner—Jerome Grant, II
(74) *Attorney, Agent, or Firm*—John Victor Pezdak

(57) **ABSTRACT**

An electronic representation of the range of color combinations that can be displayed on the monitor is created and printed. The printed representation is then scanned by the scanner. A relationship is then identified between the monitor and the scanner. This relationship is then used as the basis to determine a relationship between the scanner and the monitor. Compensation is made for errors and the relationship is further developed. Finally, a scanner RGB to printer CMYK lookup table is built.

9 Claims, 4 Drawing Sheets

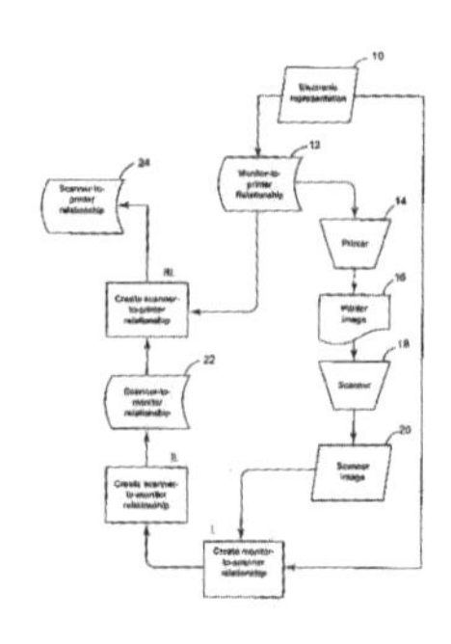

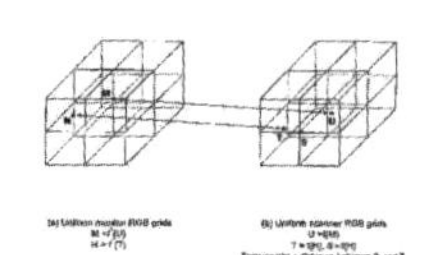

Louisiana

(12) **United States Patent**
Burch

(10) **Patent No.: US 7,007,792 B1**
(45) **Date of Patent: Mar. 7, 2006**

(54) **ANGLED-ROLLER ARTICLE-ROTATING BELT CONVEYOR**

(75) Inventor:

(73) Assignee: **Laitram, L.L.C.**, Harahan, LA (US)

(*) Notice: Subject to any disclaimer, the term of this patent is extended or adjusted under 35 U.S.C. 154(b) by 0 days.

(21) Appl. No.: **10/906,832**

(22) Filed: **Mar. 8, 2005**

(51) **Int. Cl.**
B65G 47/26 (2006.01)

(52) **U.S. Cl.** **198/457.02**; 198/779; 198/370.03

(58) **Field of Classification Search** 198/850–853, 198/597, 779, 457.02, 370.03, 370.09
See application file for complete search history.

(56) **References Cited**

U.S. PATENT DOCUMENTS

3,047,123 A 7/1962 McKay 198/33
3,866,739 A 2/1975 Sikorski 198/30
4,143,756 A * 3/1979 Chorlton 198/457.02
4,676,361 A 6/1987 Heisler 198/394
4,901,842 A 2/1990 Lemboke et al. 198/415
5,191,962 A 3/1993 Wegscheider et al. 198/415
5,400,896 A 3/1995 Loomer 198/415
5,551,543 A 9/1996 Mattingly et al. 198/370.09
5,660,262 A 8/1997 Landrum et al. 198/411
5,769,204 A 6/1998 Okada et al. 198/443
5,836,439 A 11/1998 Coyette 198/415
5,924,548 A 7/1999 Francioni 198/415
6,073,747 A 6/2000 Takino et al. 198/370.09
6,401,936 B1 6/2002 Isaacs et al. 209/656
6,494,312 B1 12/2002 Costanzo 198/779
6,571,937 B1 6/2003 Costanzo et al. 198/779
6,758,323 B1 * 7/2004 Costanzo 198/457.02

* cited by examiner

Primary Examiner—James R. Bidwell
(74) *Attorney, Agent, or Firm*—James T. Cronvich

(57) **ABSTRACT**

A conveyor system including a pair of roller-top belts arranged side by side, each individually controllable to cause articles conveyed atop both to rotate into preferred orientations. The roller-top belts include rollers that rotate about axes oblique to the direction of belt travel. The rollers on each belt direct conveyed articles toward the other belt as the belts advance in the direction of belt travel and the rollers contact an underlying bearing surface in rolling contact. The rotation of the rollers exerts a force on articles conveyed atop the rollers. The force has a component directed toward the other belt and another component directed downstream. A sensor array senses the footprint of an article at an upstream location and sends signals to a controller that determines the size and orientation of the article to selectively stop one or the other roller-top belt if necessary to rotate the article to change its orientation.

35 Claims, 6 Drawing Sheets

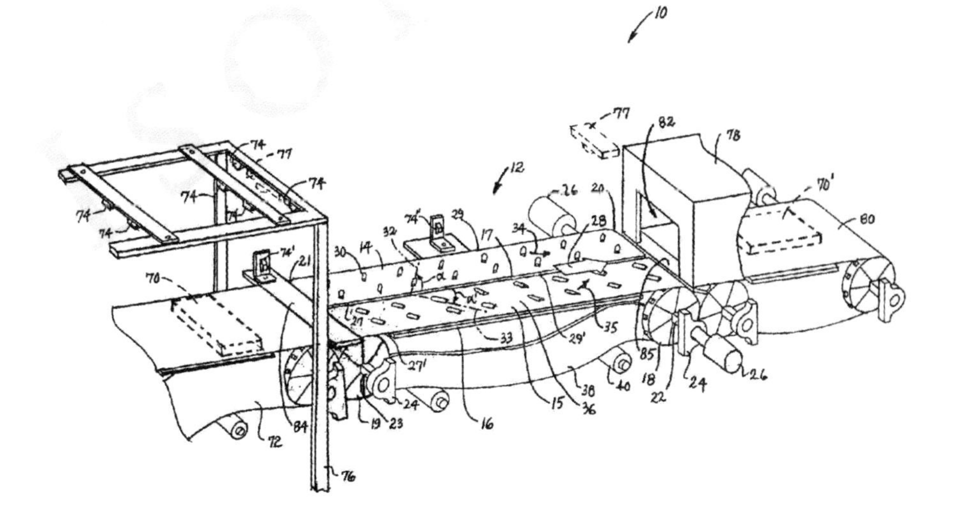

Maine

US007055461B2

(12) **United States Patent**
Harris, Jr. et al.

(10) **Patent No.:** **US 7,055,461 B2**
(45) **Date of Patent:** ***Jun. 6, 2006**

(54) METHODS FOR GROWING AND IMPRINTING FISH USING AN ODORANT

(75) Inventors:

(73) Assignee: **MariCal, Inc.**, Portland, ME (US)

(*) Notice: Subject to any disclaimer, the term of this patent is extended or adjusted under 35 U.S.C. 154(b) by 0 days.

This patent is subject to a terminal disclaimer.

(21) Appl. No.: **10/851,047**

(22) Filed: **May 21, 2004**

(65) **Prior Publication Data**

US 2004/0244714 A1 Dec. 9, 2004

Related U.S. Application Data

(63) Continuation of application No. 10/268,051, filed on Oct. 8, 2002, now Pat. No. 6,748,900.

(60) Provisional application No. 60/328,464, filed on Oct. 11, 2001.

(51) **Int. Cl.**
A01K 61/00 (2006.01)

(52) **U.S. Cl.** **119/231**; 426/2

(58) **Field of Classification Search** 119/231, 119/215, 204, 212, 230, 242, 234; 435/375; 426/2
See application file for complete search history.

(56) **References Cited**

U.S. PATENT DOCUMENTS

3,406,662 A * 10/1968 Karstein et al. 119/217
3,777,709 A * 12/1973 Anderson et al. 119/217
4,172,124 A 10/1979 Koprowski et al.
4,683,202 A 7/1987 Mullis
5,351,651 A 10/1994 Ushio et al.
5,651,651 A * 7/1997 Spencer 411/372.6
5,763,569 A 6/1998 Brown et al.
5,823,142 A * 10/1998 Cardinale et al. 119/212
5,827,551 A * 10/1998 Prochnow et al. 426/1
5,837,490 A 11/1998 Jacobs et al.
5,858,684 A 1/1999 Nemeth et al.
5,962,314 A 10/1999 Brown et al.
5,981,599 A 11/1999 Moe et al.
6,001,884 A 12/1999 Nemeth et al.
6,016,770 A * 1/2000 Fisher 119/215
6,269,586 B1 * 8/2001 Jones 43/42.06
6,337,391 B1 1/2002 Harris, Jr. et al.
6,410,249 B1 * 6/2002 Ngai et al. 435/7.21
6,463,882 B1 * 10/2002 Harris et al. 119/230
6,463,883 B1 * 10/2002 Harris et al. 119/230
6,475,792 B1 * 11/2002 Harris et al. 435/375
6,481,379 B1 * 11/2002 Harris et al. 119/230
6,564,747 B1 * 5/2003 Harris et al. 119/230

FOREIGN PATENT DOCUMENTS

SU 1784152 A1 12/1992
WO WO97/35977 A1 10/1997
WO WO02/30182 A2 4/2002
WO WO02/30215 A2 4/2002

OTHER PUBLICATIONS

Nearing, J., et al., "Polyvalent cation receptor proteins (CaRs) are salinity sensors in fish," *PNAS*, 99(14): 9231-9236 (2002).
Moore-Clark (a nutreco company) Fish Feed Sales Brochure, "USA Freshwater Product Summary," Summer/Fall 2002.
Brown, E. M., et al., "Neomycin Mimics the Effects of High Extracellular Calcium Concentrations on Parathyroid Function in Dispersed Bovine Parathyroid Cells," *Endocrinology*, 128(6):3047-3054 (1991).
Quinn, S. J. , et al., "The Ca^{2+}-sensing receptor: a target for polyamines," *American Journal of Physiology*, 273(4): C1315-C1323 (1997).
Conigrave, A.D., et al., "L-Amino acid sensing by the extracellular Ca^{2+}-sensing receptor." *Proc Natl Acad Sci*, 97(9) 4419-4819 (2000).
Nemeth, E.F., et al., "Calcimimetics with potent and selective activity on the parathyroid calcium receptor," *Proc. Natl. Acad. Sci.*, 95: 4040-4045 (1998).

(Continued)

Primary Examiner—Yvonne R. Abbott
(74) *Attorney, Agent, or Firm*—Hamilton, Brook, Smith & Reynolds, P.C.

(57) **ABSTRACT**

The present invention relates to methods of imprinting fish in freshwater with at least one odorant for the fish and causing the imprinted fish to react to the odorant in seawater, wherein the fish are maintained in freshwater prior to transfer to seawater. The method includes: adding at least one Polyvalent Cation Sensing Receptor (PVCR) modulator to the freshwater in an amount sufficient to modulate expression and/or sensitivity of at least one PVCR, the PVCR modulator being one which alters olfactory sensing of the fish to the odorant; adding feed for fish consumption to the freshwater, the feed containing the odorant and an amount of NaCl sufficient to contribute to a significantly increased level of the PVCR modulator in serum of the fish upon consumption of the feed, whereby the fish are imprinted with the odorant; transferring the imprinted fish to seawater; and providing a source of said odorant in the seawater to which the imprinted fish are transferred, thereby causing the imprinted fish to react to said odorant. The present invention also includes methods of homing or attracting fish, as well as methods for repelling fish by modulating the expression and/or sensitivity of the PVCR in the olfactory apparatus of the fish.

7 Claims, 22 Drawing Sheets

Maryland

US006992070B2

(12) **United States Patent**
Donahue et al.

(10) **Patent No.: US 6,992,070 B2**
(45) **Date of Patent: Jan. 31, 2006**

(54) METHODS AND COMPOSITIONS FOR NUCLEIC ACID DELIVERY

(75) Inventors:

(73) Assignee: **The Johns Hopkins University**, Baltimore, MD (US)

(*) Notice: Subject to any disclaimer, the term of this patent is extended or adjusted under 35 U.S.C. 154(b) by 326 days.

(21) Appl. No.: **09/977,865**

(22) Filed: **Oct. 15, 2001**

(65) **Prior Publication Data**

US 2002/0094326 A1 Jul. 18, 2002

Related U.S. Application Data

(60) Provisional application No. 60/240,231, filed on Oct. 13, 2000.

(51) **Int. Cl.**
A01N 43/04 (2006.01)
A61K 31/70 (2006.01)
C12N 15/00 (2006.01)
C12N 5/00 (2006.01)
C07H 21/04 (2006.01)

(52) **U.S. Cl.** **514/44**; 435/320.1; 435/325; 536/23.1

(58) **Field of Classification Search** 514/44; 435/320.1, 325; 536/23.1
See application file for complete search history.

(56) **References Cited**

U.S. PATENT DOCUMENTS

5,516,651 A * 5/1996 Goldring et al. 435/69.1
6,100,270 A 8/2000 Campbell
6,214,620 B1 4/2001 Johns et al.
2002/0082240 A1 * 6/2002 Linden et al. 514/46
2002/0094326 A1 7/2002 Donahue et al.

FOREIGN PATENT DOCUMENTS

WO WO 96/16657 6/1996
WO WO 99/19792 4/1999
WO WO 99/31982 7/1999

OTHER PUBLICATIONS

Carson CC Sildenafil: a 4-year update in the treatment of 20 million erectile dysfunction patients, Curr. Urol. Rep. Dec.; 2003 4(6) pp488-496.*

Crocker I.C. therapeutic potential of phosphodiesterase 4 inhibitors in allergic diseases Drugs Today. Jul.; 1999 35(7): pp519-3 and DeKorte C.J. Current and emerging therapies for the management of chronic inflammation in asthma Am J. Health Syst. Pharm.*

Netherton S.J. et al. Altered PDE-e mediated cAMP hydrolysis contributes to a hypermotile phenotype in obese JCR:LA-cp rat aortic vascular smooth muscle cells:implications for diabitets associated cardiovasular diease. Diabetes. Apr.; 2002 51(4):1194 200.*

Anderson Human Gene Therapy Nature vol. 392, supp 1998 pp 25-30.*

Verma et al. Gene therapy—promises, problems and prospects Nature vol. 389 1997 pp239-242.*

Palu et al. In pursuit of new deveopments for gene therapy of human diseases J. of Biotech. vol. 68 1999 pp1-12.*

DeKorte C.J. Current and emerging therapies for the management of chronic inflammation in asthma Am J. Health Syst. Pharm Oct. 1. 2003; (60)19: pp1949-59.*

Lamping, K, et al., "Agonist-specific impairment of coronary vascular function in genetically altered, hyperlipidemic mice," *Am. J. of Phys.*, (Apr. 1999), pp. R1023-R1029, vol. 276, No. 4.

Kuwahara, A. et al., "5-HT activates nitric oxide-generating neurons to stimulate chloride secretion in guinea pig distal colon," *Am. J. Phys.*, (Oct. 1998), pp. G829-G834, vol. 275, No. 4.

Frieden, M. et al., "Effect of 5-hydroxytriptamine on the membrane potential of endothial and smooth muscle cells in the pig coronary artery," *British Journal of Pharmacology*, (1995), pp. 95-100, vol. 115, No. 1.

Reiser, G., "Nitric oxide formation caused by calcium release from internal stores in neuronal cell line in enhanced by cyclic amp," *European Journal of Pharmacology Molecular Pharmacology Section*, (1992), pp. 89-93, vol. 9, No. 1.

Katz, S. et al., "Acute type 5 phosphodiesterase inhibition with slidenafil enhances flow-mediated vasodilation in patients with chronic heart failure," *Journal of the American College of Cardiology*, (Sep. 2000), pp. 845-851, vol. 36, No. 3.

Lai, N.C. et al., "Sodium nitroprusside facilitates intracoronary gene transfer in pigs," *Faseb Journal*, (Mar. 2001), p. A100. vol. 15, No. 4.

Bilbao, R. et al., "A blood-tumor barrier limits gene transfer to the experimental liver cancer: The effect of vasoactive compounds," *Gene Therapy*, (Nov. 2000), pp. 1824-1832, vol. 7, No. 21.

(Continued)

Primary Examiner—James Ketter
Assistant Examiner—Konstantina M. Katcheves
(74) *Attorney, Agent, or Firm*—Banner & Witcoff, Ltd.

(57) **ABSTRACT**

The present provides methods and compositions that enable effective delivery of nucleic acids to desired cells, including to a solid organ such as a mammalian heart. The methods and compositions enable effective gene transfer and subsequent expression to a majority of cells throughout a solid organ such as the heart. Methods and compositions of the invention preferably provide enhanced vascular permeability that enables increased gene transfer to targeted cells, but without significant degradation or injury to endothelial cell layers.

37 Claims, 3 Drawing Sheets

Massachusetts

US006994962B1

(12) **United States Patent**
Thilly

(10) **Patent No.: US 6,994,962 B1**
(45) **Date of Patent: Feb. 7, 2006**

(54) **METHODS OF IDENTIFYING POINT MUTATIONS IN A GENOME**

(75) Inventor:

(73) Assignee: **Massachusetts Institute of Technology**, Cambridge, MA (US)

(*) Notice: Subject to any disclaimer, the term of this patent is extended or adjusted under 35 U.S.C. 154(b) by 0 days.

(21) Appl. No.: **09/503,758**

(22) Filed: **Feb. 14, 2000**

Related U.S. Application Data

(63) Continuation of application No. PCT/US99/29379, filed on Dec. 9, 1999.

(60) Provisional application No. 60/111,457, filed on Dec. 9, 1998.

(51) **Int. Cl.**
C12Q 1/68 (2006.01)
C12P 19/34 (2006.01)
C07H 21/02 (2006.01)
C07H 21/04 (2006.01)

(52) **U.S. Cl.** **435/6**; 435/91.2; 536/23.1; 536/24.3

(58) **Field of Classification Search** 435/6, 435/91.1; 536/23.1, 24.3
See application file for complete search history.

(56) **References Cited**

U.S. PATENT DOCUMENTS

5,045,450 A	9/1991	Thilly et al.	435/6
5,633,129 A	5/1997	Karger et al.	
5,837,832 A	11/1998	Chee et al.	536/22.1
5,976,842 A	11/1999	Wurst	
2002/0132244 A1	9/2002	Li-Sucholeiki	
2003/0092021 A1	5/2003	Thilly	
2003/0143584 A1	7/2003	Li-Sucholeiki	

FOREIGN PATENT DOCUMENTS

WO	WO91/00925	1/1991
WO	WO95/21268	8/1995
WO	WO 00/34652	6/2000

OTHER PUBLICATIONS

Khrapko, K. et al., "Constant denaturant capillary electrophoresis (CDCE): a high resolution approach to mutational analysis", Nucl. Acids Res. vol. 22, pp. 364-369 (1994).*
Davies, K. E. et al., "Molecular Basis of Inherited Disease", IRL Press, New York, pp, 21-25 (1992).*
Crow, J.F. et al., "Mutation in Human Population", Adv. Hum. Genet., vol. 14, pp. 59-77 (1985).*
Conneally, P. M., "Human Genetic Polymorphisms", Dev. Biol. Stand., vol. 83, pp. 107-110 (1994).*
de la Chapelle, A., "Disease gene mapping in isolated human populations; the example of Finland", J. Med. Genet., vol. 30, pp 857-865 (1993).*
Cavalli-Sforza, L. L. et al. "The Genetics of Human Populations", W. H. Freeman and Company, San Francisco, pp. 71-110 (1971).*
Hardelin, J-P. et al., "Heterogeneity in the mutations responsible for X-chromosome linked Kallman syndrome", Hum. Mol. Genetics, vol. 2, pp. 373-377 (1993).*
Cooper, D. N. et al., "The mutational spectrum of single base-pair substitutions causing human genetic disease: patterns and predictions", Hum. Genetics, vol. 85, pp. 55-74 (1990).*
Maraglione, M. et al., "Prevalence of Apolipoprotein E Alleles in Healthy Subjects and Survivors of Ischemic Stroke", Stroke, vo 29, pp. 399-403 (Feb. 1998).*
Paik, Y-K. et al., "Nucleotide sequence and structure of the human apolipoprotein E gene", PNAS, vol. 82, pp. 3445-3449 (1985).*
de Knijff, P. et al., "Genetic Heterogeneity of Apolipoprotein E and Its Influence on Plasma Lipid and Lipoprotein Levels", Hum. Mut., pp. 178-194 (1994).*
Bjorheim J. et al., "Mutations analyses of KRAS exon 1 comparing three different techniques: temporal temperature gradient electrophoresis, constant denaturant capillary electrophoresis and allele specific polymerase chain reaction." *Mut. Res.,* 403:103-12 (Jul. 1998).
Ekstrom P.O., et al., "Detection of low-frequency mutations in exon 8 of the TP53 gene by constant denaturant capillary electrophesis (CDCE) ." *Biotechniques*, 27:128-34 (Jul. 1999).
Ekstrom, P.O., et al., "Two-point fluorescence detection and automated fraction collection applied to constant denaturant capillary electrophoresis." *Biotechniques*, 29:582-4, 586-9 (Sep. 2000).
Falt S., et al., "Identification of in vivo mutations in exon 5 of the human HPRT gene in a set of pooled T-cell mutants by constant denaturant capillary electrophoresis (CDCE) ." *Mutat. Res.*, 452:57-66 (Jul. 2000).
Fischer & Lerman, "Separation of random fragments of DNA according to properties of their sequences." *Proc. Natl. Acad. Sci. USA*, 77:4420-4424 (1980).

(Continued)

Primary Examiner—Jeffrey Fredman
Assistant Examiner—Teresa Strzelecka
(74) *Attorney, Agent, or Firm*—Hamilton, Brook, Smith & Reynolds, P.C.

(57) **ABSTRACT**

The invention relates to a method for identifying inherited point mutations in a targeted region of the genome in a large population of individuals and determining which inherited point mutations are deleterious, harmful or beneficial. Deleterious mutation are identified directly by a method of recognition using the set of point mutations observed in a large population of juveniles. Harmful mutations are identified by comparison of the set of point mutation observed in a large set of juveniles and a large set of aged individuals of the same population. Beneficial mutations are similarly identified.

11 Claims, 24 Drawing Sheets

Michigan

US007220200B2

(12) **United States Patent**
Sowul et al.

(10) **Patent No.: US 7,220,200 B2**
(45) **Date of Patent: *May 22, 2007**

(54) **ELECTRICALLY VARIABLE TRANSMISSION ARRANGEMENT WITH SPACED-APART SIMPLE PLANETARY GEAR SETS**

(75) Inventors:

(73) Assignee: **General Motors Corporation**, Detroit, MI (US)

(*) Notice: Subject to any disclaimer, the term of this patent is extended or adjusted under 35 U.S.C. 154(b) by 149 days.

This patent is subject to a terminal disclaimer.

(21) Appl. No.: **11/071,405**

(22) Filed: **Mar. 3, 2005**

(65) **Prior Publication Data**

US 2006/0025263 A1 Feb. 2, 2006

Related U.S. Application Data

(60) Provisional application No. 60/591,995, filed on Jul. 29, 2004.

(51) **Int. Cl.**
F16H 3/72 (2006.01)

(52) **U.S. Cl.** **475/5**; 475/282; 475/283; 180/65.2; 192/48.5; 192/48.6

(58) **Field of Classification Search** .... 475/5, 475/282, 283; 477/3, 4, 5; 701/22; 180/65.2, 180/65.3; 290/40 C; 192/48.5, 48.6
See application file for complete search history.

(56) **References Cited**

U.S. PATENT DOCUMENTS

4,653,348 A * 3/1987 Hiraiwa 475/276
6,478,705 B1 * 11/2002 Holmes et al. 475/5
6,527,658 B2 3/2003 Holmes et al. 475/5
2005/0227801 A1 * 10/2005 Schmidt et al. 475/5
2005/0227803 A1 * 10/2005 Holmes 475/204
2006/0019785 A1 * 1/2006 Holmes et al. 475/5
2006/0025264 A1 * 2/2006 Sowul et al. 475/5
2006/0025265 A1 * 2/2006 Sowul et al. 475/5
2006/0046886 A1 * 3/2006 Holmes et al. 475/5

FOREIGN PATENT DOCUMENTS

EP 1247679 A2 * 10/2002

* cited by examiner

Primary Examiner—David D. Le
(74) *Attorney, Agent, or Firm*—Christopher DeVries

(57) **ABSTRACT**

An electrically variable transmission is provided including an input member to receive power from an engine; an output member connected to a transfer gear; first and second motor/generators; and first and second simple planetary gear sets each having first, second and third members. The input member is continuously connected to the first member of the first gear set, and the output member is continuously connected to the first member of the second gear set. The first motor/generator is continuously connected to the second member of the first gear set. The second motor/generator is continuously connected with the third member of the first or second gear set. The transfer gear and first and second torque transfer devices are positioned between the first and second gear sets.

12 Claims, 5 Drawing Sheets

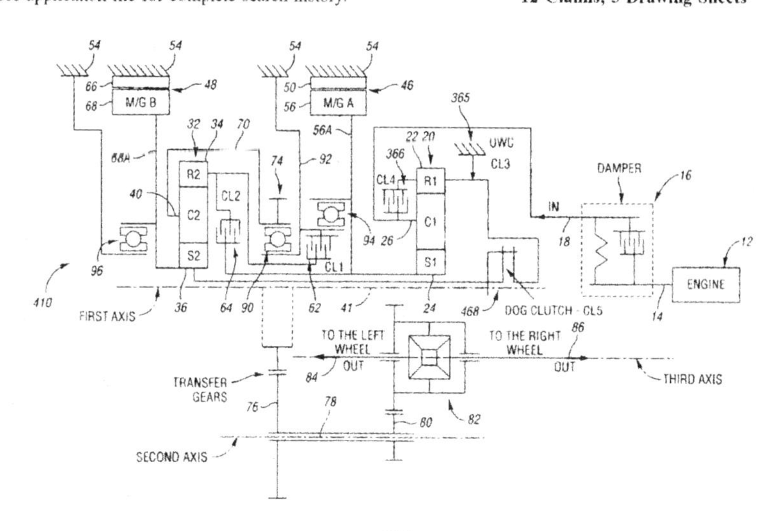

Minnesota

US007018500B2

(12) **United States Patent**
Eaton et al.

(10) **Patent No.:** **US 7,018,500 B2**
(45) **Date of Patent:** **Mar. 28, 2006**

(54) **APPARATUS AND METHOD FOR SINGULATING POROUS FUEL CELL LAYERS USING ADHESIVE TAPE PICK HEAD**

(75) Inventors:

(73) Assignee: **3M Innovative Properties Company**, Saint Paul, MN (US)

(*) Notice: Subject to any disclaimer, the term of this patent is extended or adjusted under 35 U.S.C. 154(b) by 0 days.

(21) Appl. No.: **10/980,010**

(22) Filed: **Nov. 3, 2004**

(65) **Prior Publication Data**

US 2005/0194102 A1 Sep. 8, 2005

Related U.S. Application Data

(62) Division of application No. 10/115,556, filed on Apr. 3, 2002.

(51) **Int. Cl.**
B32B 31/00 (2006.01)

(52) **U.S. Cl.** **156/247**; 156/570; 156/573; 156/584

(58) **Field of Classification Search** .... 156/247, 156/344, 564, 570, 573, 584, DIG. 29, DIG. 30, 156/DIG. 31
See application file for complete search history.

(56) **References Cited**

U.S. PATENT DOCUMENTS

2,710,234 A 6/1955 Hansen
3,178,041 A 4/1965 Wheat et al.
3,285,112 A 11/1966 Dale et al.
3,359,046 A 12/1967 Dryden
3,380,788 A 4/1968 Wilcock
3,477,558 A 11/1969 Fleischauer
3,539,177 A 11/1970 Schwenk et al.
3,861,259 A 1/1975 Hitch
3,946,920 A 3/1976 Jordan et al.

(Continued)

FOREIGN PATENT DOCUMENTS

AT 314 323 B 3/1974

(Continued)

OTHER PUBLICATIONS

U.S. Appl. No. 10/446,485, filed May 28, 2003, Roll-Good Fuel Cell Fabrication Processes, Equipment, and Articles Produced From Same (58669US002, David R. Mekala et al., pp. 1-55, 16 sheets drawings).

(Continued)

Primary Examiner—Mark A Osele
(74) *Attorney, Agent, or Firm*—Mark A. Hollingsworth; Philip Y. Dahl

(57) **ABSTRACT**

An apparatus and method provide for singulating thin and substantially porous material layers arranged in a stack. A pick head is positioned above the stack of material layers. An adhesive tape is stabilized against the pick head through use of a vacuum between the adhesive tape and the pick head. Contact is effected between the stabilized adhesive tape and the top material layer. The pick head is moved to move the top material layer from the stack to a predetermined location. While at the predetermined location, the adhesive tape is detached from the top material layer. The singulation apparatus and method are particularly well suited for destacking individual porous fluid transport layers (FTLs) from a magazine of FTLs during automated fuel cell assembly.

22 Claims, 7 Drawing Sheets

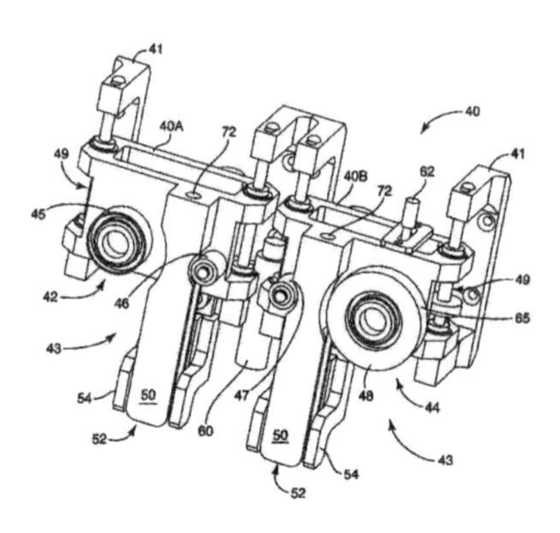

Mississippi

US007228612B2

(12) **United States Patent**
Lyvers

(10) **Patent No.: US 7,228,612 B2**
(45) **Date of Patent: Jun. 12, 2007**

(54) **METHODS OF MAKING REFRIGERATOR STORAGE ASSEMBLIES**

(75) Inventor:

(73) Assignee: **Viking Range Corporation**, Greenwood, MS (US)

(*) Notice: Subject to any disclaimer, the term of this patent is extended or adjusted under 35 U.S.C. 154(b) by 427 days.

(21) Appl. No.: **10/806,979**

(22) Filed: **Mar. 23, 2004**

(65) **Prior Publication Data**

US 2005/0210655 A1 Sep. 29, 2005

(51) **Int. Cl.**
B22D 11/126 (2006.01)
B23P 17/00 (2006.01)
A47B 96/04 (2006.01)

(52) **U.S. Cl.** **29/527.1**; 29/527.6; 264/177.1; 264/177.13; 264/177.17; 264/211.12; 264/239; 312/404; 312/406; 312/406.2; 312/408

(58) **Field of Classification Search** 29/527.1, 29/527.2, 527.4, 527.6, 557; 264/176.1, 264/177.1, 177.13, 177.16, 177.17, 211.12, 264/129, 239, 148; 312/401, 402, 404, 406, 312/406.2, 408; 62/377
See application file for complete search history.

(56) **References Cited**

U.S. PATENT DOCUMENTS

2,253,971 A * 8/1941 Dodge 312/404
2,528,807 A 11/1950 Whitney
3,375,936 A 4/1968 Kessler
3,476,852 A * 11/1969 Shattuck 264/261
3,478,138 A * 11/1969 Friesner 264/145
3,937,537 A * 2/1976 Dietterich 312/204
4,960,308 A * 10/1990 Donaghy 312/404
5,040,856 A * 8/1991 Wilkins et al. 312/402
5,044,704 A * 9/1991 Bussan et al. 312/402
5,048,247 A * 9/1991 Weldy 52/255
5,212,962 A * 5/1993 Kang et al. 62/382
RE34,377 E 9/1993 Wilkins et al.
5,322,213 A * 6/1994 Carter et al. 229/166
5,346,299 A * 9/1994 Werkmeister et al. ... 312/405.1
D358,403 S * 5/1995 Edman et al. D15/89
5,437,503 A * 8/1995 Baker et al. 312/404
5,494,630 A * 2/1996 Eraybar et al. 264/138
D384,681 S * 10/1997 Martin D15/89
5,730,516 A * 3/1998 Vismara 312/406
5,890,785 A * 4/1999 Vismara 312/406
5,918,959 A * 7/1999 Lee 312/404
5,947,573 A * 9/1999 Tovar et al. 312/404

(Continued)

FOREIGN PATENT DOCUMENTS

JP 06174363 A * 6/1994

Primary Examiner—Essama Omgba
(74) *Attorney, Agent, or Firm*—Womble Carlyle Sandridge & Rice, PLLC

(57) **ABSTRACT**

A method of making a removable refrigerator storage assembly is provided. The method includes feeding a polymeric or metal material into an extruder and directing the material through a die having an aperture with a first and second leg formed therein, thereby forming an intermediate shelf extrusion. The intermediate shelf extrusion is cut into a predetermined length to form the shelf extrusion having a bottom wall integrally formed with a side wall. The shelf extrusion is painted or plated to a specific finish. The method also encompasses attaching to both ends of the shelf extrusion end walls having brackets formed thereon for removably engaging supports formed on the door of a refrigerator.

13 Claims, 7 Drawing Sheets

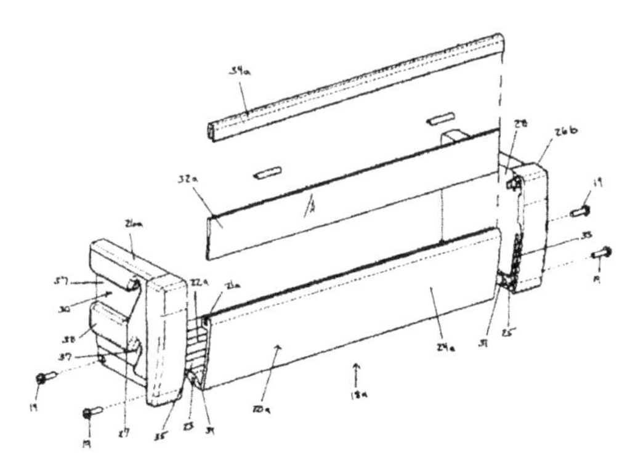

Missouri

(12) **United States Patent**
Strutz

(10) **Patent No.:** **US 7,066,415 B2**
(45) **Date of Patent:** **Jun. 27, 2006**

(54) **TOUCH PAD CONTROL INFORMATION SYSTEM FOR A FOOD WASTE DISPOSER**

(75) Inventor:

(73) Assignee: **Emerson Electric Co.**, St. Louis, MO (US)

(*) Notice: Subject to any disclaimer, the term of this patent is extended or adjusted under 35 U.S.C. 154(b) by 17 days.

(21) Appl. No.: **10/458,099**

(22) Filed: **Jun. 10, 2003**

(65) **Prior Publication Data**

US 2004/0251339 A1 Dec. 16, 2004

(51) **Int. Cl.**
B02C 25/00 (2006.01)

(52) **U.S. Cl.** **241/36**; 241/101.3; 241/46.013

(58) **Field of Classification Search** 241/46.013, 241/46.014, 46.015, 30, 36, 101.3; 4/DIG. 4
See application file for complete search history.

(56) **References Cited**

U.S. PATENT DOCUMENTS

2,678,775 A * 5/1954 Simmons 241/32.5
3,420,455 A * 1/1969 Chorney 241/36
3,545,684 A * 12/1970 Hilmanowski et al. 241/33
5,971,303 A 10/1999 Pugh-Gottlieb
6,198,079 B1 3/2001 Essig
6,364,226 B1 * 4/2002 Kubicko 241/36
6,481,652 B1 11/2002 Strutz 241/46.013
6,610,942 B1 8/2003 Anderson et al.
6,636,135 B1 * 10/2003 Vetter 335/205
6,854,673 B1 2/2005 Strutz et al.
6,904,926 B1 * 6/2005 Aylward et al. 137/119.01
2004/0178288 A1 9/2004 Berger et al.
2004/0178289 A1 9/2004 Jara-Almonte et al.

FOREIGN PATENT DOCUMENTS

DE 3233516 * 3/1984
GB 2234191 A 1/1991
JP 11028381 2/1999

OTHER PUBLICATIONS

Invention Disclosure document signed by inventor William F. Strutz on Oct. 31, 1989.
Technical Disclosure of prior art food waste disposer system manufactured by Toto (date unknown).
International Search Report, PCT/US2004/017807, European Patent Office, Nov. 5, 2004.
Written Opinion of the International Searching Authority, PCT/US2004/017807, European Patent Office, Nov. 5, 2004.

* cited by examiner

Primary Examiner—Mark Rosenbaum
(74) *Attorney, Agent, or Firm*—Locke Liddell & Sapp LLP

(57) **ABSTRACT**

A touch pad control information system for a food waste disposer is disclosed. The touch pad is mountable to a wall or countertop near the food waste disposer. The touch pad preferably includes switches which allow the user to select from a plurality of disposer functions, and light emitting diodes (LEDs) or other graphic display to indicate one of a plurality of statuses for the disposer. The touch pad is coupled to the disposer by a wire bus or by wireless means.

23 Claims, 3 Drawing Sheets

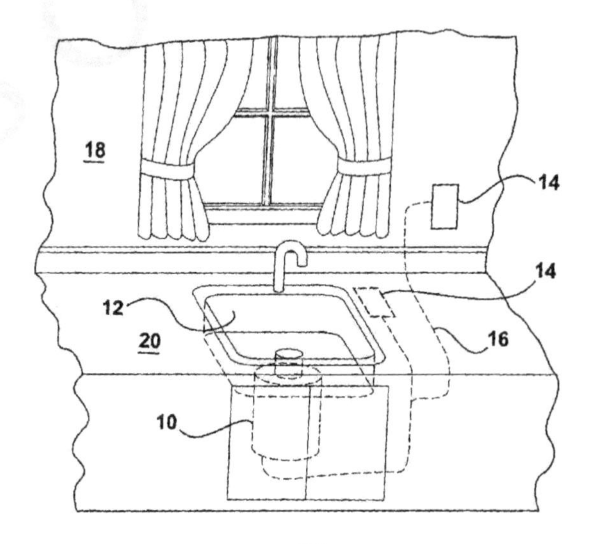

Montana

(12) **United States Patent**
Faure

(10) **Patent No.:** **US 7,048,417 B1**
(45) **Date of Patent:** **May 23, 2006**

(54) **ADJUSTABLE DESK LAMP**

(75) Inventor:

(73) Assignee: **Matthew C. Faure**, Bozeman, MT (US)

(*) Notice: Subject to any disclaimer, the term of this patent is extended or adjusted under 35 U.S.C. 154(b) by 60 days.

(21) Appl. No.: **10/857,131**

(22) Filed: **May 28, 2004**

(51) **Int. Cl.**
F21S 8/08 (2006.01)
F16M 13/00 (2006.01)

(52) **U.S. Cl.** **362/413**; 362/270; 362/418; 362/419; 362/427; 362/431; 248/124.1; 248/278.1

(58) **Field of Classification Search** 362/413, 362/418, 427, 431, 428, 419, 270; 248/124.1, 248/231.31, 295.11, 447.1
See application file for complete search history.

(56) **References Cited**

U.S. PATENT DOCUMENTS

1,474,304 A * 11/1923 Weber 362/413
1,625,100 A * 4/1927 Runge 248/295.11
2,616,029 A * 10/1952 Osowski 362/59
4,238,818 A 12/1980 Gindel 362/413
4,327,402 A 4/1982 Aubrey 362/288
4,386,393 A * 5/1983 Pike 362/427
4,605,995 A * 8/1986 Pike 362/287
4,932,620 A * 6/1990 Foy 248/124.1
5,169,226 A 12/1992 Friedman 362/190
5,203,622 A 4/1993 Sottile 362/109
5,325,278 A 6/1994 Tortola et al. 362/109
5,369,560 A 11/1994 Friedman 362/396
5,379,201 A 1/1995 Friedman 362/191
5,615,945 A 4/1997 Tseng 362/226
5,655,833 A * 8/1997 Raczynski 362/419
5,669,694 A 9/1997 Morton, Sr. 362/33
5,722,754 A 3/1998 Langner 362/1
5,868,487 A 2/1999 Polley et al. 362/33
6,089,724 A 7/2000 Shore et al. 362/85
6,152,578 A 11/2000 Hoffman et al. 362/285
6,168,292 B1 1/2001 Sherman 362/287
6,467,936 B1 10/2002 Golemba 362/413
6,481,871 B1 11/2002 Jamison 362/287

* cited by examiner

Primary Examiner—Alan Cariaso
Assistant Examiner—James W Cranson, Jr.
(74) *Attorney, Agent, or Firm*—Shane P. Coleman; Holland & Hart

(57) **ABSTRACT**

An adjustable lamp is disclosed having a rod and a light assembly that slideably connects to the rod at a connection assembly. The light assembly includes a housing and a light socket contained within the housing. The light assembly further includes an extension member that connects the housing to the connection assembly. The connection assembly moves along the rod laterally along a longitudinal axis. The connection assembly may be locked into a position along the rod. For example, the rod may be connected to a vertical surface and a light source contained in the housing may be directed downward, toward a desk or table or other horizontal surface. The slideable connection allows the light assembly to move up or down relative to the horizontal surface in this use.

34 Claims, 10 Drawing Sheets

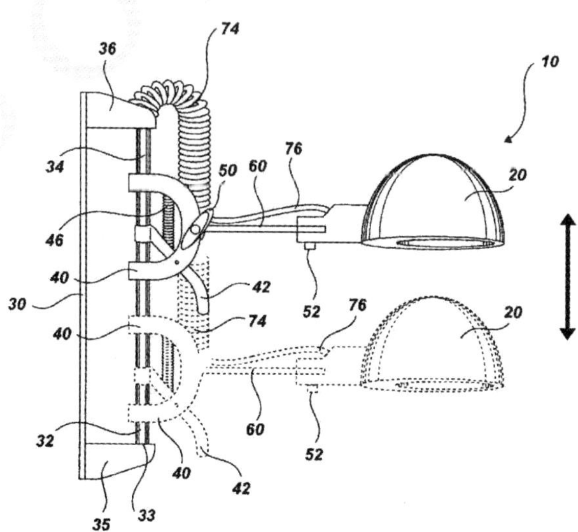

Nebraska

US007105657B2

(12) **United States Patent**
Batra et al.

(10) **Patent No.: US 7,105,657 B2**
(45) **Date of Patent: Sep. 12, 2006**

(54) **COMPOSITIONS AND METHODS FOR INHIBITING PANCREATIC CANCER METASTASIS**

(75) Inventors:

(73) Assignee: **Board of Regents of the University of Nebraska**, Lincoln, NE (US)

(*) Notice: Subject to any disclaimer, the term of this patent is extended or adjusted under 35 U.S.C. 154(b) by 500 days.

(21) Appl. No.: **10/291,151**

(22) Filed: **Nov. 8, 2002**

(65) **Prior Publication Data**

US 2004/0091869 A1 May 13, 2004

(51) **Int. Cl.**
C07H 21/04 (2006.01)

(52) **U.S. Cl.** **536/24.5**; 536/23.1

(58) **Field of Classification Search** .... None
See application file for complete search history.

(56) **References Cited**

U.S. PATENT DOCUMENTS

5,530,114 A *	6/1996	Bennett et al.	....	536/24.3
5,624,803 A *	4/1997	Noonberg et al.	....	435/6
5,981,279 A *	11/1999	Weiss	....	435/375
6,001,651 A *	12/1999	Bennett et al.	....	435/375
6,716,627 B1 *	4/2004	Dobie	....	435/375
2003/0077568 A1 *	4/2003	Gish et al.	....	435/4

OTHER PUBLICATIONS

Moniaux et al. (1999) Biochem. J. 338:325-333.*
Andrianifahanana, M., et al., "Mucin (Muc) gene expression in human pancreatic adenocarcinoma and chronic pancreatitis: A potential role of MUC4 as a tumor marker of diagnostic significance"; Clin. Cancer. Res., 7:4033-4040 (2001).
Balague, C., et al., "Altered Expression of MUC2, MUC4, and MUC5 mucin genes in pancreas tissues and cancer cell lines"; Gastroenterology 106:1054-1061 (1994).
Caldas, C., et al., "Detection of K-ras mutations in the stool of patients with pancreatic adenocarcinoma and pancreatic ductal hyperplasia" Cancer Res.; 54: 3568-3573 (1994).
Choudhury, A., et al., "MUC4 Mucin Expression in Human Pancreatic Tumors is Affected by Organ Environment: Possible Role of TGFbeta2"; Abstract, Br. J. Cancer, 90(3): 657-64 (2004).
Day, J.D., et. al., "Immunohistochemical evaluation of HER-2/neu oncogene expression in pancreatic adenocarcinoma and pancreatic intraepithelial neoplasms"; Hum. Pathol. 27: 119-124 (1996).
Dimagno, E.P., et. al., "AGA technical review on the epidemiology, diagnosis, and treatment of pancreatic ductal adenocarcinoma"; American Gastroenterological Association, Gastroenterology 117: 1464-1484 (1999).
Hameed, M., et. al., "Expression of p53 nucleophosphoprotein in in situ pancreatic ductal adenocarcinoma: an immunohistochemical analysis of 100 cases"; Lab. Invest. 70: 132A (1994).
Hollingsworth, M.A., et. al., "Expression of MUC1, MUC2, MUC3, and MUC4 mucin mRNAs in human pancreatic and intestinal tumor cell lines"; Int. J. Cancer 57: 198-203 (1994).
Khorrami, A.M., "Purification and characterization of a human pancreatic adenocarcinoma mucin"; J. Biochem. Jan; 131(1):21-9 (2002).
Komatsu, M., et al., "Muc4/sialomucin complex, an intramembrane modulator of ErbB2/HER2/Neu, potentiates primary tumor growth and suppresses apoptosis in a xenotransplanted tumor"; Oncogene 20(4): 461-470 (2001).
Komatsu, M., et al., "Potentiation of metastasis by cell surface sialomucin complex (rat MUC4), a multifunctional anti-adhesive glycoprotein"; Int. J. Cancer, 87:480-486 (2000).
Komatsu, M., et al., "Overexpression of sialomucin complex, a rat homologue of MUC4, inhibits tumor killing by lymphokine-activated killer cells"; Cancer Research, 59: 2229-2236 (1999).
Luttges, J., et. al., "The K-ras mutation pattern in pancreatic ductal adenocarcinoma usially is identical to that in associated normal hyperplastic, metaplastic ductal epithelium"; Cancer 85: 1703-1710 (1999).
Nollet, S., et. al., "Human mucin gene MUC4: Organization of its 5'-region and polymorphism of its central tandem repeat array"; Biochem. J. 15: 739-748 (1998).
Parker, S.L., et. al., "Cancer statistics, 1996"; Cancer J. Clin. 46:5-27 (1996).
Schwartz, M.J., et al., "MUC4 expression increases progressively in pancreatic intraepithelial neoplasia"; Am. J. Clin. Pathol., 117:791-796 (2002).
Tada, M., et. al., "Analysis of K-ras gene mutation in hyperplastic duct cells of the pancreas without pancreatic disease"; Gastroenterology 110: 227-231 (1996).
Walsh, M.D., et. al., "Expression of MUC2 epithelial mucin in breast carcinoma"; J. Clin. Pathol. 46: 922-925 (1993).
Warshaw, A.L., et. al., "Pancreatic carcinoma"; N. Engl. J. Med. 326: 455-465 (1992).
Wilentz, R.E., et. al., "Immunohistochemistry labeling for Dpc4 mirrors genetic status in pancreatic and peripancreatic adenocarcinomas: a new marker of DPC4 inactivation"; Am. J. Pathol. 156: 37-43 (2000).
Terris, B., et al., "Mucin gene expression in intraductal papillary-mucinous pancreatic tumours and related lesions"; J. Pathol. 197(5): 632-637 (2002).

* cited by examiner

Primary Examiner—Andrew Wang
Assistant Examiner—Louis V. Wollenberger
(74) *Attorney, Agent, or Firm*—Dann Dorfman Herrell and Skillman; Kathleen D. Rigaut

(57) **ABSTRACT**

Methods are provided for the inhibition of metastasis of cancer cells expressing MUC4, metastatic pancreatic cancer cells.

18 Claims, 3 Drawing Sheets

Nevada

(12) **United States Patent**
Baerlocher et al.

(10) **Patent No.: US 7,112,137 B2**
(45) **Date of Patent: *Sep. 26, 2006**

(54) **GAMING DEVICE HAVING AN INDICATOR SELECTION WITH PROBABILITY-BASED OUTCOME**

(75) Inventors:

(73) Assignee: **IGT**, Reno, NV (US)

(*) Notice: Subject to any disclaimer, the term of this patent is extended or adjusted under 35 U.S.C. 154(b) by 276 days.

This patent is subject to a terminal disclaimer.

(21) Appl. No.: **10/734,307**

(22) Filed: **Dec. 12, 2003**

(65) **Prior Publication Data**

US 2005/0027384 A1 Feb. 3, 2005

Related U.S. Application Data

(63) Continuation of application No. 09/990,693, filed on Nov. 9, 2001, now Pat. No. 6,676,516, which is a continuation-in-part of application No. 09/605,809, filed on Jun. 28, 2000, now Pat. No. 6,315,664.

(51) **Int. Cl.**
A63F 13/00 (2006.01)

(52) **U.S. Cl.** **463/20**; 463/19; 463/21; 273/138.1; 273/139

(58) **Field of Classification Search** .... 463/16–21, 463/25; 273/138.1, 138.2, 139, 143 R
See application file for complete search history.

(56) **References Cited**

U.S. PATENT DOCUMENTS

4,182,515 A 1/1980 Nemeth
4,624,459 A 11/1986 Kaufman
4,991,848 A 2/1991 Greenwood et al.
5,116,055 A 5/1992 Tracy
5,456,465 A 10/1995 Durham
5,542,669 A 8/1996 Charron et al.
5,711,525 A 1/1998 Breeding
5,833,538 A 11/1998 Weiss
5,851,147 A 12/1998 Stupak et al.

(Continued)

FOREIGN PATENT DOCUMENTS

EP 0 874 337 A1 10/1998

(Continued)

OTHER PUBLICATIONS

Welcome To Video Reality Brochure written by Atronic Casino Technology Ltd. published in 1995.

(Continued)

Primary Examiner—Xuan M. Thai
Assistant Examiner—Alex F. R. P. Rada, II
(74) *Attorney, Agent, or Firm*—Bell, Boyd & Lloyd LLC

(57) **ABSTRACT**

A gaming device which presents a plurality of indicators to the player. Each indicator may be a success indicator or a failure indicator based on a pre-determined probability. Upon or prior to the selection of the indicator, the processor in the gaming device determines, based on that probability, if the indicator is a success indicator or a failure indicator. When a player selects an indicator, the gaming device displays if the selected indicator is a failure indicator or a success indicator and a value associated with the success indicator. In one embodiment, the player selects indicators until the player selects all of the success indicators or the player selects a failure indicator. Accordingly, based on chance and the pre-determined probability, a game may include all success indicators and no failure indicators to increase player excitement and enjoyment.

18 Claims, 5 Drawing Sheets

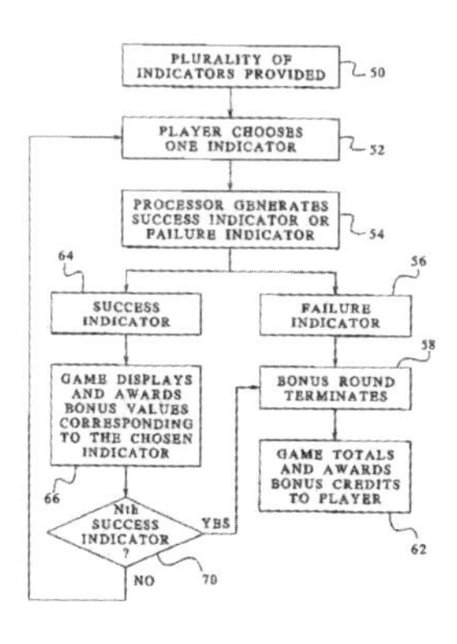

New Hampshire

(12) **United States Patent**
Marshall et al.

(10) **Patent No.: US 6,996,443 B2**
(45) **Date of Patent: Feb. 7, 2006**

(54) RECONFIGURABLE DIGITAL PROCESSING SYSTEM FOR SPACE

(75) Inventors:

(73) Assignee: **Bae Systems Information and Electronic Systems Integration Inc.**, Nashua, NH (US)

(*) Notice: Subject to any disclaimer, the term of this patent is extended or adjusted under 35 U.S.C. 154(b) by 498 days.

(21) Appl. No.: **10/334,317**

(22) Filed: **Dec. 31, 2002**

(65) **Prior Publication Data**

US 2004/0078103 A1 Apr. 22, 2004

Related U.S. Application Data

(60) Provisional application No. 60/347,670, filed on Jan. 11, 2002.

(51) **Int. Cl.**
G06N 5/00 (2006.01)

(52) **U.S. Cl.** **700/87**; 712/1

(58) **Field of Classification Search** 700/87; 712/1; 385/58; 341/120
See application file for complete search history.

(56) **References Cited**

U.S. PATENT DOCUMENTS

5,960,191 A 9/1999 Sample et al.
6,175,940 B1 1/2001 Saunders
6,462,684 B1 * 10/2002 Medelius et al. 341/120
6,811,320 B1 * 11/2004 Abbott 385/58

OTHER PUBLICATIONS

Bergmann et al., Adaptive Interfacing With Reconfigurable Computers, Australian Computer Science Communications, Proceedings of the 6th Australasian conference on Computer systems architecture ACSAC '01, Jan. 2001, vol. 23 Issue 4, pp. 11-18.*

PCT International Search Report dated Sep. 23, 2003 International Application No. PCT/US03/00568 filed Jan. 9, 2003.

* cited by examiner

Primary Examiner—Wilbert Starks
(74) *Attorney, Agent, or Firm*—Daniel J. Long; Robert K. Tendler

(57) **ABSTRACT**

A reconfigurable digital processing system for space includes the utilization of field programmable gate arrays utilizing a hardware centric approach to reconfigure software processors in a space vehicle through the reprogramming of multiple FPGAs such that one obtains a power/performance characteristic for signal processing tasks that cannot be achieved simply through the use of off-the-shelf processors. In one embodiment, for damaged or otherwise inoperable signal processors located on a spacecraft, the remaining processors which are undamaged can be reconfigured through changing the machine language and binary to the field programmable gate arrays to change the core processor while at the same time maintaining undamaged components so that the signal processing functions can be restored utilizing a RAM-based FPGA as a signal processor. In one embodiment, multiple FPGAs are connected together by a data bus and are also provided with data pipes which interconnect selected FPGAs together to provide the necessary processing function. Flexibility in reconfiguration includes the utilizing of a timing and synchronization block as well as a common configuration block which when coupled to an interconnect block permits reconfiguration of a customizable application core, depending on the particular signal processing function desired. The result is that damaged or inoperable signal processing components can be repaired in space without having to physically attend to the hardware by transmitting to the spacecraft commands which reconfigure the particular FPGAs thus to alter their signal processing function. Also mission changes can be accomplished by reprogramming the FPGAs.

1 Claim, 6 Drawing Sheets

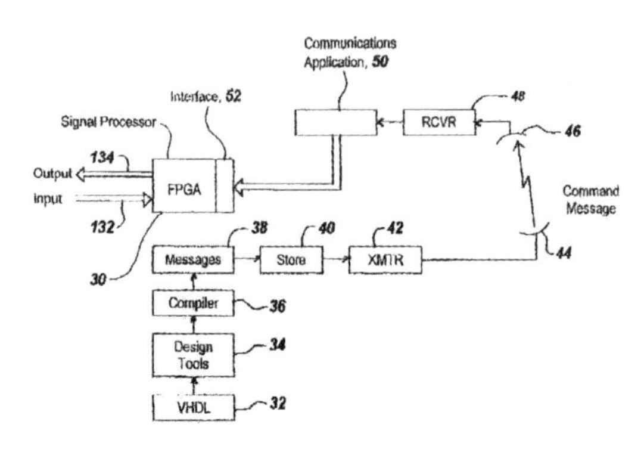

New Jersey

US006993591B1

(12) **United States Patent**
Klemm

(10) **Patent No.: US 6,993,591 B1**
(45) **Date of Patent: Jan. 31, 2006**

(54) **METHOD AND APPARATUS FOR PREFETCHING INTERNET RESOURCES BASED ON ESTIMATED ROUND TRIP TIME**

(75) Inventor:

(73) Assignee: **Lucent Technologies Inc.**, Murray Hill, NJ (US)

(*) Notice: Subject to any disclaimer, the term of this patent is extended or adjusted under 35 U.S.C. 154(b) by 577 days.

(21) Appl. No.: **09/164,509**

(22) Filed: **Sep. 30, 1998**

(51) **Int. Cl.**
G06F 15/16 (2006.01)

(52) **U.S. Cl.** **709/232**; 709/219; 709/223; 707/2; 707/10

(58) **Field of Classification Search** 709/106, 709/202, 219, 213, 214, 215, 216, 223, 235, 709/232, 224; 707/2, 10
See application file for complete search history.

(56) **References Cited**

U.S. PATENT DOCUMENTS

5,864,863 A * 1/1999 Burrows
5,918,002 A * 6/1999 Klemets et al.
5,918,017 A * 6/1999 Attanasio et al. 709/224
5,935,210 A * 8/1999 Strak
5,961,603 A * 10/1999 Kunkel et al.
6,006,265 A * 12/1999 Rangan et al.
6,012,096 A * 1/2000 Link et al. 709/233
6,021,439 A * 2/2000 Turek et al. 709/224
6,026,452 A * 2/2000 Pitts 710/56
6,029,182 A * 2/2000 Nehab et al.
6,067,565 A * 5/2000 Horvitz 709/223
6,078,956 A * 6/2000 Bryant et al. 709/224
6,115,752 A * 9/2000 Chauhan
6,119,235 A * 9/2000 Vaid et al.
6,122,661 A * 9/2000 Stedman et al.
6,134,584 A * 10/2000 Chang et al. 709/219
6,173,318 B1 * 1/2001 Jackson et al. 709/219

(Continued)

OTHER PUBLICATIONS

Venkata N. Padmanabhan and Jeffrey C. Mogul, Using Predictive Prefetching to Improve World Wide Web Latency, Proc. of ACM SIGCOMM '96, (Jul. 1996).

(Continued)

Primary Examiner—Andrew Caldwell
Assistant Examiner—Stephan Willett

(57) **ABSTRACT**

A method and apparatus are disclosed for prefetching Internet resources based on the estimated round trip time of the resources. Whenever a user clicks on an embedded hyperlink, the prefetching strategy aims to ensure that the corresponding document has been prefetched or can be fetched very quickly from its origin server. Web access time as perceived by the user is reduced, while also minimizing the network, server and local resource overhead due to prefetching. The estimated round trip time is obtained or approximated for all referenced documents. The "round trip" time or access time of a resource is the time interval between the sending of the first byte of an HTTP request for the resource until the last byte of the server response has arrived at the requesting Web client. Documents with the longest access times are prefetched first and prefetching generally continues until the estimated round trip time falls below a predefined threshold. An HTTP HEAD request may be used to determine the estimated round trip time of a Web resource. The prefetching agent can be configured to prevent prefetching of those documents that are quickly fetchable, dynamically generated or non-HTTP based resources, or those documents whose size exceed a certain limit, to minimize the network, server and local resource overhead due to prefetching. The thresholds applied to the list of documents to be prefetched can be dynamically adjusted by the agent, based on changing network and server conditions.

29 Claims, 5 Drawing Sheets

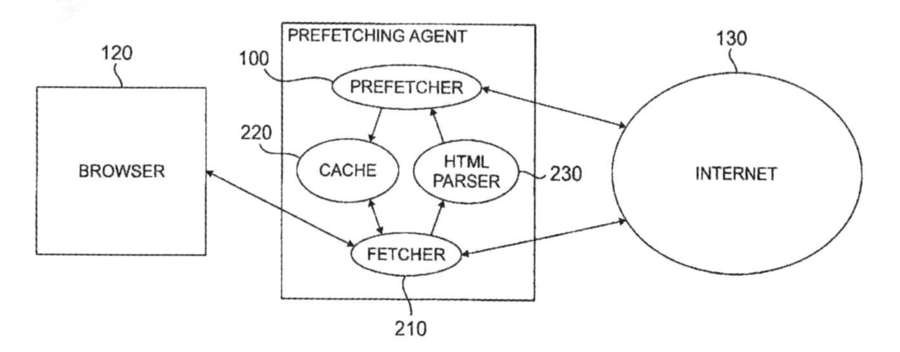

New Mexico

US006985867B1

(12) **United States Patent**
Pryor et al.

(10) **Patent No.:** **US 6,985,867 B1**
(45) **Date of Patent:** **Jan. 10, 2006**

(54) METHOD OF PREDICTING A CHANGE IN AN ECONOMY

(75) Inventors:

(73) Assignee: **Sandia Corporation**, Albuquerque, NM (US)

(*) Notice: Subject to any disclaimer, the term of this patent is extended or adjusted under 35 U.S.C. 154(b) by 973 days.

(21) Appl. No.: **08/791,724**

(22) Filed: **Jan. 29, 1997**

(51) **Int. Cl.**
G06F 17/60 (2006.01)

(52) **U.S. Cl.** **705/1**; 703/6

(58) **Field of Classification Search** 705/1, 705/10, 35, 36, 7; 395/200.3, 200.34, 200.36, 395/200.37, 925; 709/201, 202, 204, 205, 709/206, 207; 703/6
See application file for complete search history.

(56) **References Cited**

U.S. PATENT DOCUMENTS

5,406,477 A *	4/1995	Harhen		703/6
5,444,819 A *	8/1995	Negishi		705/7
5,486,995 A *	1/1996	Krist et al.		700/29
5,652,717 A *	7/1997	Miller et al.		703/6
5,689,650 A *	11/1997	McClelland et al.		705/36
5,737,581 A *	4/1998	Keane		703/6
5,774,878 A *	6/1998	Marshall		705/35
5,812,988 A *	9/1998	Sandretto		705/36
5,818,737 A *	10/1998	Orr et al.		703/6
5,884,283 A *	3/1999	Manos		705/30
5,946,667 A *	8/1999	Tull et al.		705/36
6,212,649 B1 *	4/2001	Yalowitz et al.		709/201 R

FOREIGN PATENT DOCUMENTS

DE	3830326 A1 *	3/1990
JP	08314892 A *	11/1996
WO	WO9744741 *	11/1997

OTHER PUBLICATIONS

Leijonhufvud, Axel, "Towards a Not-Too-Rational Macroeconomics", Southern Economic Journal v. 60 n. 1 p1(13), Jul. 1993.*
Jurik, Mark "Trading Techniques: The Care and Feeding of a Neural Network", Futures: The Magazine of Commodities and Options v. 21 n. 12 p. 40-44, Oct. 1992.*
Coats, Pamela K, "Will Economic System Work for Economic Forecasting?" Journal of Business Forecasting v. 6 n. 1 p. 23-27, 1987.*
Pryor et al., "Development of Aspen: A Microanalytic Simulation Model of the US Economy" Report Number SAND-96-0434, Feb. 1996 (abstract).*
Bull et al., "Monetry Implications of Tax Reforms", National Tax Journal, Sep. 1996, v. 49, n.3, pp. 359-379.*
Basu et al., "Aspen: A Microsimulation Model of the Economy" Sandia Report SAND96-2459, Oct. 1996.*
Goonatilake et al, "Intelligent Systems for Finance and Business", 1995, John Wiley and Sons, chapters 14-16, pp 251-311.*
Pindyck et al, "Microeconomics", 1989, Macmillan Publishing Company, pp 649-657.*

(Continued)

Primary Examiner—James W. Myhre

(57) **ABSTRACT**

An economy whose activity is to be predicted comprises a plurality of decision makers. Decision makers include, for example, households, government, industry, and banks. The decision makers are represented by agents, where an agent can represent one or more decision makers. Each agent has decision rules that determine the agent's actions. Each agent can affect the economy by affecting variable conditions characteristic of the economy or the internal state of other agents. Agents can communicate actions through messages. On a multiprocessor computer, the agents can be assigned to processing elements.

16 Claims, 3 Drawing Sheets

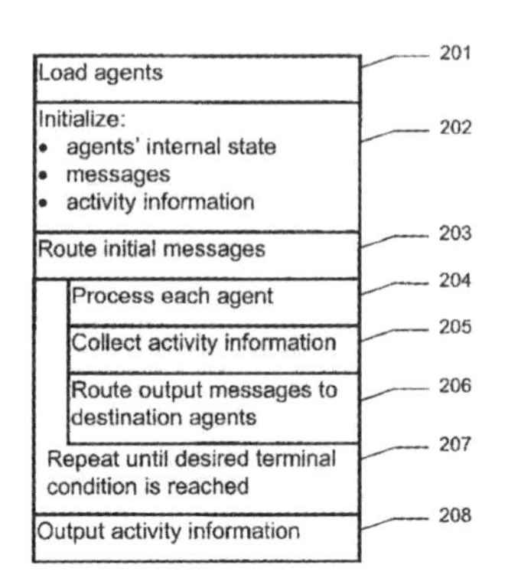

New York

US006989766B2

(12) **United States Patent**
Mese et al.

(10) **Patent No.:** **US 6,989,766 B2**
(45) **Date of Patent:** **Jan. 24, 2006**

(54) SMART TRAFFIC SIGNAL SYSTEM

(75) Inventors:

(73) Assignee: **International Business Machines Corporation**, Armonk, NY (US)

(*) Notice: Subject to any disclaimer, the term of this patent is extended or adjusted under 35 U.S.C. 154(b) by 203 days.

(21) Appl. No.: **10/745,260**

(22) Filed: **Dec. 23, 2003**

(65) **Prior Publication Data**

US 2005/0134478 A1 Jun. 23, 2005

(51) **Int. Cl.**
G08G 1/095 (2006.01)

(52) **U.S. Cl.** **340/907**; 701/118

(58) **Field of Classification Search** 340/905
See application file for complete search history.

(56) **References Cited**

U.S. PATENT DOCUMENTS

3,188,927 A * 6/1965 Woods 404/9
3,750,099 A * 7/1973 Proctor 340/932
3,872,423 A * 3/1975 Yeakley 340/932
5,134,393 A * 7/1992 Henson 340/933
5,539,398 A 7/1996 Hall et al. 340/907
5,673,039 A * 9/1997 Pietzsch et al. 340/905
5,926,113 A 7/1999 Jones et al. 340/906
6,243,026 B1 6/2001 Jones et al. 340/906
6,246,954 B1 6/2001 Berstis et al. 701/117
6,385,531 B2 5/2002 Bates et al. 701/117
2002/0008637 A1 1/2002 Lemelson et al. 340/907
2003/0016143 A1 1/2003 Ghazarian 340/901

* cited by examiner

Primary Examiner—Jeffery Hofsass
Assistant Examiner—George A Bugg
(74) *Attorney, Agent, or Firm*—Jeanine S. Ray-Yarletts

(57) **ABSTRACT**

Traffic signal data is broadcast, for receipt by vehicles traversing the roadways controlled by the traffic signals. If desired, traffic lights are provided with the capability to broadcast their location, status, changing cycles and timing data continuously. A receiving system in a vehicle is configured to receive the traffic signal data and display, to a user of the vehicle, visual display information and/or audible information informing the user of a speed range which, if followed, optimizes the use of the highway and minimizes the number of starts and stops that must be made.

18 Claims, 7 Drawing Sheets

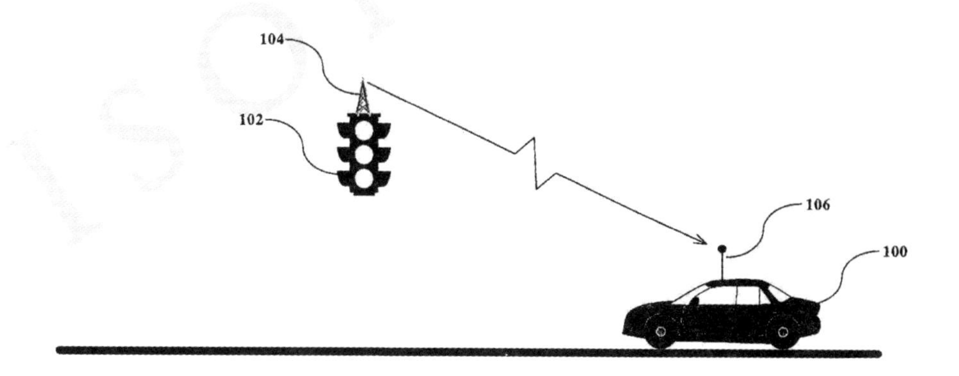

North Carolina

(12) **United States Patent**
Kem et al.

(10) **Patent No.: US 7,087,106 B2**
(45) **Date of Patent: Aug. 8, 2006**

(54) MATERIALS AND METHODS FOR INHIBITING FOULING OF SURFACES EXPOSED TO AQUATIC ENVIRONMENTS

(75) Inventors:

(73) Assignees: **University of Florida**, Gainesville, FL (US); **Duke University**, Durham, NC (US)

(*) Notice: Subject to any disclaimer, the term of this patent is extended or adjusted under 35 U.S.C. 154(b) by 0 days.

(21) Appl. No.: **10/783,312**

(22) Filed: **Feb. 19, 2004**

(65) **Prior Publication Data**

US 2004/0235901 A1 Nov. 25, 2004

Related U.S. Application Data

(60) Provisional application No. 60/449,098, filed on Feb. 20, 2003.

(51) **Int. Cl.**
A01N 43/40 (2006.01)
A01N 443/42 (2006.01)

(52) **U.S. Cl.** **106/18.32**; 106/15.05; 514/277; 514/290; 514/298; 514/299; 514/332; 514/334

(58) **Field of Classification Search** 106/15.05, 106/18.32; 424/78.09; 514/277, 290, 298, 514/299, 332, 334
See application file for complete search history.

(56) **References Cited**

U.S. PATENT DOCUMENTS

5,358,749 A * 10/1994 Fears 427/397.7
5,695,552 A * 12/1997 Taylor 106/15.05
5,945,171 A * 8/1999 Cook 427/456
5,977,144 A 11/1999 Meyer et al.
5,989,323 A * 11/1999 Taylor 106/15.05
6,221,374 B1 * 4/2001 Ghosh et al. 424/405
6,753,397 B1 * 6/2004 Nakamura et al. 528/7
6,900,394 B1 * 5/2005 Itabashi et al. 174/262

FOREIGN PATENT DOCUMENTS

EP 1 033 392 A 9/2000
GB 1 514 651 A 6/1978
JP 51-63835 A * 6/1976
JP 04-337369 A * 11/1992
JP 05-65433 A * 3/1993
WO WO 2004/058901 A 7/2004

OTHER PUBLICATIONS

Kem W., Abbott B., Coates R., "Isolation and structure of a hoplonemertine toxin" *Toxicon*(1971), vol. 9, pp. 15-22, no month.

Kem W., Abbott B., Coates R., "Isolation and structure of a hoplonemertine toxin" *Toxicon*(1971), vol. 9, pp. 15-22, no month.

Kem W., "A study of the occurrence of anabaseine in Paranemertes and other nemertines" *Toxicon* (1971), vol. 9, pp. 23-32, no month.
Kem W., Scott K., Duncan J., "Hoplonemertine worms—a new source of pyridine neurotoxins" *Experientia* (1976), vol. 32, pp. 684-686, no month.
Zoltewicz, J. A., Bloom, L. B., Kem, W. R., "N-Mehtylated 2,3'-Bipyridinium Ion. First Synthesis of the More Sterically Hindered Isomer" *J. Org. Chem.* (1992), vol. 57, pp. 2392-2395, no month.
Hatt, H. and I. Schmiedel-Jakob (1984) "Electrophysiological studies of pyridine-sensitive units on the crayfish walking leg" *J. Comp. Physiol.* 154:855-863.
Kem, William R. (2002) In: Handbook of Neurotoxicology, E.J. Massaro, Ed. vol. 1 Humana Press, Totowa, NJ, pp. 161-193.
Kem, William R. (1988) "Pyridine alkaloid distribution in the hoplonemertines" *Hydrobiologia* 156:145-151.
Matsushima, Sanji, Tomoko Ohsumi, Shiro Sugawara (1983) "Composition of Trace Alkaloids in Tobacco Leaf Lamina" *Agric. Biol. Chem.* 47(3):507-510.
Strausfeld, Nicholas J. and John G. Hildebrand (1999) "Olfactory systems: common design, uncommon origins?" *Curr. Rev. Neurobiol.* 9:634-639.
Warfield, A.H., W.D. Galloway, A.G. Kallianos (1972) "Some New Alkaloids from Burley Tobacco" *Phytochemistry* 11:3371-3375.
Zoltewicz, John A., Linda B. Bloom, William R. Kem (1990) "Hydrolysis of Cholinergic Anabaseine and N-Methylanabaseine: Influence of Cosolvents on the Position of the Ring-Chain Equilibrium-Compensatory Changes" *J. Bioorg. Chem.* 18:395-412.
Zoltewicz, John A., Linda B. Bloom, William R. Kem (1989) "Quantitative Determination of the Ring-Chain Hydrolysis Equilibrium Constant for Anabaseline and Related Tobacco Alkaloids" *J. Org. Chem.* 54:4462-4468.
Patent Abstracts of Japan vol. 0173, No. 82 (C-1085), Jul. 19, 1993 & JP 5065433 A (Suzuki Sogyo Co. Ltd.), Mar. 19, 1993, abstract.
Database WPI, Section Ch, Week 197629, Derwent Publications Ltd., London, GB; AN 1976-55047X & JP 51 063835 A (Kamata T) Jun. 2, 1976, Abstract.
Database WPI, Section Ch, Week 199140, Derwent Publications Ltd., London, GB; AN 1991-290870 & JP 03 192167 A (Chugoku Toryo) Aug. 22, 1991, Abstract.

* cited by examiner

Primary Examiner—Anthony J. Green
(74) *Attorney, Agent, or Firm*—Saliwanchik, Lloyd & Saliwanchik

(57) **ABSTRACT**

The subject invention provides materials and methods for inhibiting the biofouling of surfaces exposed to aquatic environments. In one embodiment, the subject invention provides additives for marine paints and surface treatments. The subject invention further provides repellants and selective inhibitors for aquatic and/or terrestrial crustacean pests.

4 Claims, 2 Drawing Sheets

North Dakota

(12) **United States Patent**
Tormaschy et al.

(10) **Patent No.: US 7,306,719 B2**
(45) **Date of Patent: Dec. 11, 2007**

(54) **WATER CIRCULATION SYSTEMS FOR PONDS, LAKES, AND OTHER BODIES OF WATER**

(75) Inventors:

(73) Assignee: **PSI-ETS, a North Dakota partnership**, Dickinson, ND (US)

(*) Notice: Subject to any disclaimer, the term of this patent is extended or adjusted under 35 U.S.C. 154(b) by 420 days.

(21) Appl. No.: **11/067,135**

(22) Filed: **Feb. 25, 2005**

(65) **Prior Publication Data**

US 2005/0155922 A1 Jul. 21, 2005

Related U.S. Application Data

(63) Continuation-in-part of application No. 10/749,064, filed on Dec. 30, 2003.

(60) Provisional application No. 60/437,217, filed on Dec. 31, 2002.

(51) **Int. Cl.**
C02F 1/00 (2006.01)

(52) **U.S. Cl.** **210/170.05**; 417/61

(58) **Field of Classification Search** 210/170.05, 210/170.08, 170.09; 417/61
See application file for complete search history.

(56) **References Cited**

U.S. PATENT DOCUMENTS

1,530,326 A		3/1925	Prindle	
2,827,268 A		3/1958	Staaf	
3,204,768 A		9/1965	Daniel	
3,512,375 A		5/1970	Madarasz	
3,794,303 A		2/1974	Hirshon	
3,856,272 A	*	12/1974	Ravitts	417/61
4,030,859 A		6/1977	Henegar	
4,179,243 A		12/1979	Aide	
4,764,313 A		8/1988	Cameron et al.	
5,021,154 A	*	6/1991	Haegeman	210/221.2
5,122,266 A		6/1992	Kent	
6,273,402 B1		8/2001	Cheng	
6,432,302 B1		8/2002	Obritsch	
6,439,853 B2		8/2002	Tormaschy	

FOREIGN PATENT DOCUMENTS

CA	1262052	10/1989
JP	11-47794	2/1999

* cited by examiner

Primary Examiner—Michael Koczo, Jr.
(74) *Attorney, Agent, or Firm*—W. Scott Carson

(57) **ABSTRACT**

Circulation systems for ponds, lakes, or other bodies of water using a flotation platform, dish, and impeller. One embodiment has a connecting arrangement between the drive motor and the impeller that permits the two to be easily and quickly coupled and uncoupled. The connecting arrangement also is designed to accommodate slight misalignments between the shafts. An arrangement to adjust and calibrate the depth of the inlet to the draft tube is disclosed. The system further includes solar panels that can be pivotally swung outwardly to open positions, angularly adjusted about a horizontal axis, and mounted to face toward the central axis of the flotation platform rather than away from it. Arrangements are further provided to monitor and control the components of the system including remotely from shore.

7 Claims, 32 Drawing Sheets

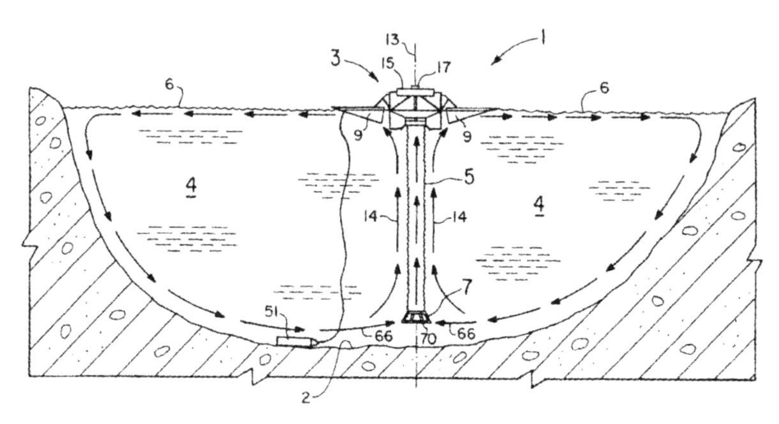

Ohio

US007307125B2

(12) **United States Patent**
Chundury et al.

(10) **Patent No.: US 7,307,125 B2**
(45) **Date of Patent: Dec. 11, 2007**

(54) **THERMOPLASTIC OLEFIN COMPOSITIONS AND INJECTION MOLDED ARTICLES MADE THEREOF**

(75) Inventors:

(73) Assignee: **Ferro Corporation**, Cleveland, OH (US)

(*) Notice: Subject to any disclaimer, the term of this patent is extended or adjusted under 35 U.S.C. 154(b) by 37 days.

(21) Appl. No.: **11/012,897**

(22) Filed: **Dec. 15, 2004**

(65) **Prior Publication Data**

US 2006/0128860 A1 Jun. 15, 2006

(51) **Int. Cl.**
C08L 23/10 (2006.01)
C08K 3/34 (2006.01)

(52) **U.S. Cl.** **525/240**; 524/451; 524/584; 524/579

(58) **Field of Classification Search** 525/240; 524/451, 584, 579
See application file for complete search history.

(56) **References Cited**

U.S. PATENT DOCUMENTS

5,106,696 A 4/1992 Chundury et al. 428/517
5,264,280 A 11/1993 Chundury et al. 428/330
5,321,081 A 6/1994 Chundury et al. 525/98
5,360,868 A 11/1994 Mosier et al. 525/89
5,723,527 A * 3/1998 Sadatoshi et al. 524/451
5,747,592 A * 5/1998 Huff et al. 525/191
5,969,027 A 10/1999 Chundury et al. 524/436
5,985,973 A * 11/1999 Sumitomo et al. 524/451
6,046,264 A 4/2000 Muller et al. 524/407
6,245,856 B1 6/2001 Kaufman et al. 525/240
6,306,972 B1 * 10/2001 Ohkawa et al. 525/240
6,441,081 B1 * 8/2002 Sadatoshi et al. 524/451
6,593,005 B2 * 7/2003 Tau et al. 428/516
6,620,891 B2 9/2003 Yu et al. 525/240
6,656,987 B2 12/2003 Takashima et al. 524/127
6,803,421 B2 * 10/2004 Joseph 525/240
6,815,490 B2 * 11/2004 Seelert et al. 524/451
2002/0039630 A1 * 4/2002 Rousselet et al. 428/35.7
2004/0092631 A1 5/2004 Joseph 524/230
2004/0242776 A1 * 12/2004 Strebel et al. 525/100
2005/0009991 A1 * 1/2005 Meka et al. 525/240

OTHER PUBLICATIONS

KR 2004/099865 (Byun et al.), Dec. 2, 2004; abstract in English.*
STRUKTOL TR106 technical data sheet.*

* cited by examiner

Primary Examiner—Ling-Sui Choi
(74) *Attorney, Agent, or Firm*—Rankin, Hill, Porter & Clark LLP

(57) **ABSTRACT**

A thermoplastic polymer composition that exhibits substantially isotropic post injection molding shrinkage. The thermoplastic polymer composition includes a blend of highly crystalline polypropylene homopolymer, an ethylene-C_{4-8} α-olefin plastomer and talc having a D_{50} of about 2.0 μm or less. The thermoplastic polymer composition exhibits post injection molding shrinkage, low temperature impact strength and tensile strength that is similar to or better than relatively expensive engineering resins and blends. The thermoplastic polymer composition according to the invention is particularly useful for forming injection molded parts for motor vehicles.

8 Claims, No Drawings

Oklahoma

US006992048B2

(12) **United States Patent**
Reddy et al.

(10) **Patent No.: US 6,992,048 B2**
(45) **Date of Patent: *Jan. 31, 2006**

(54) **METHODS OF GENERATING GAS IN WELL TREATING FLUIDS**

(75) Inventors:

(73) Assignee: **Halliburton Energy Services, Inc.**, Duncan, OK (US)

(*) Notice: Subject to any disclaimer, the term of this patent is extended or adjusted under 35 U.S.C. 154(b) by 0 days.

This patent is subject to a terminal disclaimer.

(21) Appl. No.: **10/792,999**

(22) Filed: **Mar. 4, 2004**

(65) **Prior Publication Data**

US 2004/0168801 A1 Sep. 2, 2004

Related U.S. Application Data

(63) Continuation of application No. 10/159,588, filed on May 31, 2002, now Pat. No. 6,722,434.

(51) **Int. Cl.**
C09K 3/00 (2006.01)

(52) **U.S. Cl.** **507/202**; 507/211; 507/217; 166/309

(58) **Field of Classification Search** 507/102, 507/202, 110, 115, 211, 217; 166/309
See application file for complete search history.

(56) **References Cited**

U.S. PATENT DOCUMENTS

3,117,699 A * 1/1964 Epstein 222/389
3,591,394 A 7/1971 Digglemann et al. 106/87
3,958,638 A 5/1976 Johnston 166/294
3,977,470 A 8/1976 Chang 166/273
4,099,912 A * 7/1978 Ehrlich 8/137
4,121,674 A 10/1978 Fischer et al. 175/66
4,142,909 A 3/1979 Gaines 106/87
4,201,678 A 5/1980 Pye et al. .... 252/8.5 A
4,219,083 A 8/1980 Richardson et al. 166/300
4,232,741 A 11/1980 Richardson et al. 166/281
4,289,633 A 9/1981 Richardson et al. ... 252/8.55 B
4,304,298 A 12/1981 Sutton 166/293
4,333,764 A 6/1982 Richardson 106/87
4,340,427 A 7/1982 Sutton 106/87
4,367,093 A 1/1983 Burkhalter et al. 106/87
4,450,010 A * 5/1984 Burkhalter et al. 106/673
4,452,898 A * 6/1984 Richardson 436/2
4,565,578 A 1/1986 Sutton et al. 106/87
4,692,269 A 9/1987 Kmiec et al. 252/350
4,728,675 A * 3/1988 Pressman 521/92
4,741,401 A 5/1988 Walles et al. 166/300
4,758,003 A * 7/1988 Goldstein et al. 277/314
4,813,484 A 3/1989 Hazlett 166/270
4,832,123 A 5/1989 Abou-Sayed et al. 166/281
4,844,163 A 7/1989 Hazlett et al. 166/270
4,848,465 A 7/1989 Hazlett 166/270
4,899,819 A 2/1990 Hazlett et al. 166/285
5,373,901 A 12/1994 Norman et al. 166/300
5,413,178 A 5/1995 Walker et al. 166/300
5,495,891 A 3/1996 Sydansk 166/295
5,613,558 A 3/1997 Dillenbeck, III 166/293
5,658,380 A 8/1997 Dillenbeck, III 106/823
5,669,446 A 9/1997 Walker et al. 166/300
5,789,352 A 8/1998 Carpenter et al. 507/209
5,950,731 A 9/1999 Shuchart et al. 166/300
5,962,808 A 10/1999 Lundstrom 149/19.1
5,972,103 A 10/1999 Mehta et al. 106/728
5,996,693 A 12/1999 Heathman 166/291
6,035,933 A * 3/2000 Khalil et al. 166/263
6,063,738 A 5/2000 Chatterji et al. 507/269
6,138,760 A 10/2000 Lopez et al. 166/300
6,162,839 A * 12/2000 Klauck et al. 521/83
6,187,720 B1 2/2001 Acker et al. 507/238
6,209,646 B1 4/2001 Reddy et al. 166/300
6,270,565 B1 8/2001 Heathman et al. 106/696
6,354,381 B1 * 3/2002 Habeeb et al. 166/400
6,364,945 B1 4/2002 Chatterji et al. 106/677
6,419,016 B1 7/2002 Reddy 166/293
6,444,316 B1 9/2002 Reddy et al. 428/407
6,460,632 B1 10/2002 Chatterji et al. 175/66
6,592,660 B2 7/2003 Nguyen et al. 106/677
6,605,304 B1 * 8/2003 Wellinghoff et al. 424/489
6,681,857 B2 * 1/2004 Habeeb et al. 166/299
6,715,553 B2 * 4/2004 Reddy et al. 166/309
6,722,434 B2 * 4/2004 Reddy et al. 166/292
2002/0035951 A1 3/2002 Chatterji et al. 106/677
2004/0110643 A1 6/2004 Zavallos

OTHER PUBLICATIONS

Paper entitled "Specification for Materials And Testing for Well Cements" by American Petroleum Institute, dated Jul. 1990.

* cited by examiner

Primary Examiner—Philip C. Tucker
(74) *Attorney, Agent, or Firm*—Craig W. Roddy; Baker Botts L.L.P.

(57) **ABSTRACT**

The present invention relates to methods of generating gas in and foaming well treating fluids during pumping of the treating fluids or after the treating fluids are placed in a subterranean zone, or both. A method of the present invention provides a method of making a foamed well fluid that comprises a gas comprising the steps of combining an aqueous fluid, a surfactant, an encapsulated activator, and a gas generating chemical, the gas generating chemical being present in an amount in the range of from about 0.1% to 100% of a water component in the aqueous well fluid; and allowing the gas generating chemical and the encapsulated activator to react so that gas is generated in the aqueous fluid to form a foamed well fluid. Methods of cementing, fracturing, cementing compositions, fracturing fluid compositions, and foamed well fluid compositions also are provided.

28 Claims, No Drawings

Oregon

(12) **United States Patent**
Dua

(10) **Patent No.: US 6,986,269 B2**
(45) **Date of Patent: Jan. 17, 2006**

(54) **FOOTWEAR WITH KNIT UPPER AND METHOD OF MANUFACTURING THE FOOTWEAR**

(75) Inventor:

(73) Assignee: **Nike, Inc.,** Beaverton, OR (US)

(*) Notice: Subject to any disclaimer, the term of this patent is extended or adjusted under 35 U.S.C. 154(b) by 0 days.

(21) Appl. No.: **11/024,480**

(22) Filed: **Dec. 30, 2004**

(65) **Prior Publication Data**

US 2005/0115284 A1 Jun. 2, 2005

Related U.S. Application Data

(62) Division of application No. 10/323,608, filed on Dec. 18, 2002, now Pat. No. 6,931,762.

(51) **Int. Cl.**
D04B 11/00 (2006.01)

(52) **U.S. Cl.** **66/177**

(58) **Field of Classification Search** .... 66/170, 66/171, 177–188; 36/9 R, 45

See application file for complete search history.

(56) **References Cited**

U.S. PATENT DOCUMENTS

1,597,934 A *	8/1926	Stimpson	36/3 A
2,092,616 A *	9/1937	Ruckel	66/182
2,147,197 A	2/1939	Glidden	
2,675,631 A *	4/1954	Doughty	36/9 R
4,211,806 A	7/1980	Civardi et al.	
4,216,662 A *	8/1980	Harris et al.	66/186
4,255,949 A	3/1981	Thorneburg	
4,373,361 A	2/1983	Thorneburg	
4,447,967 A	5/1984	Zaino	
4,607,439 A	8/1986	Harada	
4,785,558 A	11/1988	Shiomura	
4,813,158 A	3/1989	Brown	

OTHER PUBLICATIONS

Internet publication entitled "Acorn Footwear—Slipper Sock," from Northland Marine, which shows products that were on sale in this country at least one year prior to the filing date of the present application, 1 page.

Internet publication entitled "Welcome to Arcopedico Shoe, " from Arcopedico Shoes, which shows products that were on sale in this country at least one year prior to the filing date of the present application, 4 pages.

Leaflet entitled "X machine," from Sangiacomo S.p.A., which was on sale in this country at least one year prior to the filing date of the present application, 1 page.

Advertising material entitled "Still Crazy After All These Years," which shows a product entitled "Sock Racer," and was sold in this country in 1986 by NIKE, Inc., 3 pages.

* cited by examiner

Primary Examiner—Danny Worrell
(74) *Attorney, Agent, or Firm*—Banner & Witcoff, Ltd.

(57) **ABSTRACT**

An article of footwear with a knit upper and a method of manufacturing the footwear are disclosed. The upper is formed through a knitting process to include a plurality of sections formed of different yarns and knits to provide the sections with different physical properties. In portions of the upper where sections formed of different yarns are in adjacent wales, a tuck stitch is utilized to join the sections. The method utilizes a circular knitting machine having multiple feeds that work together to knit the upper into a unitary, seamless structure. The multiple feeds, each of which provide multiple types of yarns, produce the sections to have varying physical properties.

27 Claims, 3 Drawing Sheets

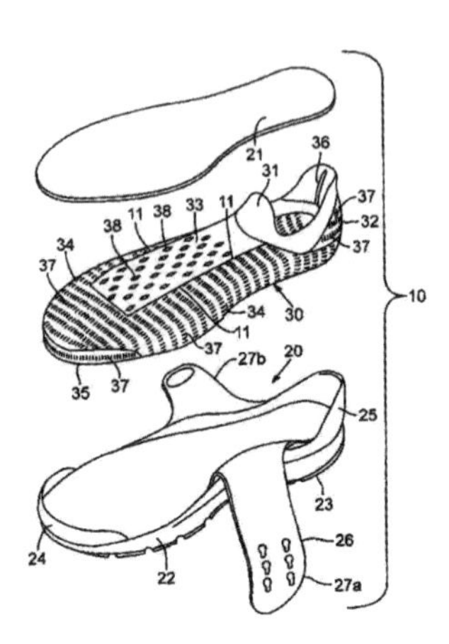

Pennsylvania

(12) **United States Patent**
Gabara et al.

(10) **Patent No.:** **US 7,024,000 B1**
(45) **Date of Patent:** **Apr. 4, 2006**

(54) **ADJUSTMENT OF A HEARING AID USING A PHONE**

(75) Inventors:

(73) Assignee: **Agere Systems Inc.**, Allentown, PA (US)

(*) Notice: Subject to any disclaimer, the term of this patent is extended or adjusted under 35 U.S.C. 154(b) by 963 days.

(21) Appl. No.: **09/589,391**

(22) Filed: **Jun. 7, 2000**

(51) **Int. Cl.**
H04B 29/00 (2006.01)
A61B 5/00 (2006.01)
H04M 11/00 (2006.01)

(52) **U.S. Cl.** **381/60**; 600/559; 379/52

(58) **Field of Classification Search** 381/60; 73/585; 600/559; 379/52, 102.01, 102.02, 379/102.07
See application file for complete search history.

(56) **References Cited**

U.S. PATENT DOCUMENTS

4,284,847 A * 8/1981 Besserman 73/585
5,226,086 A * 7/1993 Platt 381/58
5,479,522 A 12/1995 Lindemann et al. 381/68.2
5,825,894 A 10/1998 Shennib 381/60
5,835,611 A 11/1998 Kaiser et al. 381/321
5,852,668 A 12/1998 Ishige et al. 381/312
5,870,481 A 2/1999 Dymond et al. 381/60
5,923,764 A 7/1999 Shennib 381/60
6,086,541 A * 7/2000 Rho 600/559
6,118,877 A * 9/2000 Lindemann et al. 381/60
6,212,496 B1 * 4/2001 Campbell et al. 704/221
6,522,988 B1 * 2/2003 Hou 702/122
6,549,633 B1 * 4/2003 Westermann 381/312
6,741,712 B1 * 5/2004 Bisgaard 381/312

* cited by examiner

Primary Examiner—Vivian Chin
Assistant Examiner—Devona E Faulk

(57) **ABSTRACT**

A system and method for using a telephone to reconfigure or readjust the performance characteristics of a hearing aid or to check whether a user has a hearing problem. The telephone is used to generate one or more frequency tests covering the audible spectrum using a DSP contained in the phone, an external computer and/or a hearing aid. The keypad of the phone or keyboard of an attached computer is used as a feedback mechanism. The generated frequencies can be used to test the hearing of a user and the quality (or fit) of a hearing aid while being worn by the user. A local memory may be used to store the results of the tests for future reference or for transmission over the network for analysis at a later time. Once the hearing response of a user wearing the hearing aid has been measured, an updated compensation configuration (audiogram) can be downloaded into the hearing aid via an infra-red link, via a physical connection or a direct audio transmission from the telephone to the DSP in the hearing aid.

45 Claims, 3 Drawing Sheets

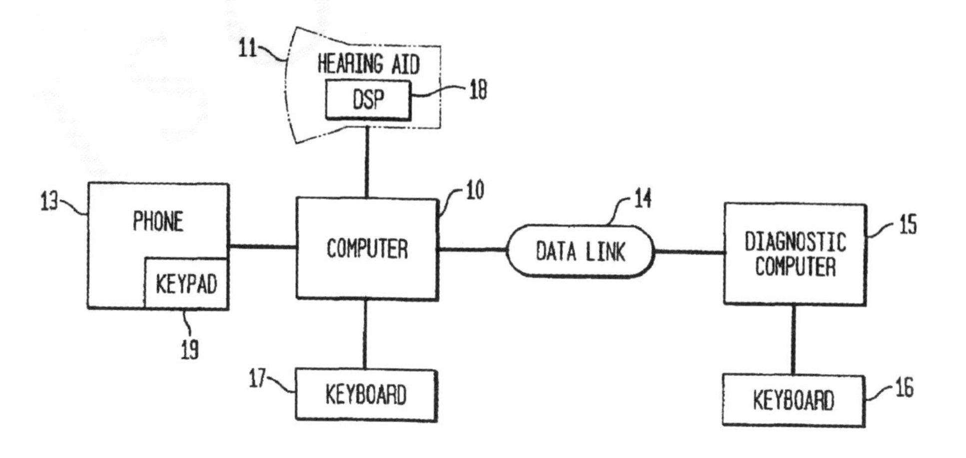

Rhode Island

(12) **United States Patent**
Wilk et al.

(10) **Patent No.:** **US 7,104,543 B2**
(45) **Date of Patent:** **Sep. 12, 2006**

(54) **ACTION FIGURE GAME PIECE AND METHOD OF PLAYING ACTION FIGURE GAME**

(75) Inventors:

(73) Assignee: **Hasbro, Inc.**, Pawtucket, RI (US)

(*) Notice: Subject to any disclaimer, the term of this patent is extended or adjusted under 35 U.S.C. 154(b) by 0 days.

(21) Appl. No.: **10/643,526**

(22) Filed: **Aug. 19, 2003**

(65) **Prior Publication Data**

US 2005/0040598 A1 Feb. 24, 2005

(51) **Int. Cl.**
A63F 3/00 (2006.01)

(52) **U.S. Cl.** **273/288**; 446/272; 446/297

(58) **Field of Classification Search** 446/236, 446/230, 231, 232, 102, 103, 83, 84, 268, 446/269, 397, 54, 265, 272, 297; 273/288
See application file for complete search history.

(56) **References Cited**

U.S. PATENT DOCUMENTS

1,222,551	A	5/1917	MacPherson	
2,052,035	A *	8/1936	Potter	273/317.3
2,610,060	A	9/1952	Powell	
2,735,221	A *	2/1956	Fields	446/435
3,222,068	A	12/1965	Cowels, Jr.	
3,387,778	A *	6/1968	Althaus	235/95 R
3,945,640	A	3/1976	Denmark	
4,005,543	A *	2/1977	McKay	446/65
4,039,188	A	8/1977	Goldfarb et al.	
4,083,564	A	4/1978	Matsumoto	
4,211,408	A	7/1980	Tickle	
4,280,300	A *	7/1981	Kulesza et al.	446/414
4,317,570	A *	3/1982	Brunton	273/246
4,783,080	A	11/1988	Rosenwinkel et al.	

(Continued)

FOREIGN PATENT DOCUMENTS

EP 0 850 671 7/1998

(Continued)

OTHER PUBLICATIONS

International Search Report, International Application No. PCT/US2004/011557, search completed Oct. 4, 2004.

(Continued)

Primary Examiner—Vishu Mendiratta
(74) *Attorney, Agent, or Firm*—Marshall, Gerstein & Borun LLP

(57) **ABSTRACT**

An action figure game piece is disclosed. The game piece comprises a game figure character adapted for movement a distance over a surface and a mechanism for indicating the distance of movement over the surface. A method of playing an action figure game on the surface is also disclosed. The method comprises providing for each player a plurality of the game pieces, each game piece having an indicium indicating a point value, a mechanism for indicating a distance the game piece has moved over the surface, and a mechanism for attacking an opponent player's game pieces. The method also comprises selecting for each player a group of the game pieces having point values summing to a predetermined value and selecting a number of actions each player can take per player's turn. An action comprises either a move, comprising moving a game piece a distance up to the particular game piece's point value, or an attack, comprising actuation of the particular game piece's attacking mechanism. Players alternate taking turns, selectively making a move or an attack, until the game is determined to have ended.

7 Claims, 4 Drawing Sheets

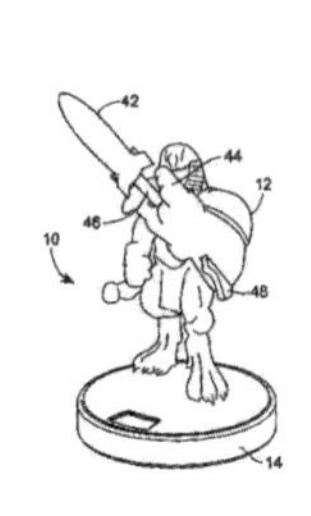

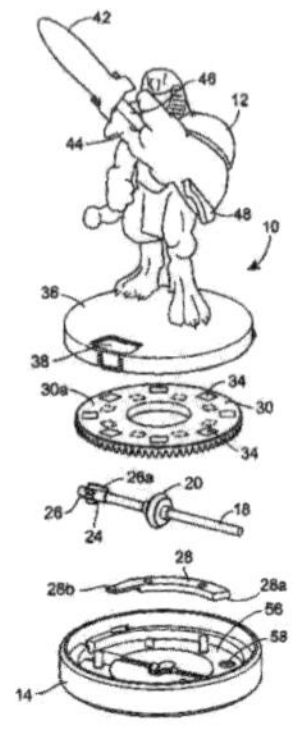

South Carolina

(12) **United States Patent**
Brown et al.

(10) **Patent No.:** **US 6,990,703 B2**
(45) **Date of Patent:** **Jan. 31, 2006**

(54) METHOD OF REPRODUCING, RECOLORING AND/OR RECYCLING CARPET TILES

(75) Inventors:

(73) Assignee: **Milliken & Company**, Spartanburg, SC (US)

(*) Notice: Subject to any disclaimer, the term of this patent is extended or adjusted under 35 U.S.C. 154(b) by 477 days.

(21) Appl. No.: **09/920,152**

(22) Filed: **Aug. 1, 2001**

(65) **Prior Publication Data**

US 2002/0074075 A1 Jun. 20, 2002

Related U.S. Application Data

(60) Provisional application No. 60/223,450, filed on Aug. 4, 2000.

(51) **Int. Cl.**
D06B 1/02 (2006.01)

(52) **U.S. Cl.** .... **8/150**; 68/200; 68/205 R

(58) **Field of Classification Search** .... 8/150, 8/148; 68/28, 200, 205 R
See application file for complete search history.

(56) **References Cited**

U.S. PATENT DOCUMENTS

3,010,859 A * 11/1961 Stephens et al.
3,402,094 A * 9/1968 Levitch
3,969,779 A 7/1976 Stewart, Jr. 8/149
4,877,669 A * 10/1989 Endrenyi, Jr. et al.
4,926,520 A * 5/1990 Watson
4,991,307 A 2/1991 Higgins 33/526
5,096,747 A 3/1992 Scholla et al. 427/299
5,116,439 A * 5/1992 Raus
5,324,562 A * 6/1994 Millinax et al.
5,380,561 A 1/1995 Dorn 427/430
5,381,592 A * 1/1995 Higgins
5,457,845 A 10/1995 Higgins et al. 15/302
5,658,430 A * 8/1997 Drake, Jr. et al.
5,763,001 A * 6/1998 Brown

FOREIGN PATENT DOCUMENTS

CA	50162/54		6/1990
DE	19603951		8/1997
GB	1338030		11/1973
JP	4988512		8/1974
JP	5196177		8/1976
JP	52148970		12/1977
JP	8-92880	*	4/1996

* cited by examiner

Primary Examiner—Frankie L. Stinson
(74) *Attorney, Agent, or Firm*—Terry T. Moyer; Daniel R. Alexander

(57) **ABSTRACT**

Reproducing, recoloring and/or recycling of used carpet tiles is provided. In accordance with one embodiment, used carpet tiles, which are recovered, are subjected to a choosing step according to the degree of stains, etc. The chosen carpet tiles are washed with a high-pressure fluid, and entangling of piles is removed and piles raised. The resultant carpet tiles are subjected to a choosing step once again according to the degree of stains, etc. The chosen carpet tiles are treated with such a design and color for recycling as to make less visible the stains or non-uniformity in color remaining after the washing, using a design computer, etc. Thus, reproduced carpet tiles are produced and subjected a choosing step once again, and the chosen ones are shipped.

19 Claims, 10 Drawing Sheets

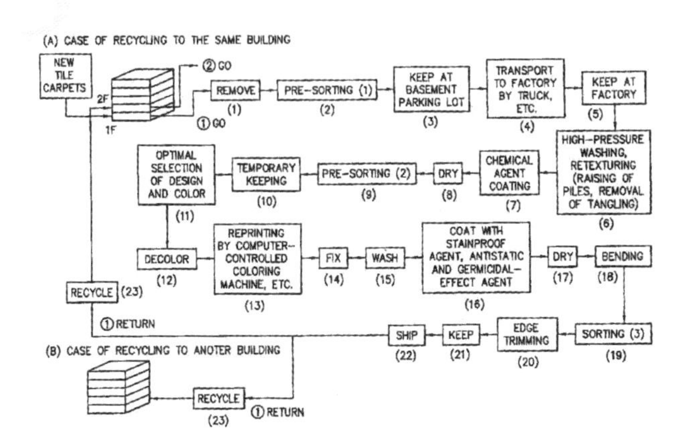

South Dakota

US007004065B2

(12) **United States Patent**
Roth

(10) **Patent No.: US 7,004,065 B2**
(45) **Date of Patent: *Feb. 28, 2006**

(54) APPARATUS FOR TREATING AMMONIATED MEATS

(75) Inventor:

(73) Assignee: **Freezing Machines, Inc.**, Dakota Dunes, SD (US)

(*) Notice: Subject to any disclaimer, the term of this patent is extended or adjusted under 35 U.S.C. 154(b) by 119 days.

This patent is subject to a terminal disclaimer.

(21) Appl. No.: **10/662,616**

(22) Filed: **Sep. 15, 2003**

(65) **Prior Publication Data**

US 2004/0067287 A1 Apr. 8, 2004

Related U.S. Application Data

(60) Continuation of application No. 10/173,955, filed on Jun. 18, 2002, which is a continuation-in-part of application No. 09/579,783, filed on May 26, 2000, now Pat. No. 6,406,728, which is a division of application No. 09/286,699, filed on Apr. 6, 1999, now Pat. No. 6,142,067.

(51) **Int. Cl.**
A23L 1/00 (2006.01)

(52) **U.S. Cl.** **99/534**; 99/510

(58) **Field of Classification Search** 99/325–331, 99/348, 516, 467, 468, 472, 484, 485–492, 99/493, 510, 534–536; 241/38–43, 62, 95, 241/301, 292, 299; 426/231, 646, 319, 332, 426/518, 630, 632–634, 656, 807; 366/132, 366/140–142

See application file for complete search history.

(56) **References Cited**

U.S. PATENT DOCUMENTS

2,711,373	A	6/1955	Coleman et al.
3,023,109	A	2/1962	Hines
3,119,696	A	1/1964	Williams
3,681,851	A	8/1972	Fleming
3,711,392	A	1/1973	Metzger
3,875,310	A	4/1975	Rawlings et al.
4,107,262	A	8/1978	Leuders et al.
4,419,414	A	12/1983	Fischer
4,919,955	A	4/1990	Mitchell
5,082,679	A	1/1992	Chapman
5,393,547	A	2/1995	Balaban et al.
5,405,630	A	4/1995	Ludwig
5,433,142	A	7/1995	Roth
5,558,774	A	9/1996	Tonelli et al.

(Continued)

FOREIGN PATENT DOCUMENTS

GB	1223159	2/1971

(Continued)

OTHER PUBLICATIONS

Author unknown, "Mott Spager Application; pH Control—Neutralizing Alkaline Solutions," Mott Industrial, Division of Mott Corporation, Feb. 1996.

(Continued)

Primary Examiner—Timothy F. Simone
(74) *Attorney, Agent, or Firm*—Russell D. Culbertson; The Culbertson Group, P.C.

(57) **ABSTRACT**

An initial comminuted meat (**21**) is exposed to ammonia to produce an ammoniated meat product. The ammoniated meat product is further comminuted in a comminuting device (**12**) to produce a further comminuted meat product (**22**). The ammonia exposure to the original comminuted meat product (**21**) is controlled to result in a pH of at least around 6.0 in the further comminuted meat product (**22**).

20 Claims, 3 Drawing Sheets

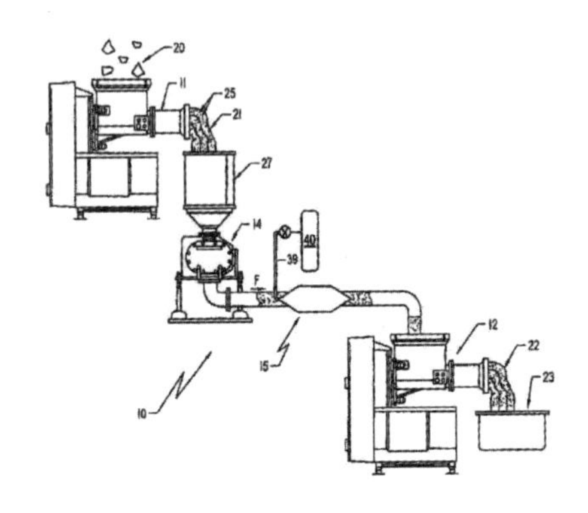

Tennessee

US007072705B2

(12) **United States Patent**
Miga et al.

(10) **Patent No.: US 7,072,705 B2**
(45) **Date of Patent: Jul. 4, 2006**

(54) **APPARATUS AND METHODS OF BRAIN SHIFT COMPENSATION AND APPLICATIONS OF THE SAME**

(75) Inventors:

(73) Assignee: **Vanderbilt University**, Nashville, TN (US)

(*) Notice: Subject to any disclaimer, the term of this patent is extended or adjusted under 35 U.S.C. 154(b) by 28 days.

(21) Appl. No.: **10/988,982**

(22) Filed: **Nov. 15, 2004**

(65) **Prior Publication Data**

US 2005/0101855 A1 May 12, 2005

Related U.S. Application Data

(63) Continuation-in-part of application No. 10/936,339, filed on Sep. 8, 2004.

(60) Provisional application No. 60/501,514, filed on Sep. 8, 2003.

(51) **Int. Cl.**
A61B 5/055 (2006.01)

(52) **U.S. Cl.** **600/411**; 600/427

(58) **Field of Classification Search** 600/410, 600/425, 437, 411, 427; 382/128–131, 168–170; 324/309, 318
See application file for complete search history.

(56) **References Cited**

U.S. PATENT DOCUMENTS

6,006,126 A * 12/1999 Cosman 600/426
6,272,370 B1 * 8/2001 Gillies et al. 600/411
2003/0071194 A1 * 4/2003 Mueller et al. 250/208.1
2004/0013289 A1 * 1/2004 Labudde 382/128
2004/0046557 A1 * 3/2004 Karmarkar et al. 324/322

OTHER PUBLICATIONS

C. R. Maurer, J. M. Fitzpatrick, M. Y. Wang, R. L. Galloway, R. J. Maciunas, and G. S. Allen, "Registration of head volume images using implantable fiducial markers," *IEEE Trans. Med. Imag.*, vol. 16, pp. 447-462, Apr. 1997.

(Continued)

Primary Examiner—Eleni Mantis-Mercader
(74) *Attorney, Agent, or Firm*—Morris Manning & Martin; Tim Tingkang Xia, Esq.

(57) **ABSTRACT**

A method of compensation for intra-operative brain shift of a living subject. In one embodiment, the method includes the steps of pro-operatively acquiring brain images of the living subject, constructing a statistical atlas of brain displacements of the living subject from the pro-operatively acquired brain images, intra-operatively measuring brain displacements of the living subject, deriving an intra-operative displacement atlas from the intra-operatively measured brain displacements and the statistical atlas, obtaining intra-operative brain shift at least from the intra-operative displacement atlas, and compensating for the intra-operative brain shift.

33 Claims, 6 Drawing Sheets

110 — Pro-operatively acquiring brain images of the living subject.

120 — Constructing a statistical atlas of brain displacements of the living subject from the pro-operatively acquired brain images.

130 — Intra-operatively measuring brain displacements of the living subject.

140 — Deriving an intra-operative displacement atlas from the intra-operatively measured brain displacements and the statistical atlas.

150 — Obtaining intra-operative brain shift at least from the intra-operative displacement atlas.

160 — Compensating for the intra-operative brain shift.

Texas

(12) **United States Patent**
Wolf

(10) **Patent No.:** **US 6,993,704 B2**
(45) **Date of Patent:** **Jan. 31, 2006**

(54) **CONCURRENT MEMORY CONTROL FOR TURBO DECODERS**

(75) Inventor:

(73) Assignee: **Texas Instruments Incorporated**, Dallas, TX (US)

(*) Notice: Subject to any disclaimer, the term of this patent is extended or adjusted under 35 U.S.C. 154(b) by 466 days.

(21) Appl. No.: **10/141,416**

(22) Filed: **May 8, 2002**

(65) **Prior Publication Data**

US 2003/0208716 A1 Nov. 6, 2003

Related U.S. Application Data

(60) Provisional application No. 60/293,014, filed on May 23, 2001.

(51) **Int. Cl.**
H03M 13/03 (2006.01)

(52) **U.S. Cl.** **714/794**; 714/795

(58) **Field of Classification Search** 714/794, 714/795; 375/262, 341
See application file for complete search history.

(56) **References Cited**

U.S. PATENT DOCUMENTS

6,516,444 B1 * 2/2003 Maru 714/795
6,598,204 B1 * 7/2003 Giese et al. 714/795

OTHER PUBLICATIONS

J. Dielissen, et al.; *Power-Efficient Layered Turbo Decoder Processor*, IEEE Proc. of Conf. Design, Automation and Test in Europe; Los Alamitos, US, 13 Mar. 2001, pp. 246-251.
C. Schurgers, et al.; *Energy Efficient Data Transfer and Storage Organization for a MAP Turbo Decoder Module*, Proc. 1999 Int'l Symp. On Low Power Electronics and Design, San Diego, CA Aug. 16-17, 1999, pp. 76-81.
J. Vogt, et al.; *Comparision of Different Turbo Decoder Realization for LMT-2000*, Proc. of the Global Telecommunications Conf. 1999, GLOBECOM '99, vol. 5, Dec. 5, 1999, pp. 2704-2708.

* cited by examiner

Primary Examiner—Joseph Torres
(74) *Attorney, Agent, or Firm*—Robert D. Marshall, Jr.; W. James Brady, III; Frederick J. Telecky, Jr.

(57) **ABSTRACT**

The concurrent memory control turbo decoder solution of this invention uses a single port main memory and a simplified scratch memory. This approach uses an interleaved forward-reverse addressing which greatly relieves the amount of memory required. This approach is in marked contrast to conventional turbo decoders which employ either a dual port main memory or a single port main memory in conjunction with a complex ping-ponged scratch memory. In the system of this invention, during each cycle accomplishes one read and one write operation in the scratch memories. If a particular location in memory, has been read, then that location is free. The next write cycle can use that location to store its data. Similarly a simplified beta RAM is implemented using a unique addressing scheme which also obviates the need for a complex ping-ponged beta RAM.

1 Claim, 6 Drawing Sheets

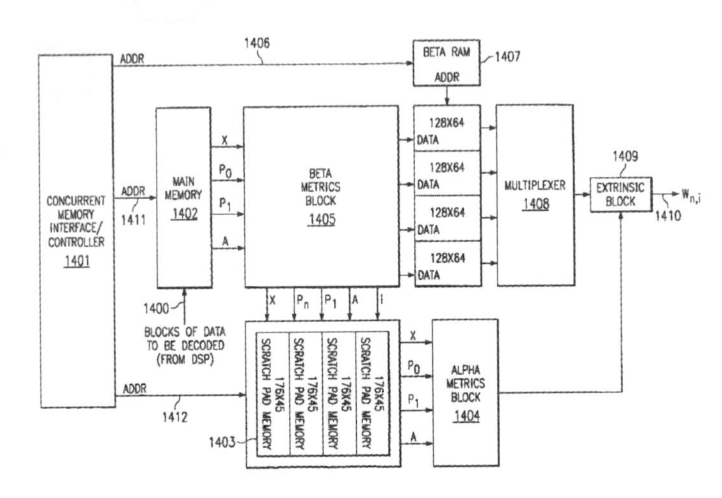

Utah

US007001290B2

(12) **United States Patent**
Mower et al.

(10) **Patent No.: US 7,001,290 B2**
(45) **Date of Patent: Feb. 21, 2006**

(54) **BLOW MOLDED BASKETBALL BACKBOARD FRAME**

(75) Inventors:

(73) Assignee: **Lifetime Products, Inc.**, Clearfield, UT (US)

(*) Notice: Subject to any disclaimer, the term of this patent is extended or adjusted under 35 U.S.C. 154(b) by 0 days.

(21) Appl. No.: **10/352,940**

(22) Filed: **Jan. 29, 2003**

(65) **Prior Publication Data**

US 2003/0158005 A1 Aug. 21, 2003

Related U.S. Application Data

(60) Provisional application No. 60/357,404, filed on Feb. 15, 2002.

(51) **Int. Cl.**
A63B 63/08 (2006.01)

(52) **U.S. Cl.** **473/481**; D21/701

(58) **Field of Classification Search** .... 473/476, 473/472, 479, 484, 485, 486, 488, FOR 10, 473/415, 433, 447, 448, 481; D21/302, 305, D21/314, 355, 698, 699, 701–704, 781
See application file for complete search history.

(56) **References Cited**

U.S. PATENT DOCUMENTS

4,478,415 A 10/1984 Shaffer et al.
5,082,261 A * 1/1992 Pelfrey 473/483
5,507,484 A 4/1996 van Nimwegen et al.
D402,338 S 12/1998 Davis
5,902,198 A 5/1999 Martin et al.
6,004,231 A * 12/1999 Schickert et al. 473/481
6,007,437 A 12/1999 Schickert et al.
D438,920 S * 3/2001 Matteucci et al. D21/701
6,367,948 B1 * 4/2002 Branson 362/234
6,468,373 B1 * 10/2002 Grinwald et al. 156/82

OTHER PUBLICATIONS

WebPage, Backboard Ad,www.dickssportinggoods.com/product/index.jsp, Download Date Aug. 27, 2004, 2 pages.*
Communication Pursuant to Article 96(2), European Patent Application No. 03250923.4, issued Dec. 3, 2003, 5 pages.
Communication Pursuant to Article 96(2), European Patent Application No. 03 250 923.4, issued May 11, 2004, 2 pages.

* cited by examiner

Primary Examiner—Gregory Vidovich
Assistant Examiner—M. Chambers
(74) *Attorney, Agent, or Firm*—Workman Nydegger

(57) **ABSTRACT**

A basketball goal system having a lightweight blow molded support frame and a transparent acrylic backboard is disclosed. The support frame includes an outer periphery and a support structure that is disposed within the outer periphery. The support structure divides the frame into two or more sections and at least a portion of the two or more sections are preferably covered by the acrylic backboard. Advantageously, the support structure and frame can create a lightweight basketball backboard with consistent rebounding characteristics.

49 Claims, 13 Drawing Sheets

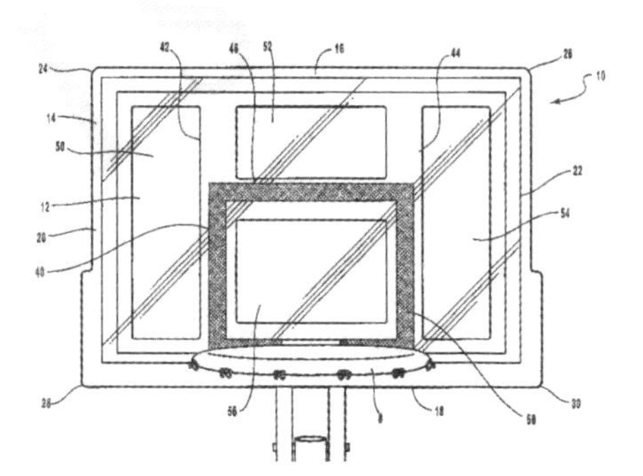

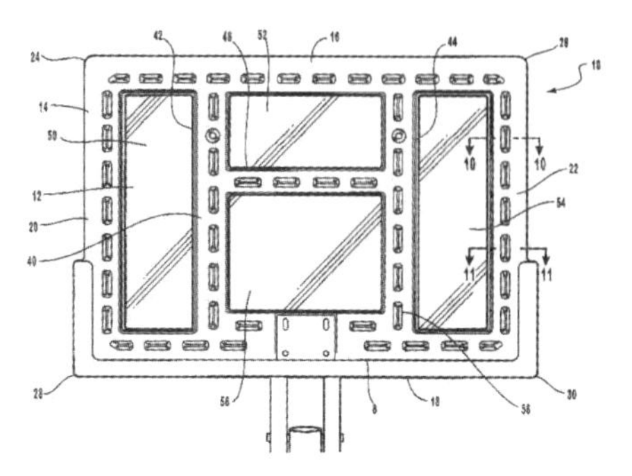

Vermont

US007075192B2

(12) **United States Patent**
Bywaters et al.

(10) **Patent No.: US 7,075,192 B2**
(45) **Date of Patent: Jul. 11, 2006**

(54) **DIRECT DRIVE WIND TURBINE**

(75) Inventors:

(73) Assignee: **Northern Power Systems, Inc.**, Waitsfield, VT (US)

(*) Notice: Subject to any disclaimer, the term of this patent is extended or adjusted under 35 U.S.C. 154(b) by 79 days.

(21) Appl. No.: **10/709,176**

(22) Filed: **Apr. 19, 2004**

(65) **Prior Publication Data**

US 2005/0230979 A1   Oct. 20, 2005

(51) **Int. Cl.**
F03D 9/00 (2006.01)
F03D 3/00 (2006.01)

(52) **U.S. Cl.** **290/55**; 290/44

(58) **Field of Classification Search** 290/44, 290/55; 416/44, 47
See application file for complete search history.

(56) **References Cited**

U.S. PATENT DOCUMENTS

2,153,523 A * 4/1939 Edmonds et al. 290/55
5,281,094 A * 1/1994 McCarty et al. 416/147
5,289,041 A * 2/1994 Holley 290/44
5,990,568 A * 11/1999 Hildingsson et al. 290/55
6,285,090 B1 9/2001 Brutsert et al.
6,400,039 B1 6/2002 Wobben
6,452,287 B1 9/2002 Looker
6,870,281 B1 * 3/2005 Weitkamp 290/55
6,888,262 B1 * 5/2005 Blakemore 290/44
6,921,243 B1 * 7/2005 Canini et al. 415/4.3
6,945,752 B1 * 9/2005 Wobben .... 416/170 R
2003/0194310 A1 10/2003 Canini et al.
2004/0041409 A1 * 3/2004 Gabrys 290/55

FOREIGN PATENT DOCUMENTS

BE	902092	7/1985
DE	261 395 A1	10/1988
DE	4402 184 A1	8/1995
DE	4402184 A1	8/1995
DE	4402184 A1	8/1995
DE	101 02 255 A1	8/2001
EP	0037002 A1	3/1980
EP	0 811 764 B1	5/2000
EP	1371845 A2	12/2003
ES	2156706 A1	1/2001
WO	WO 00/70219	10/2000
WO	WO-00/70219 A	11/2000
WO	WO 01/21956 A1	3/2001
WO	WO-02/057624	7/2002
WO	WO 02/057624 A1	7/2002
WO	WO-03/023943 A	3/2003
WO	WO 03/023943 A2	3/2003

OTHER PUBLICATIONS

Jean-Marc Fernandez, International Search Report Form PCT/ISA/206, Dated Jul. 27,2005.

* cited by examiner

Primary Examiner—Nicholas Ponomarenko
(74) *Attorney, Agent, or Firm*—Dave S. Christensen

(57) **ABSTRACT**

A wind turbine is provided that minimizes the size of the drive train and nacelle while maintaining the power electronics and transformer at the top of the tower. The turbine includes a direct drive generator having an integrated disk brake positioned radially inside the stator while minimizing the potential for contamination. The turbine further includes a means for mounting a transformer below the nacelle within the tower.

14 Claims, 3 Drawing Sheets

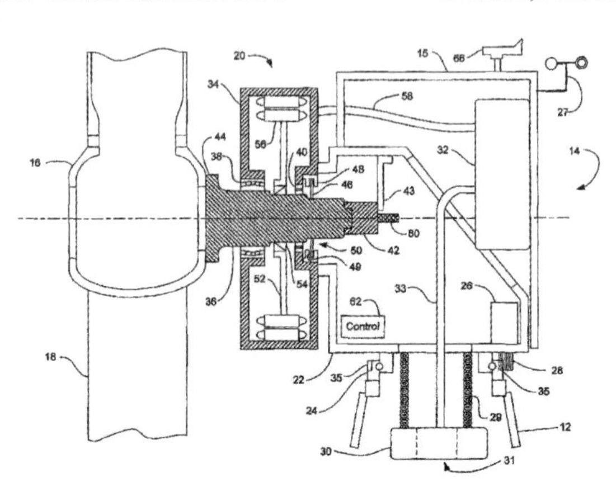

Virginia

(12) **United States Patent**
Bellesfield et al.

(10) **Patent No.: US 6,993,430 B1**
(45) **Date of Patent: Jan. 31, 2006**

(54) **AUTOMATED TRAVEL PLANNING SYSTEM**

(75) Inventors:

(73) Assignee: **America Online, Inc.**, Dulles, VA (US)

(*) Notice: Subject to any disclaimer, the term of this patent is extended or adjusted under 35 U.S.C. 154(b) by 0 days.

(21) Appl. No.: **10/273,889**

(22) Filed: **Oct. 21, 2002**

Related U.S. Application Data

(63) Continuation of application No. 09/901,082, filed on Jul. 10, 2001, now Pat. No. 6,498,982, which is a continuation of application No. 08/069,161, filed on May 28, 1993, now Pat. No. 6,282,489.

(51) **Int. Cl.**
G21C 21/26 (2006.01)

(52) **U.S. Cl.** **701/202**; 701/208; 701/209; 701/211; 340/995.14; 340/995.19; 340/995.24

(58) **Field of Classification Search** .... 701/23–26, 701/200, 201, 202, 207, 208, 209; 340/995.13, 340/995.14, 995.15, 995.17, 995.22, 995.25, 340/995.26, 988, 990, 995.19, 995.1, 995.16, 340/995.24; 342/450–452, 350.01, 357.13, 342/357.17, 457; 73/178 R
See application file for complete search history.

(56) **References Cited**

U.S. PATENT DOCUMENTS

4,689,747 A * 8/1987 Kurose et al. 701/200
5,067,081 A * 11/1991 Person 701/202
5,353,034 A * 10/1994 Sato et al. 342/457
5,557,524 A * 9/1996 Maki 701/35
5,608,635 A * 3/1997 Tamai 701/209
6,282,489 B1 * 8/2001 Bellesfield et al. 701/201

* cited by examiner

Primary Examiner—Jacques H. Louis-Jacques
(74) *Attorney, Agent, or Firm*—Fish & Richardson P.C.

(57) **ABSTRACT**

A list of places of interest geographically located near a travel route may be generated by accessing a routing database storing shape points capable of defining a travel route, determining a travel route from the shape points stored in the routing database, accessing a place of interest database including geographic centers, selecting from the places of interest database at least one geographic center that is geographically proximate to the travel route, and generating a list of places of interest associated with at least one of the selected geographic centers.

28 Claims, 7 Drawing Sheets

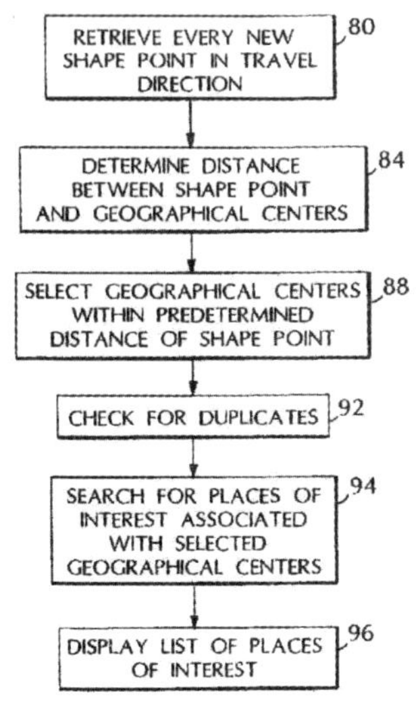

Washington

US006993563B2

(12) **United States Patent**
Lytle et al.

(10) **Patent No.: US 6,993,563 B2**
(45) **Date of Patent: *Jan. 31, 2006**

(54) **SYSTEM AND METHOD FOR COMPOSING, PROCESSING, AND ORGANIZING ELECTRONIC MAIL MESSAGE ITEMS**

(75) Inventors:

(73) Assignee: **Microsoft Corporation**, Redmond, WA (US)

(*) Notice: Subject to any disclaimer, the term of this patent is extended or adjusted under 35 U.S.C. 154(b) by 59 days.

This patent is subject to a terminal disclaimer.

(21) Appl. No.: **10/360,181**

(22) Filed: **Feb. 7, 2003**

(65) **Prior Publication Data**

US 2003/0120737 A1 Jun. 26, 2003

Related U.S. Application Data

(60) Continuation of application No. 09/019,245, filed on Feb. 5, 1998, now Pat. No. 6,549,950, which is a division of application No. 08/658,840, filed on May 31, 1996, now Pat. No. 5,923,848.

(51) **Int. Cl.**
G06F 15/16 (2006.01)

(52) **U.S. Cl.** **709/206**; 709/202; 709/246

(58) **Field of Classification Search** 709/202, 709/205, 206, 207, 217, 246, 203, 219, 214, 709/228, 236, 245
See application file for complete search history.

(56) **References Cited**

U.S. PATENT DOCUMENTS

4,803,619 A 2/1989 Bernstein et al. 364/200
(Continued)

FOREIGN PATENT DOCUMENTS

WO WO 9608779 A1 * 3/1996

OTHER PUBLICATIONS

H. Lee Murphy, "Computers Encourage Participation, Fast Results" Crain's Chicago Business, Jan. 18, 1993, 3 pp.
(Continued)

Primary Examiner—Paul H. Kang
(74) *Attorney, Agent, or Firm*—Merchant & Gould LLC

(57) **ABSTRACT**

In an electronic mail system environment, a system and method for automatically checking recipients' names, providing message flags, providing custom forms, and providing an autoresponse feature. Recipients' names are resolved in the background, while the user of the e-mail system is composing the message. The user easily resolves ambiguous names by using a context menu. The resolved ambiguous names are automatically used to create nicknames, which are used to resolve ambiguous names in the future. Message flags allow a sender or recipient to identify required follow-up action and a deadline. The recipient may use the message flags to quickly determine which messages require follow-up action. The e-mail system notifies a recipient when a due date is approaching or when a follow-up action is past due. A custom forms feature allows a user to create and share custom forms without requiring the form to be published or installed by other user. The custom form's attributes are transmitted to the recipient as an element of the e-mail message. An autoresponse feature allows a sender to create a message that includes voting buttons corresponding to the possible responses to a query. A recipient replies by selecting one of the voting buttons. The recipient's vote is automatically tallied in the sender's copy of the message, thus allowing the sender to view a vote tally, a list of the recipients, and their response.

19 Claims, 32 Drawing Sheets

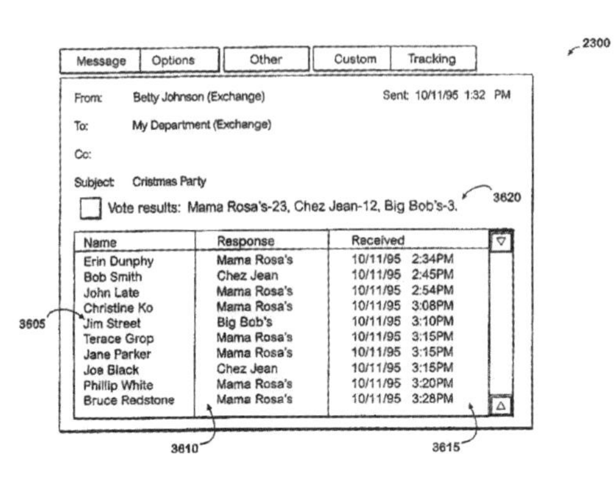

West Virginia

(12) **United States Patent**
Kerns

(10) **Patent No.: US 7,014,236 B2**
(45) **Date of Patent: Mar. 21, 2006**

(54) **PICKUP TRUCK RECREATIONAL EQUIPMENT RACK**

(75) Inventor:

(73) Assignee: **Christopher L. Kerns**, Martinsburg, WV (US)

(*) Notice: Subject to any disclaimer, the term of this patent is extended or adjusted under 35 U.S.C. 154(b) by 0 days.

(21) Appl. No.: **10/847,252**

(22) Filed: **May 17, 2004**

(65) **Prior Publication Data**

US 2004/0232718 A1 Nov. 25, 2004

Related U.S. Application Data

(60) Provisional application No. 60/471,586, filed on May 20, 2003.

(51) **Int. Cl.**
B60P 3/00 (2006.01)

(52) **U.S. Cl.** **296/3**

(58) **Field of Classification Search** 296/3, 296/180.1, 26.04; 49/358; 135/158; 52/127.2; 211/182, 189, 175, 187, 60.1; 248/218.4; 224/405, 403, 402, 531, 532, 917.5; 414/462
See application file for complete search history.

(56) **References Cited**

U.S. PATENT DOCUMENTS

2,947,566 A * 8/1960 Pierce 296/3
3,263,692 A * 8/1966 Questi et al. 135/158
4,057,281 A * 11/1977 Garrett 296/3
4,152,020 A * 5/1979 Brown et al. 296/3
4,215,894 A * 8/1980 Sidlinger 296/3
4,267,948 A * 5/1981 Lewis 296/3
4,278,175 A * 7/1981 Jackson 296/3
4,405,170 A * 9/1983 Raya 296/3
4,449,656 A 5/1984 Wouden
4,509,787 A * 4/1985 Knaack et al. 296/3
4,877,169 A 10/1989 Grim
5,002,324 A * 3/1991 Griffin 296/3
5,108,141 A * 4/1992 Anderson 296/3
5,139,375 A * 8/1992 Franchuk 410/105
5,143,415 A * 9/1992 Boudah 296/3
5,152,570 A * 10/1992 Hood 296/3
5,439,152 A * 8/1995 Campbell 224/405
5,806,905 A * 9/1998 Moore 296/3
5,836,635 A * 11/1998 Dorman 296/3
5,848,743 A 12/1998 Derecktor
5,927,782 A * 7/1999 Olms 296/3
6,089,795 A * 7/2000 Booth 296/3
D436,915 S * 1/2001 Burger D12/406
6,186,571 B1 * 2/2001 Burke 296/3
6,332,637 B1 * 12/2001 Chambers 296/3
6,347,731 B1 * 2/2002 Burger 224/405
6,513,849 B1 * 2/2003 Carter 296/3
6,520,723 B1 * 2/2003 Christensen 410/100
6,634,689 B1 * 10/2003 Soto 296/3
6,676,220 B1 * 1/2004 Mistler .... 298/1 A
6,786,522 B1 * 9/2004 Kench et al. 296/3

* cited by examiner

Primary Examiner—Kiran B. Patel

(57) **ABSTRACT**

A pickup truck bed load carrying rack is described. A pickup truck bed load carrying rack is supported by the bedrails of the pickup truck bed. It is designed to carry items that are too long to fit in the bed of a pickup or to carry items so as to free up space that is needed in the pickup bed. The rack includes two tubular horizontal crossbars (**4**), two tubular braces (**6**), tubular brace connecting straps (**6*a***) four tubular vertical supporting braces (**8**), and braces (**12**). All of the above are supported by two horizontal brackets (**10**) which in turn are supported by bedrails (**18**).

4 Claims, 10 Drawing Sheets

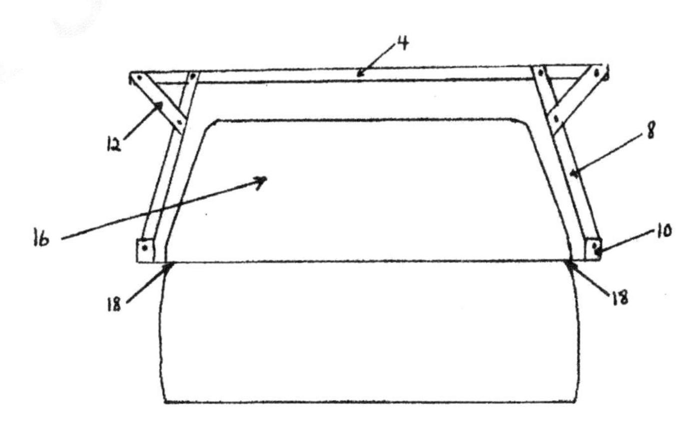

Wisconsin

(12) **United States Patent**
Behnke et al.

(10) **Patent No.: US 7,117,816 B2**
(45) **Date of Patent: Oct. 10, 2006**

(54) **HEATED PET BED**

(75) Inventors:

(73) Assignee: **Kimberly-Clark Worldwide, Inc.**, Neenah, WI (US)

(*) Notice: Subject to any disclaimer, the term of this patent is extended or adjusted under 35 U.S.C. 154(b) by 35 days.

(21) Appl. No.: **10/880,759**

(22) Filed: **Jun. 30, 2004**

(65) **Prior Publication Data**

US 2006/0000415 A1 Jan. 5, 2006

(51) **Int. Cl.**
A01K 29/00 (2006.01)

(52) **U.S. Cl.** **119/28.5**; 5/421

(58) **Field of Classification Search** 119/28.5; 5/421

See application file for complete search history.

(56) **References Cited**

U.S. PATENT DOCUMENTS

3,125,663 A 3/1964 Hoffman
4,064,835 A 12/1977 Rabenbauer
4,303,074 A 12/1981 Bender
4,332,214 A 6/1982 Cunningham
4,366,804 A * 1/1983 Abe 126/263.02
4,516,564 A * 5/1985 Koiso et al. 126/263.02
4,591,694 A 5/1986 Phillips
4,899,693 A 2/1990 Arnold
4,995,126 A * 2/1991 Matsuda 5/421
5,025,777 A 6/1991 Hardwick
5,092,271 A 3/1992 Kleinsasser
5,324,911 A 6/1994 Cranston
5,371,340 A 12/1994 Stanfield
5,425,975 A 6/1995 Koiso et al.
5,469,592 A * 11/1995 Johnson 5/654
5,653,741 A 8/1997 Grant
5,702,375 A * 12/1997 Angelillo et al. 604/358
5,715,772 A 2/1998 Kamrath et al.
5,724,911 A 3/1998 McAlister
5,730,721 A * 3/1998 Hyatt et al. 604/500
5,975,074 A * 11/1999 Koiso et al. 126/204
6,078,026 A 6/2000 West
6,084,209 A 7/2000 Reusche et al.

(Continued)

FOREIGN PATENT DOCUMENTS

CA 1261908 A 9/1989

(Continued)

Primary Examiner—Teri Pham Luu
Assistant Examiner—Kimberly S. Smith
(74) *Attorney, Agent, or Firm*—Schwegman, Lundberg, Woessner, & Kluth, P.A.

(57) **ABSTRACT**

The present invention relates to a pet bed that includes a heating layer which is formed of an enclosure and a heating composition sealed inside the enclosure. The enclosure includes a gas-permeable section such that the heating composition generates heat when a gas is received through the gas-permeable section. The pet bed further includes (i) a liner attached to the heating layer to engage a pet as the pet lies on the pet bed; (ii) a conforming layer attached to the heating layer such that the conforming layer conforms to the shape of the pet as the pet lies on the pet bed; and/or (iii) a backing layer attached to the heating layer such that the backing layer restricts movement of the pet bed. The enclosure may further include a cover over the gas-permeable section such that the heating composition generates heat when the cover is removed from the enclosure.

20 Claims, 5 Drawing Sheets

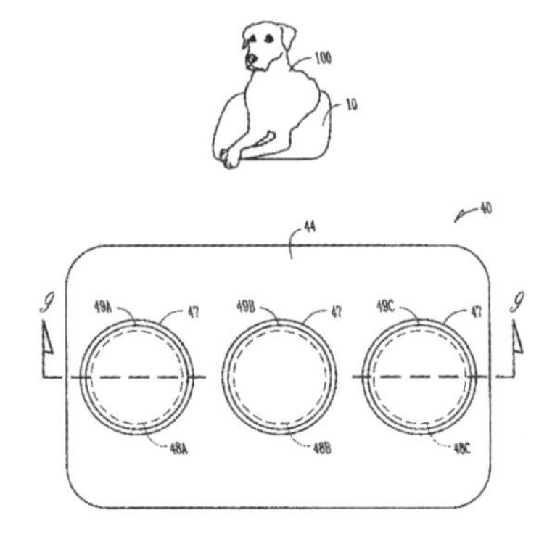

Wyoming

(12) **United States Patent**
Gomez

(10) **Patent No.:** **US 7,025,362 B1**
(45) **Date of Patent:** **Apr. 11, 2006**

(54) **MANUAL LARGE GAME CARRYING DEVICE**

(75) Inventor:

(73) Assignee: **Matthew L. Gomez**, Rock Springs, WY (US)

(*) Notice: Subject to any disclaimer, the term of this patent is extended or adjusted under 35 U.S.C. 154(b) by 0 days.

(21) Appl. No.: **10/717,423**

(22) Filed: **Nov. 19, 2003**

(51) **Int. Cl.**
B62B 3/10 (2006.01)

(52) **U.S. Cl.** **280/47.11**; 280/263; 280/63

(58) **Field of Classification Search** 280/263, 280/264, 267–269, 63, 47.11, 47.131, 47.17, 280/281.1; 482/66, 68
See application file for complete search history.

(56) **References Cited**

U.S. PATENT DOCUMENTS

89,443 A * 4/1869 Smith 280/230
94,056 A * 8/1869 Allen 280/267
404,562 A * 6/1889 Reynolds 280/257
486,056 A * 11/1892 Saladee 280/261
2,190,779 A * 2/1940 Fogle .... 33/1 H
2,284,333 A * 5/1942 McGirl et al. 280/261
2,992,834 A * 7/1961 Tidwell et al. 280/47.3
3,907,323 A 9/1975 Knapp et al.
D257,587 S * 12/1980 Doyich D34/12
4,506,902 A * 3/1985 Maebe 280/266
4,934,724 A * 6/1990 Allsop et al. 280/281.1
5,195,394 A * 3/1993 Latta 74/551.8
5,328,192 A 7/1994 Thompson
5,645,292 A * 7/1997 McWilliams et al. 280/494
5,897,131 A 4/1999 Brown et al.
6,341,787 B1 * 1/2002 Mason 280/47.26
2001/0004148 A1 6/2001 Darling, III
2003/0080538 A1 5/2003 Watts et al.

* cited by examiner

Primary Examiner—Jeff Restifo

(57) **ABSTRACT**

A manual large game carrying device of the type having an substantially elongated upper receiving bar (**30**) that accommodates and secures large game at points of securement (**32**). The manual large game carrying device can be turned by a handlebar configuration (**22**) which engages a front fork axle assembly (**20**) at the frame/front fork/handlebar junction (**44**). This front fork axle assembly (**20**) is connected to the front rotatable wheel (**10**A) by the releasable front axle (**12**). In addition, the manual large game carrying device's speed is controlled by braking mechanism levers (**26**). The braking mechanism levers (**26**) engage the braking mechanism linkages (**28**) which engage the braking mechanisms (**42**) attached to the front brake rotor (**16**) and rear brake rotor (**18**). In addition to steering and stopping the manual large game carrying device, the handlebar configuration (**22**) has handle grips (**24**) conveniently placed for efficient transport.

2 Claims, 4 Drawing Sheets

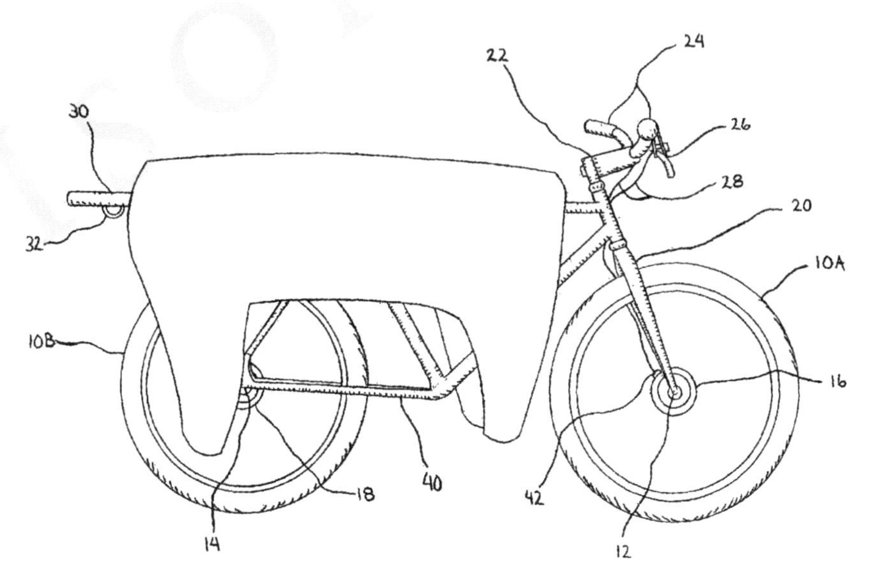

Chapter 19
Conclusion

Innovation is the Wave of the Future

The generation of wealth through innovation requires an inherently valuable innovation, solid patent protection and a sound business plan. In today's marketplace, a strategy not founded on innovation and consisting solely of, for example, getting to market fast and developing brand recognition may be unrealistic and suboptimal, particularly for a start-up company or any business in a highly-competitive market.[73] Product information is now easy to obtain, consumers are sophisticated, and competitors nimble. Assume that without meaningful innovation and protection, competitors can and will copy every valuable aspect of your product or service. You will be left to compete based solely on price and delivery time, and potentially not have the opportunity or resources to develop market penetration or brand awareness. Not only will profit opportunities be lost, but so will the ability to attract investors and buyers.

Innovation is the new business model – there is no going back to the days of commodity manufacturing and slow-moving competitors. Embrace innovation, develop your approach and adjust it to compensate for market conditions. If you make innovation a cornerstone of your business strategy, the results may astound you.

We Want to Hear from You

Thanks for taking the time to read *Business Success Through Innovation.* We welcome your comments, questions and feedback about the book. Additionally, if you have any experiences with innovations and patents you would like to share with us, please contact us at memberservices@isopatent.com.

Endnotes

1 Throckmorton, Ray, United States: Royalty Audit Checklists (January 30, 2008).

2 Bigelow, Bruce V., Patent Payoff, The San Diego Union-Tribune (May 14, 2006).

3 35 U.S.C. § 112.

4 International Monetary Fund, Trilateral 2006 Report, EPO, JPO, USPTO (2007).

5 Central Intelligence Agency World Factbook (2007).

6 International Monetary Fund (in current prices, U.S. dollars.), World Economic Database (October, 2008).

7 Id.

8 As used in this book "concepts" and "innovations" refer to inventions that have a new or improved function and that can be protected by utility patents. Trademarks, copyrights and designs can also be valuable assets, but are not addressed here.

9 "Intellectual property" includes patents, trademarks, service marks, trade dress, copyrights and trade secrets. See 15 U.S.C. § 1051 et seq. (trademarks, service marks and trade dress); 17 U.S.C. § 101 at seq. (copyrights); 35 U.S.C. § 1 et seq. (patents).

10 House Committee Notes, Industrial Espionage Act of 1996 (S. 1556) (1996).

11 That number is estimated to be $368 billion for 2007. About 72% of which was contributed by businesses, 13% contributed by universities and colleges, 11% by the federal government and the remainder from other sources. National Science Foundation (August, 2008).

12 United States Patent and Trademark Office Press Release, USPTO Introduces New Intellectual Property Curriculum (April 14, 2008); A Market For Ideas, The Economist

(October, 2005)

13 Rivette, Kevin; Kline, David, Discovering New Value in Intellectual Property, Harvard Business Review (January-February, 2000).

14 This is for utility patents. 35 U.S.C. § 154. See also 35 U.S.C. §§ 155, 155A and 156.

15 There are four types of patents in the United States: (1) A utility patent, which protects the functional, or utilitarian, aspects of an invention and comprises about 86% of all U.S. patents. (2) A design patent, which protects new, original and ornamental (i.e., non-functional and non-utilitarian aspects) of a design for an article of manufacture. About 13% of the patents issued in the United States are design patents. (3) A plant patent, which protects any distinct and new variety of plant that is asexually reproduced including cultivated sports, mutants, hybrids, and newly found seedlings, and which comprises about 1% of all patents issued in the United States. (4) A reissue patent, which is based on an existing utility, design or plant patent. The patent is "reissued" to correct an error by the Applicant during patent examination, which may have resulted in either part of the patent being invalid or too narrow in scope. If the rules for reissue are met, the patent is reissued by the USPTO and replaces the original patent. A very small percentage of all patents are reissue patents. 35 U.S.C. §§ 101, 161, 171, 251; United States Commerce Department (2008).

16 35 U.S.C. § 101.

17 United States Commerce Department (2008).

18 This is a quote from the Supreme Court case of Diamond v. Chakrabarty, 447 U.S. 303 (1980).

19 This is for utility patents. See 35 U.S.C. § 101; Diamond v. Chakrabarty, 447 U.S. 303 (1980).

20 State Street Bank & Trust Co. v. Signature Financial Group, Inc., 149 F.3d 1368 (Fed. Cir. 1998). In re Bilski, 2008 U.S. App. LEXIS 22478 (Fed. Cir. 2008) was decided on October 30, 2008, just before this book was printed. Under the rule of this case, called "the machine or transformation test," a process must either be (1) tied to a particular machine or apparatus (such as a computer system), or (2) transform a particular article into a different state or thing.

21 The same is true for applications originally prepared in other countries. The attorneys preparing them should consider issues for filing outside of their own country, such as the "best mode" requirement in the United States.

22 35 U.S.C. § 102. Some art that is not technically publicly known can still bar you from receiving a patent, such as a secret method you commercialized prior to filing for patent protection or an earlier invention by another who did not suppress, abandon or conceal it. 35 U.S.C. §§ 102(a), (g).

23 35 U.S.C. § 103; Graham v. John Deere Co., 383 U.S. 1 (1966); see note 22 with respect to the use of the term "publicly available."

24 While we use the word "innovation," you cannot protect merely an idea (except through contractual obligations), you must have an idea and know how to make and use it to have an innovation that can be patented. 35 U.S.C. § 112.

25 United States Department of Commerce (2008). These numbers include utility, design, plant and reissue patents.

26 United States Patent and Trademark Office (2008); Bigelow, Bruce V., Patent Payoff, The San Diego Union-Tribune (May 14, 2006); United States Commerce Department (2008).

27 United States Patent and Trademark Office Press Release, USPTO 2007 Fiscal Year-End Results Demonstrate Trend of Improved Patent and Trademark Quality (November 15, 2007).

28 This procedure is called "prosecution." Quality prosecution, in addition to quality preparation of a patent application, is extremely important to obtaining strong patent rights and is a lengthy, detailed topic outside the scope of this book.

29 Subject to potential costs and attorney's fees required to protect the market segment. You may have to litigate to determine whether another is infringing your patent.

30 Innovations can also be protected through contractual means or trade secret laws, but such protection is often inapplicable or impractical, particularly if you have no contractual relationship to enforce and/or your products or processes are publicly available and can be reverse engineered.

31 35 U.S.C. § 41; 37 CFR § 1.20.

32 United States Commerce Department (2008).

33 United States Patent and Trademark Office 2008 fees. This is the small entity fee for organizations with 500 employees or less. For large entities having over 500 employees the 3½ year maintenance fee is $980.

34 See note 30, above, concerning contractual or trade secret protection.

35 The first-to-invent rule is explained in more detail in *Inventor's Notebook*, shown in the back of this book. Consult a good patent attorney if you have any questions about the rule or requirements.

36 Most businesses and individuals use a mechanism known as a Patent Cooperation Treaty ("PCT") application to delay the costs of foreign filing into individual foreign countries by 18 or more months, depending upon the country.

37 Unfortunately, Ralph Waldo Emerson's famous quote: "If a man can make a better mousetrap, though he builds his house in the woods the world will make a beaten path to his door," is no longer accurate. You need a sound, thorough business plan to profit from your innovation.

38 IBM 2007 Annual Report, p.21; Hamm, Steve, IBM Takes on Amazon, Business Week (October, 2006); Rivette, Kevin; Kline, David, Discovering New Value in Intellectual Property, Harvard Business Review (January/February, 2000).

39 Xerox Press Release, Xerox Fuels Innovation Engine with 584 New Patents in 2007, (January, 2008).

40 Qualcomm Press Release, Qualcomm Acquires Wireless Location Leader SnapTrack, (January, 2000).

41 Qualcomm, Incorporated, Qualcomm 2007 Annual Report.

42 Pollack, Andrew M., Takeda Pharmaceutical of Japan to Buy Millenium Pharmaceuticals for $8.8 Billion, International Herald Tribune (April, 2008).

43 CERM Foundation for the European Commission, Study on Evaluating the Knowledge Economy--What Are Patents Actually Worth?--The Value of Patents for Today's Economy and Society (July, 2006).

44 velcro.com Home Page, Velcro Industries (2007-2008); Massachusetts Institute of

Technology (April, 2004).

45 Massachusetts Institute of Technology (January, 2007); mars.com History Page (2008).

46 Google Press Release, Google Announces Fourth Quarter and Fiscal Year 2007 Results (January, 2008); google.com Milestones Page, Google (2008).

47 Schudel, Matt, Accomplished, Frustrated Inventor Dies, Washington Post (February, 2005).

48 NASA, Spaceport News, John F. Kennedy Space Center (July, 2004); www.cwru.edu/development/alum2/newsevents/publications/fastforward/content/walking.html (2008).

49 Design Continuum, Inc., Innovation Solutions to Age-Old Problems (2008).

50 Widman, Wendy, A Witch's Broom Spells Big Profits, Forbes (August, 2005).

51 Swiffer SweepVac Product Manual available at www.swiffer.com (2008).

52 Huston, Larry; Sakkab, Nabil, Connect and Develop, Harvard Business Review (March, 2006).

53 Anderson, Diane, When Crocs Attack, An Ugly Shoe Tale, Business 2.0 (November, 2006); Crocs, Inc. Press Release, Crocs, Inc. Reports Record Fiscal 2007 Fourth Quarter and Full Year Financial Results (February, 2008).

54 Anderson, Diane, Instant Company, Crocs Edition, Business 2.0 (November, 2006).

55 Massachusetts Institute of Technology (September, 2005).

56 http://em.wikipedia.org/wiki/thomas_edison(2008); about.com: The Inventions of Thomas Edison (2008); National Inventors Hall of Fame Foundation, Inc., Inventor Profile - Thomas Alva Edison, www.invent.org/hall_of_fame/50.html (2008).

57 about.com, Bellis, Mary, Inventors of the Modern Computer (2008).

58 hasbro.com Monopoly History Page, Hasbro (2008); Massachusetts Institute of Technology (October, 1997). An earlier game, called "The Landlord's Game," was patented in 1904.

59 Stickgold, Emma, Stanley I. Mason, 84; His Inventions Became Part of Everyday Life, The Boston Globe (January, 2006).

60 liquidpaper.com About Us Page, Sanford, a Division of Newell Rubbermaid (2007).

61 Pfeiffer, Eric, Setting Patent Traps, Forbes (June, 2002); lemelson.org, The Lemelson Foundation, (2007).

62 IBM 2007 Annual Report; United States Patent and Trademark Office (2008).

63 United States Patent and Trademark Office (2008).

64 See, e.g., 35 U.S.C. § 102.

65 www.epo.org/about.us/epo/member-states.html (2008).

66 An "unenforceable" patent is one in which the entire patent cannot be enforced because the inventor perpetrated a "fraud" on the USPTO during the procurement of the patent. The standards for, and consequences of, unenforceability are not discussed here.

67 See 35 U.S.C. § 111(b).

68 All data in Chapters 12-18 without individual footnotes was obtained from the United States Commerce Department in 2008. The data for the most recent year (which was usually 2007) available is presented.

69 International Monetary Fund, World Economic Outlook Database (October, 2008).

70 Id.

71 Population data for Chapter 15 is from the U.S. Census Bureau, Population Estimates Program (July, 2007). Age, household income and eductaion data are from the U.S. Census Bureau, 2006 American Community Survey (2006). Patent data is from the United States Commerce Department (2008).

72 The total number of United States patents listed in this chart is 93,584 for the 50 United States. If the U.S. Virgin Islands, District of Columbia and other United States Territories are included, the total number of U.S. patents issued in 2007 to U.S.-based persons and entities is 93,691.

73 Some businesses, such as utilities or heavy equipment manufacturers, with extensive infrastructure, have non-legal barriers to entry.